高等职业院校技能应用型教材·软件技术系列

Android嵌入式开发及实训

周　薇　王想实　李　昊　主　编

张　超　陈　斌　陈姣姣　副主编

電子工業出版社

Publishing House of Electronics Industry

北京·BEIJING

内 容 简 介

本书基于 Android Studio 集成开发工具，循序渐进地介绍了 Android 应用程序开发的基本内容。本书共 11 章，第 1 章为 Android 入门概述；第 2 章～第 10 章以项目的形式，分别介绍了图形界面、Activity 与 Intent、数据库与 ContentProvider、数据存储、Service、BroadcastReceiver、多线程、网络编程和串口编程。其中，大部分项目通过需求分析、界面设计、实施等环节，逐步引导读者完成项目操作，同时，在每章的后半部分，讲解了对应项目所用到的基本概念与功能，并通过章末的实训环节加深理解；第 11 章为综合实训，旨在通过一个完整的实训项目巩固前面所学的知识。

本书配有源代码、电子课件等教学资源，读者可以登录华信教育资源网（www.hxedu.com.cn）注册后免费下载。本书内容翔实、语言精练，既可以作为高等院校、高等职业院校计算机、物联网等专业的教材，也可以作为移动互联开发人员参考用书。

图书在版编目（CIP）数据

Android 嵌入式开发及实训/周薇，王想实，李昊主编. —北京：电子工业出版社，2019.9
ISBN 978-7-121-35779-4

Ⅰ. ①A… Ⅱ. ①周… ②王… ③李… Ⅲ. ①移动终端－应用程序－程序设计－高等学校－教材
Ⅳ.①TN929.53

中国版本图书馆 CIP 数据核字（2018）第 285943 号

策划编辑：薛华强（xuehq@phei.com.cn）
责任编辑：程超群 文字编辑：薛华强
印 刷：北京虎彩文化传播有限公司
装 订：北京虎彩文化传播有限公司
出版发行：电子工业出版社
北京市海淀区万寿路 173 信箱 邮编：100036
开 本：787×1 092 1/16 印张：18 字数：576 千字
版 次：2019 年 9 月第 1 版
印 次：2021 年 4 月第 4 次印刷
定 价：55.00 元

凡所购买电子工业出版社图书有缺损问题，请向购买书店调换。若书店售缺，请与本社发行部联系，联系及邮购电话：（010）88254888，88258888。

质量投诉请发邮件至 zlts@phei.com.cn，盗版侵权举报请发邮件至 dbqq@phei.com.cn。

本书咨询联系方式：（010）88254569，xuehq@phei.com.cn，QQ1140210769。

前　言

当今的世界已步入移动互联时代，各种智能手机、平板电脑等移动设备迅速普及，逐步地改变着人们的工作、生活、消费方式。各传统软件及互联网企业也在调整产品布局，向移动互联应用软件进军。总体上看，移动互联应用产品开发产业呈井喷式发展，相关企业的人才需求量巨大。从高等职业教育的服务对象来看，高等职业教育要面向行业企业，以就业为导向，走产学结合道路，培养高素质的技能型人才。

Android 是一种基于 Linux 的自由及开放源代码的操作系统，由 Google 公司和开放手机联盟领导及开发。过去，开发 Android 应用程序采用 Eclipse 集成开发工具。Android 6.0 之后，Google 公司开始主推 Android Studio。Android Studio 提供了集成的 Android 程序开发工具用于开发和调试。

Android Studio 是一个全新的 Android 开发环境，成功解决了多分辨率、多语言等诸多程序开发与运行问题，开发者可以在编写程序的同时预览在不同尺寸屏幕中的外观效果。相关企业和技能大赛也已采用 Android Studio 开发环境。

本书注重培养读者的动手能力，循序渐进地介绍了 Android 应用程序开发的基本内容。本书共 11 章，第 1 章为“Android 入门概述”，介绍了 Android 基本知识、搭建开发环境、项目结构分析等；第 2 章为“图形界面——计算器项目”，介绍了界面设计基础、事件处理、布局、常用的控件、调试技巧等；第 3 章为“Activity 与 Intent——运动会报名项目”，介绍了 Activity 的状态与生命周期、Intent、Activity 间传递数据等；第 4 章为“数据库与 ContentProvider——用户管理项目”，介绍了 SQLite 数据库管理系统、ContentProvider 和 ContentResolver、使用内置的 ContentProvider 等；第 5 章为“数据存储——简易相册项目”，介绍了数据存储概述、文件存储、共享偏好设置、通知、访问远程数据等；第 6 章为“Service——MP3 音乐播放器项目”，介绍了 Service、多媒体等；第 7 章为“BroadcastReceiver——短信过滤器项目”，介绍了 BroadcastReceiver、手机通话、手机短信等；第 8 章为“多线程——射击游戏项目”，介绍了多线程技术、绘图技术等；第 9 章为“嵌入式开发：网络编程——天气预报项目”，介绍了网络编程概述、网络编程综合项目等；第 10 章为“嵌入式开发：串口编程——读卡器项目”，介绍了串口编程的相关内容；第 11 章为“综合实训——诗词赏析项目”，旨在通过一个完整的实训项目巩固前面所学的知识。

在第 2 章～第 10 章中，大部分项目通过需求分析、界面设计、实施等环节，逐步引导读者完成项目操作，同时，在每章的后半部分，讲解了对应项目所用到的基本概念与功能，并通过章末的实训环节加深理解。其中，第 9 章和第 10 章为嵌入式开发的相关内容，适用于物联网专业学习。

本书内容翔实、语言精练，涉及的知识面广，部分项目融入了技能大赛的设计理念。本书既可以作为高等院校、高等职业院校计算机、物联网专业的教材，也可以作为移动互联开发人员参考用书。

本书的建议课时安排如下表。

章　　节	课时数	32 学时（入门）	48 学时（基本）	64 学时（完整）	64 学时（基本+嵌入式）	80 学时（完整+嵌入式）
第 1 章　Android 入门概述	8	8	8	8	8	8
第 2 章　图形界面——计算器项目	10	10	10	10	10	10
第 3 章　Activity 与 Intent——运动会报名项目	12	12	12	12	12	12
第 4 章　数据库与 ContentProvider——用户管理项目	10	0	10	10	10	10
第 5 章　数据存储——简易相册项目	6	0	6	6	6	6
第 6 章　Service——MP3 音乐播放器项目	4	0	0	4	0	4
第 7 章　BroadcastReceiver——短信过滤器项目	6	0	0	6	0	6
第 8 章　多线程——射击游戏项目	6	0	0	6	0	6
第 9 章　嵌入式开发：网络编程——天气预报项目	8	0	0	0	8	8
第 10 章　嵌入式开发：串口编程——读卡器项目	8	0	0	0	8	8
第 11 章　综合实训——诗词赏析项目	0	0	0	0	0	0
机　　动	2	2	2	2	2	2
合　　计	80	32	48	64	64	80

本书由无锡职业技术学院周薇、王想实，河南交通职业技术学院李昊担任主编；无锡科技职业技术学院张超，石家庄信息工程职业学院陈斌，合肥信息技术职业学院陈姣姣担任副主编。

本书配有源代码、电子课件等教学资源，读者可以登录华信教育资源网（www.hxedu.com.cn）注册后免费下载。

由于编者水平有限，书中的疏漏和不妥之处在所难免，敬请各界专家和读者朋友批评指正，意见和建议请反馈至编者的电子邮箱 20831065@qq.com，我们将不胜感激。

编　者

目　录

CONTENTS

第 1 章 Android 入门概述

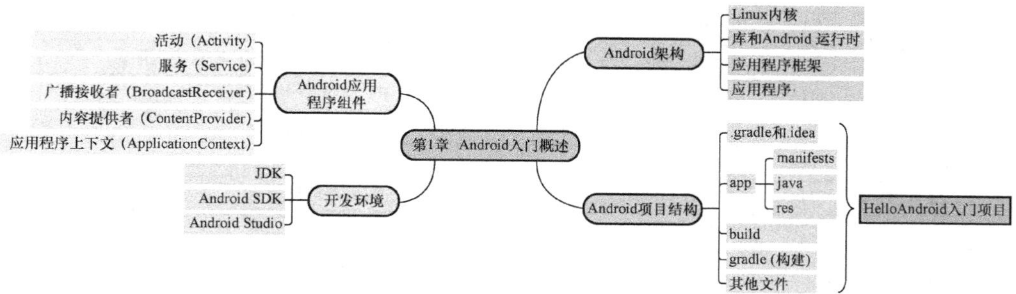

本章知识要点思维导图

Android 是一种基于 Linux 系统的自由且开放源代码的操作系统，主要用于移动设备（如智能手机和平板电脑），由 Google 公司和开放手机联盟（Open Handset Alliance，OHA）领导及开发。

Android 操作系统最初由 Andy Rubin 开发，主要用于手机。2005 年 8 月由 Google 公司收购注资。2007 年 11 月，Google 公司与 84 家硬件制造商、软件开发商及电信运营商组建开放手机联盟，共同研究并改良 Android 系统。随后 Google 公司以 Apache 开源许可证的授权方式，发布了 Android 系统的源代码。2008 年 9 月，Google 公司正式发布了 Android 1.0 系统，这也是 Android 系统最早的版本。2017 年 8 月 22 日，Google 公司发布了 Android 8.0 版本。

第一部 Android 智能手机发布于 2008 年 10 月。随后，Android 系统逐渐扩展到平板电脑及其他设备上，如电视、数码相机、游戏机等。

1.1 Android 基本知识

1.1.1 Android SDK 与 API Level 对应关系

不同的 Android SDK 版本需要对应不同的 API Level，其关系见表 1-1。

表 1-1 Android SDK 版本、发布时间与 API Level

版　本	API Level	发布时间	代　号
1.0	1	2008 年 9 月 23 日	无代号
1.1	2	2009 年 2 月 9 日	无代号
1.5	3	2009 年 4 月 27 日	Cupcake 纸杯蛋糕
1.6	4	2009 年 9 月 15 日	Donut 甜甜圈
2.0	5	2009 年 10 月 26 日	Eclair 泡芙
2.0.1	6	2009 年 12 月 3 日	Eclair 泡芙

续表

版　本	API Level	发布时间	代　号
2.1	7	2010 年 1 月 12 日	Eclair 泡芙
2.2	8	2010 年 5 月 20 日	Froyo 冻酸奶
2.3	9	2010 年 12 月 6 日	Gingerbread 姜饼
2.3.3	10	2011 年 2 月 9 日	Gingerbread 姜饼
3.0	11	2011 年 2 月 22 日	Honeycomb 蜂巢
3.1	12	2011 年 5 月 10 日	Honeycomb 蜂巢
3.2	13	2011 年 7 月 15 日	Honeycomb 蜂巢
4.0.1	14	2011 年 10 月 21 日	Ice Cream Sandwich 冰激凌三明治
4.0.3	15	2011 年 12 月 16 日	Ice Cream Sandwich 冰激凌三明治
4.1	16	2012 年 7 月 9 日	Jelly Bean 果冻豆
4.2	17	2012 年 11 月 13 日	Jelly Bean 果冻豆
4.3	18	2013 年 7 月 24 日	Jelly Bean 果冻豆
4.4	19	2013 年 10 月 31 日	KitKat 奇巧巧克力
4.4W	20	2014 年 6 月 25 日	KitKat 奇巧巧克力
5.0	21	2014 年 11 月 12 日	Lollipop 棒棒糖
5.1	22	2015 年 3 月 9 日	Lollipop 棒棒糖
6.0	23	2015 年 10 月 5 日	Marshmallow 棉花糖
7.0	24	2016 年 8 月 22 日	Nougat 牛轧糖
7.1.1	25	2016 年 12 月 5 日	Nougat 牛轧糖
8.0	26	2017 年 8 月 22 日	Oreo 奥利奥
8.1	27	2017 年 12 月 5 日	Oreo 奥利奥
9.0	28	2018 年 8 月 7 日	Pie 派

1.1.2　Android 架构

Android 架构如图 1-1 所示，由 Linux Kernel（Linux 内核）、Libraries（库）和 Android RunTime（Android 运行时）、Application Framework（应用程序框架）以及 Applications（应用程序）四个层组成，中间两层（库和 Android 运行时、应用程序框架）被称为中间件，它们充当了操作系统与应用程序之间的接口，是供开发人员使用的。

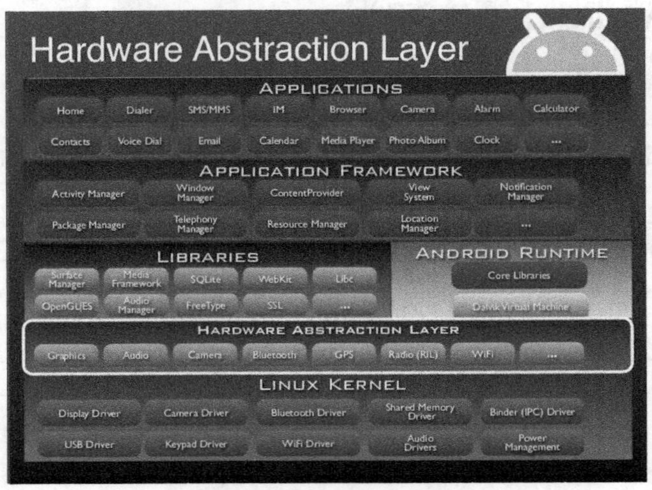

图 1-1　Android 架构

1. Linux 内核

Android 系统是基于 Linux 内核的，由 Linux 提供核心系统服务，如安全、内存管理、进程管理、网络堆栈、驱动模型等。Linux 内核也作为硬件和软件之间的抽象层，它隐藏具体硬件细节而为上层提供统一的服务。

Google 公司为了让 Linux 系统在移动设备上能良好地运行，对其进行了修改和扩充，所以 Linux 系统中的多数核心程序能在 Android 系统中运行。但是，Android 系统去除了 Linux 系统中的本地 X Window System，并且 Android 系统也不支持标准的 GNU 库，这使得 Linux 系统中的应用程序很难移植到 Android 系统。因此，多数 Linux 应用程序在 Android 系统中是无法运行的。

2. 库和 Android 运行时

Android 包含一个 C/C++库的集合，供 Android 系统的各组件使用。这些功能通过 Android 的应用程序框架对开发者开放。以下是一些核心库的简单介绍。

- 系统 C 库是标准 C 系统库（libc）的 BSD 衍生，适用于嵌入式 Linux 系统。
- 媒体库是基于 PacketVideo 的 OpenCore 的，该库支持播放和录制许多流行的音频文件、视频文件及静态图像文件，包括的文件格式有 MPEG-4、H.264、MP3、AAC、AMR、JPG、PNG 等。
- 界面管理库用于管理访问显示子系统以及无缝组合多个应用程序的二维、三维图形层。
- LibWebCore 库是新式的 Web 浏览器引擎，用于驱动 Android 浏览器和内嵌的 Web 视图。
- SGL 库是基本的 2D 图形引擎。
- 3D 库是基于 OpenGL ES 1.0 API 实现的，该库使用硬件 3D 加速或包含高度优化的 3D 软件光栅。
- FreeType 库用于位图和矢量字体渲染。
- SQLite 库是轻量级的关系数据库引擎，其功能强大，所有应用程序都可以使用。

Android 平台的 Java 虚拟机被称为 Dalvik 虚拟机，它由 Google 公司设计，是 Android 平台的重要组成部分。Dalvik 虚拟机和一般的 Java 虚拟机（JVM，即 JDK 中包含的 Java 虚拟机，一部 JVM 只能运行一个 Java Project，文件格式为.class）是不同的，虽然它们都使用 Java 语言，但是，Dalvik 虚拟机为了在移动设备上运行，进行了精简和优化，它资源消耗少并且效率较高。

Android 提供了很多在 Java 编程语言核心类库中可以使用的功能。例如，一台设备可以高效地运行多部虚拟机；一部 Dalvik 虚拟机可运行多个 Android 应用程序；每个 Android 应用程序是 Dalvik 虚拟机中的实例，并且每个应用程序运行在自己的进程中。

Dalvik 虚拟机运行的可执行文件格式为.dex，.dex 格式是专为 Dalvik 设计的一种压缩格式，适合内存和处理器速度有限的系统。

Dalvik 虚拟机依赖 Linux 内核提供的基本功能，如线程和底层内存管理。

有关 Java 虚拟机与 Android 平台的 Dalvik 虚拟机的对比情况见表 1-2。

表 1-2　Java 虚拟机与 Dalvik 虚拟机的对比情况

	Java 虚拟机	Dalvik 虚拟机
语言	Java	Java
名称	JVM	Dalvik
运行格式	.class	.dex
可运行程序数量	一个	多个

3. 应用程序框架

通过提供开放的开发平台，Android 使开发者能够开发丰富、新颖的应用程序。开发者可以自由地利用设备的硬件优势，进行访问位置信息、运行后台服务、设置闹钟、向状态栏添加通知等。

开发者可以完全使用核心应用程序所用的框架 API。应用程序的体系结构旨在简化组件的重复调用，任何应用程序都能发布功能且其他应用程序可以使用这些功能（需要服从框架执行的安全限制）。这一机制允许用户替换组件。

所有的应用程序其实是一组服务和系统，包括以下内容。

- 视图（View）指丰富的、可扩展的视图集合，可用于构建一个应用程序，包括列表、网格、文本框、按钮以及内嵌的网页浏览器。
- 内容提供者（ContentProvider）能够使某应用程序访问其他应用程序（如通信录）的数据，或共享自己的数据。
- 资源管理器（Resource Manager）提供访问非代码资源，如本地化字符串、图形和布局文件。
- 通知管理器（Notification Manager）能够使所有的应用程序在状态栏显示自定义警告。
- 活动管理器（Activity Manager）用于管理应用程序的生命周期，并提供通用的导航回退功能。

4．应用程序

Android 包含了一个核心应用程序集合，包括电子邮件客户端、SMS 程序、日历、地图、浏览器、联系人和其他设置。此外，开发人员可以开发更多应用程序。

1.1.3 Android 应用程序组件

Android 应用程序是用 Java 编程语言编写的。编译后的 Java 代码（包括应用程序要求的数据和资源文件）通过 aapt 工具捆绑为一个 Android 包，即生成.apk 格式的归档文件，该文件是分发应用程序和安装到移动设备的中介工具，用户可以将这个文件下载到设备中。我们可以这样理解：一个.apk 格式的归档文件中的所有代码构成一个应用程序。

一个 Android 应用程序通常运行在一个 Linux 进程中，但需要注意以下几点。

- 默认情况下，每个应用程序运行在自己的 Linux 进程中。当应用程序中的任何代码需要被执行时，Android 系统将启动进程；当不需要运行该应用程序且系统资源被其他应用程序请求时，Android 系统将关闭进程。
- 每个应用程序都有自己的 Dalvik 虚拟机，因此，应用程序的代码是单独运行的，独立于其他应用程序。
- 默认情况下，每个应用程序被分配唯一的 Linux 用户 ID。设置权限后，每个应用程序的文件仅对用户和应用程序本身可见，但在某些特殊情况下，采用某些方法可以将文件暴露给其他应用程序。

实际操作中，有可能将两个应用程序设置为共享一个用户 ID，则它们能够看到对方的文件。所以，为了节省系统资源，具有相同 ID 的应用程序也可以安排在同一个 Linux 进程中，共享同一个 Dalvik 虚拟机。

每个应用程序都会有一个入口，如 C/C++以及 Java 都有一个入口，即 main 方法。但是，Android 应用程序却没有这样的入口。Android 应用程序是通过实例化一些组件来启动一个应用的，这样的组件共有四种：活动（Activity）、服务（Service）、广播接收器（BroadcastReceiver）和内容提供者（ContentProvider）。

每个应用程序必须包含这四种组件中的一个或多个，并且应将所用的组件列在 AndroidManifest.xml 文件中，从而声明应用程序组件以及它们的特性和要求。

下面开始介绍四种组件、应用程序上下文（ApplicationContext）以及意图（Intent）的相关内容。

1．活动（Activity）

一个活动表示一个可视化的用户界面，关注用户所从事的事件。例如，一个文本短信应用程序有若干个活动，第一个活动的功能为显示联系人名单并发送信息；第二个活动的功能为写信息给选定的联系人；第三个活动的功能为查看旧信息或更改设置。虽然，上述活动共同工作形成一个整体的用户界面，但是，每个活动都是独立的，即每个活动都作为 Activity 基类的一个子类。

一个应用程序可能只包含一个活动，或者包含多个活动。活动的数量取决于如何设计应用程序。一般来讲，当应用程序启动时，第一个被标记的活动通常用于向用户展示内容，当完成当前活动后，开始下一个活动。

每个活动都有一个默认的窗口。一般来讲，窗口会填满整个屏幕，但是它可能比屏幕小或悬浮在其他窗口上。活动还可以使用额外的窗口，比如弹出式窗口，即当用户选择特定选项时，屏幕中弹出窗口以便提示用户。

窗口的可视内容是由继承自 View 基类的一个分层的视图（对象）提供的。每个视图控件是窗口内一个特定的矩形空间。父视图包含子视图，并组织子视图的布局；子视图（在分层的底层）绘制的矩形直接控制和响应用户的操作。因此，视图是活动发生的地方，也是产生用户交互操作的地方。例如，一个视图可以显示一张图片，并且当用户单击图片时还能发起一个行为。Android 系统有一些现成的视图可以使用，如按钮（buttons）、文本域（text fields）、滚动条（scroll bars）、菜单项（menu items）、复选框（check boxes）等。

2．服务（Service）

服务没有可视化的用户界面，而是在后台无期限地运行。例如，服务可以是用户在进行其他操作时播放背景音乐；也可以是从网络中获取数据；甚至可以是计算一些数据并将结果提供给需要的活动……每个服务都继承自 Service 基类。

举例说明，一个媒体播放器应用程序可能有一个或多个活动，允许用户选择所需的歌曲并播放，当用户离开播放器应用程序后进行其他操作时，为了让音乐持续播放，播放器活动可以启动一个服务在后台运行，系统将保持音乐播放服务运行。然而，音乐播放这一过程本身不会被当作一个活动处理。

当连接（绑定）到一个持续运行的服务（如果该服务尚未运行，可以启动服务）后，可以通过服务的接口进行交流。例如，对于音乐服务，该接口允许用户暂停、倒带、停止和重新播放等操作。

类似于活动以及其他组件，服务运行在应用程序进程中的主线程中。因此，它们不会阻止其他组件或用户界面，但是，它们往往产生一些耗时的任务（如音乐播放）。

3．广播接收器（BroadcastReceiver）

广播接收器本身不会启动进程，仅仅接收广播信息并且做出相应反应。有很多广播信息是系统发出的，如时区改变、电池电量低、用户改变了语言偏好设置等。此外，应用程序也可以发起广播，例如，A 应用程序中的部分数据已经下载到相关设备，需要让其他应用程序获知当前情况，并允许其他应用程序使用这部分数据，则由 A 应用程序发起广播。

一个应用程序可以包含任意数量的广播接收器，这些广播接收器可以对重要的信息做出反应。所有的接收器继承自 BroadcastReceiver 基类。

广播接收器一般不在用户界面单独显示，它们通常启动活动响应收到的信息，或者使用 Notification Manager 通知用户。通知的方式多种多样，包括闪光灯闪烁、设备振动、播放声音等，其原则是引起用户注意。最常见的通知方式是将信息放在状态栏中，以便用户随时获取。

4．内容提供者（ContentProvider）

内容提供者可以将应用程序的指定数据集提供给其他应用程序，这些数据可以存储在文件系统中、SQLite 数据库中以及其他合理的存储环境中。内容提供者继承自 ContentProvider 基类，并实现了一个标准的方法集，使其他应用程序可以检索和存储数据。然而，应用程序并不直接调用这些方法，而是使用 ContentResolver 对象并调用其方法。ContentResolver 能和所有内容提供者相互通信并合作，来管理进程间的通信。

5．应用程序上下文（ApplicationContext）

前面讨论了活动、服务、广播接收器和内容提供者，这四种组件构成了 Android 应用程序，也可以说它们共同存在于一个应用程序上下文（ApplicationContext，简称 Context）里。

应用程序上下文是当前应用程序所在的进程及运行环境，因为不同的组件可以共享应用程序上下文，所以，借助它能实现在不同的组件中共享数据和资源。

无论应用程序启动哪个组件，系统首先会初始化应用程序上下文。因此，应用程序上下文的生存周期就与应用程序保持一致，与其他组件（如 Activity）无关。应用程序上下文可以通过 Context.getApplicationContext()方法（或 Activity.getApplication()方法）获得引用。

注意：Activity 与 Service 都是 Context 的子类，因此继承了 Context 的所有方法。

6. 意图（Intent）

Intent 是一种应用程序组件之间传递消息的机制。通过 Intent 可以显示活动、启动或停止服务，也可以发起简单的广播。Intent 是异步的，即发送时不需要阻塞等待组件完成响应。

Intent 分为显式（explicit）和隐式（implicit）。显式的 Intent 要求发送者明确指出接收者是谁；而隐式的 Intent 只要求发送者指明接收者属于哪个类别。例如，某 Activity 发送一个“我要打开网页”的 Intent，那么此时，只要能够“打开网页”的应用程序就能收到这个 Intent。

1.2 搭建开发环境

编写 Android 程序需要掌握 Java 编程语言，因此，我们需要了解以下内容。

- JDK。JDK 是 Java 编程语言的软件开发工具包，它包含了 Java 的运行环境、工具集合、基础库等内容，本书采用 JDK8。
- Android SDK。Android SDK 是 Google 公司提供的 Android 开发工具包，在开发 Android 程序时，需要引入该工具包才能使用 Android 相关的 API。
- Android Studio。Android Studio（简称 AS）是 Google 公司推出的 Android 集成开发工具，在开发 Android 程序方面要比 Eclipse 强大且方便得多。本书采用 Android Studio2.2.2 版本。

读者可以从 Android 的官方网站（https://developer.android.google.cn/studio/index.html）下载相应的开发工具。

本书使用的 Android Studio2.2.2 版本其安装程序为 android-studio-bundle_2.2.0.0.exe。启动安装程序，弹出 Android Studio Setup 对话框，如图 1-2 所示。

图 1-2 Android Studio Setup 对话框

单击“Next”按钮，选择安装组件，如图 1-3 所示，选中 Android SDK 复选框和 Android Virtual Device 复选框。本书测试与模拟程序时使用 Android 虚拟仿真器（Android Virtual Device，AVD），读者也可以不安装 AVD，而使用 Android 模拟器（Genymotion）。

Genymotion 是一款专业的 Android 虚拟环境模拟软件，专为 Android 开发测试人员设计，可用于所有 Android 平台，如手机、电视、平板电脑等。Genymotion 是目前运行速度较快的 Android 模拟器，开机和关机速度为 5~20 秒，并且能够集成 ADT 和 Android Studio 开发工具。

Android Studio 安装完毕，弹出如图 1-4 所示的 Welcome to Android Studio 对话框。

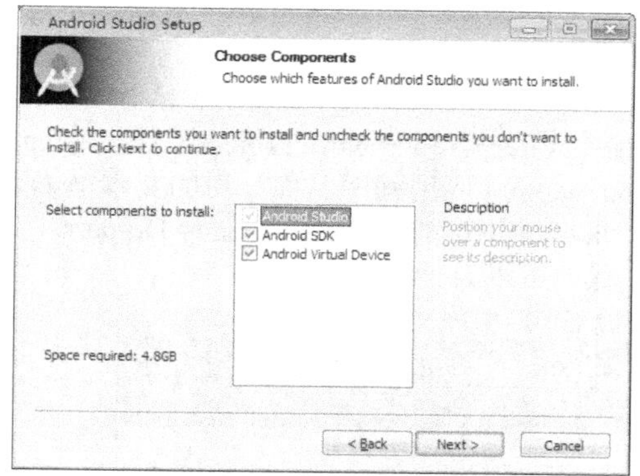

图 1-3　选择安装组件

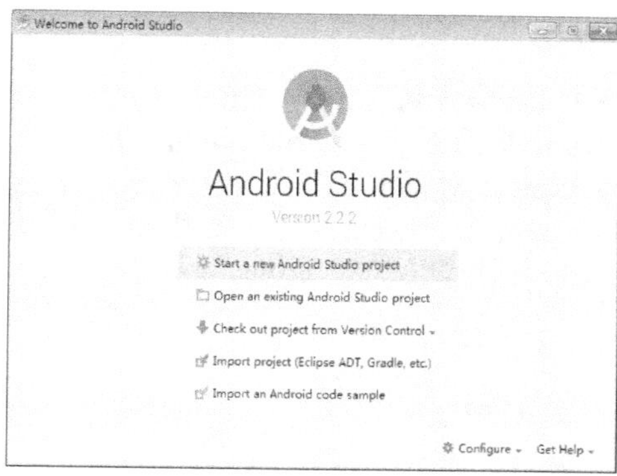

图 1-4　Welcome to Android Studio 对话框

Android Studio 程序图标如图 1-5 所示，Android Studio 启动界面如图 1-6 所示。

图 1-5　Android Studio 程序图标

图 1-6　Android Studio 启动界面

1.3　HelloAndroid 入门项目

1.3.1　需求分析

编写一个 HelloAndroid 入门项目，该项目仅需要一个界面，包括三个控件：显示文字“Hello World!”的文本框、“开始”按钮和“退出”按钮。界面设计效果如图 1-7 所示，项目运行结果如图 1-8 所示。

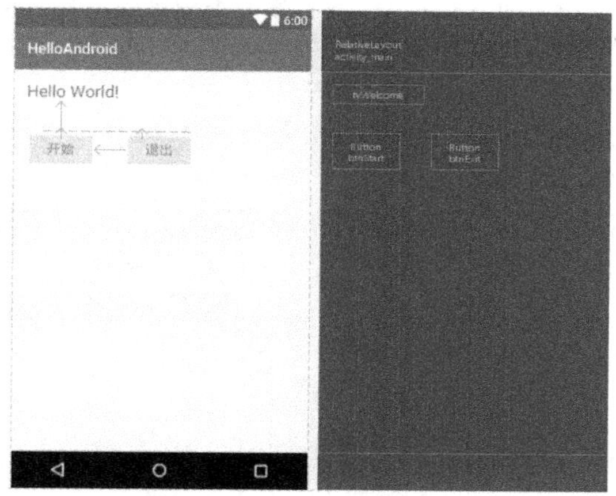

图 1-7　HelloAndroid 项目界面设计效果

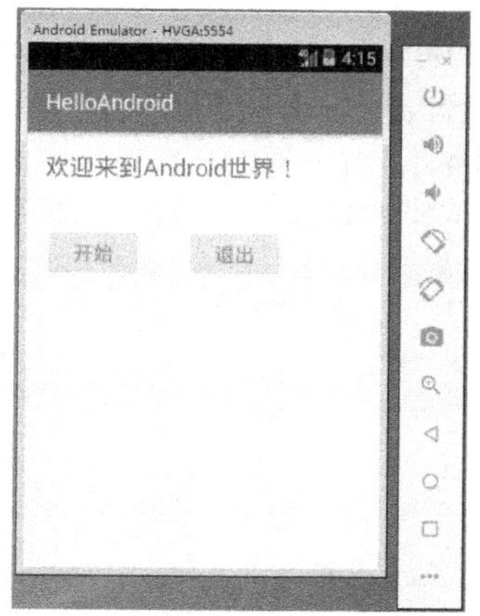

图 1-8　HelloAndroid 项目运行结果

项目运行时，文字被改为“欢迎来到 Android 世界！”。单击“开始”按钮，屏幕出现提示信息“单击了开始”，两秒后自动消失；单击“退出”按钮，程序结束运行。

本项目需要两个按钮设计动作，采用 onClick 属性和用户自定义方法实现。

1.3.2 实施

1. 创建项目

双击桌面上的Android Studio图标，执行菜单命令File→New→New Project，弹出Create New Project对话框，如图1-9所示。在Application name（应用程序名称）后的文本框输入HelloAndroid，但是要注意，应用程序名称必须符合Java标识符的规范；Company Domain（公司域名）保留默认值；Project location（项目保存位置）可以自行设定。

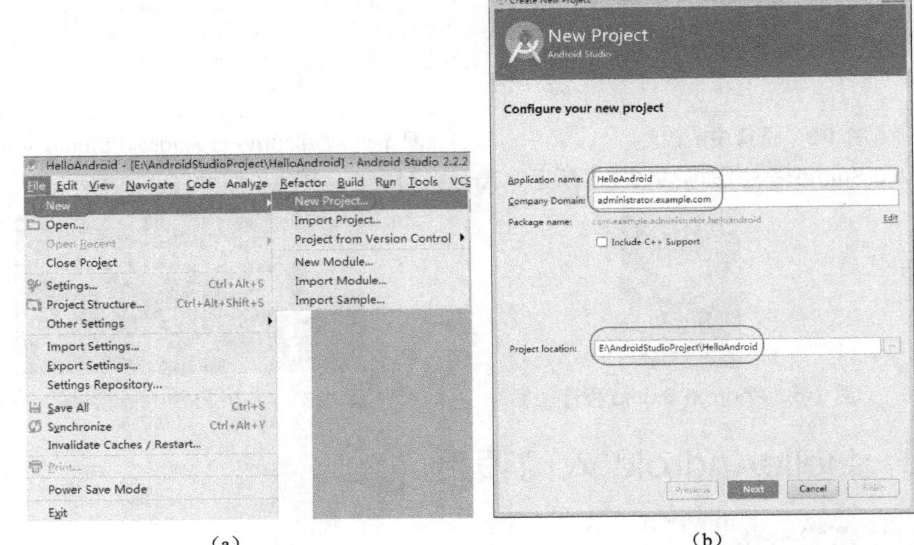

（a）　　　　（b）

图1-9 创建项目

单击“Next”按钮，进入下一个对话框，将Minimum SDK设置为API19；再单击“Next”按钮，在对话框中选择“Empty Activity”模板创建一个空的活动；继续单击“Next”按钮，对话框中的Activity Name（活动名称）和Layout Name（布局名称）不用修改，默认即可。最后单击“Finish”按钮，项目创建完成。

2. 设计界面——布局文件

如图1-10所示为默认的Android项目结构模式，但不是真实的目录结构。参照图1-11，将项目结构模式切换为Project，即真实的目录结构。

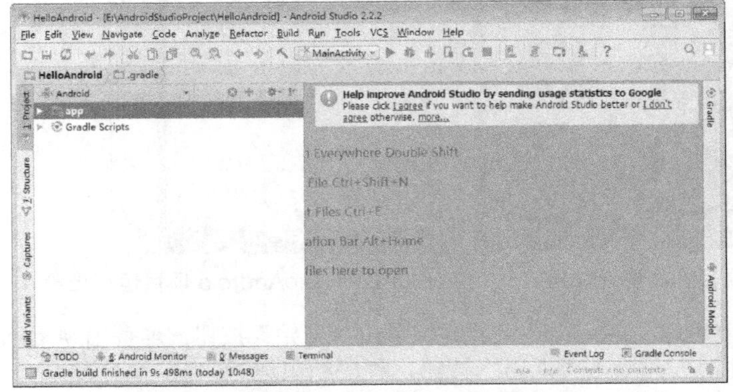

图1-10 默认的Android项目结构模式

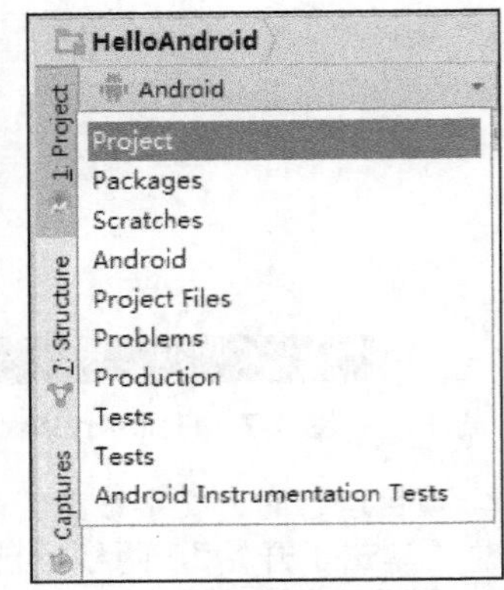

图1-11 切换项目结构模式

在如图 1-12 所示的 Project 项目结构中，选择 app→src→main→res→layout→activity_main.xml 文件，打开设计界面。该界面是一个.xml 格式的文件，可以通过图形化操作直接进行设计，也可以编辑 xml 源代码进行设计。通常，将两种设计方式结合可以明显提高设计效率。

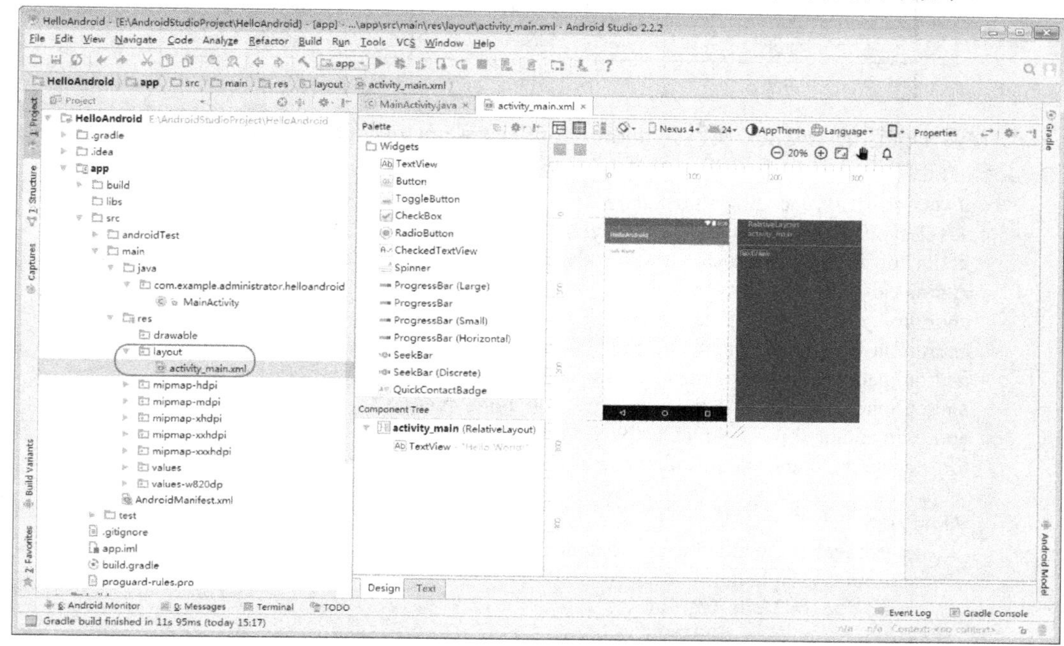

图 1-12　HelloAndroid 项目设计界面

删除 Widgets 目录下的既有 TextView 控件，再通过工具箱添加一个新的 TextView 控件和两个 Button 控件，如图 1-13 所示。TextView 控件、Button 控件的 Text 属性及 ID 属性按照表 1-3 设置。

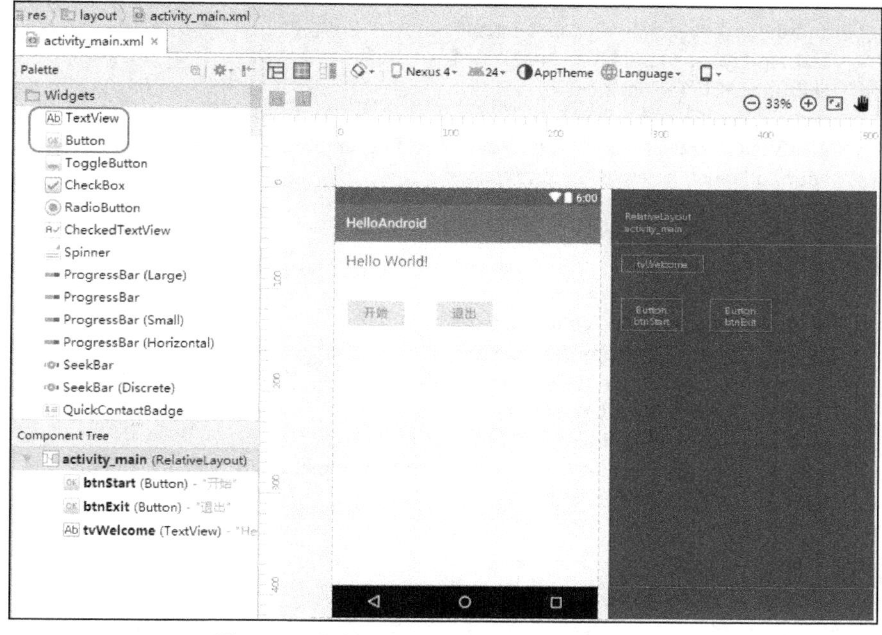

图 1-13　添加 TextView 控件和 Button 控件

表 1-3 TextView 图形控件、Button 控件的 Text 属性及 ID 属性

控 件	Text 属 性	ID 属 性
TextView	Hello World!	tvWelcome
Button	开始	btnStart
Button	退出	btnExit

设置两个按钮的 onClick 属性为 android:onClick="onClick"。

具体的布局设计及参数设置，参考 activity_main.xml 文件，代码如下：

```
<?xml version="1.0" encoding="utf-8"?>
<RelativeLayout xmlns:android="http://schemas.android.com/apk/res/android"
    xmlns:tools="http://schemas.android.com/tools"
    android:id="@+id/activity_main"
    android:layout_width="match_parent"
    android:layout_height="match_parent"
    android:paddingLeft="@dimen/activity_horizontal_margin"
    android:paddingRight="@dimen/activity_horizontal_margin"
    android:paddingTop="@dimen/activity_vertical_margin"
    tools:context="com.example.administrator.helloandroid.MainActivity">

    <TextView
        android:layout_width="wrap_content"
        android:layout_height="wrap_content"
        android:text="Hello World!"
        android:textColor="#FF0000"
        android:id="@+id/tvWelcome"
        android:textAppearance="@style/Base.TextAppearance.AppCompat.Large" />

    <Button
        android:text="开始"
        android:layout_width="wrap_content"
        android:layout_height="wrap_content"
        android:layout_marginTop="40dp"
        android:onClick="onClick"
        android:id="@+id/btnStart"
        android:textAppearance="@style/Base.TextAppearance.AppCompat.Medium"
        android:layout_below="@+id/tvWelcome" />

    <Button
        android:text="退出"
        android:layout_width="wrap_content"
        android:layout_height="wrap_content"
        android:layout_alignTop="@+id/btnStart"
        android:layout_toRightOf="@id/btnStart"
        android:layout_marginLeft="40dp"
        android:onClick="onClick"
        android:textAppearance="@style/Base.TextAppearance.AppCompat.Medium"
        android:id="@+id/btnExit" />

</RelativeLayout>
```

3．设计功能——Activity 类

在项目结构中选择 app→src→main→java→com.example.administrator.helloandroid（下文简

称包）→MainActivity 文件，进入 Java 源代码编辑区，如图 1-14 所示。在编写代码前，我们补充介绍三项内容，请读者注意。

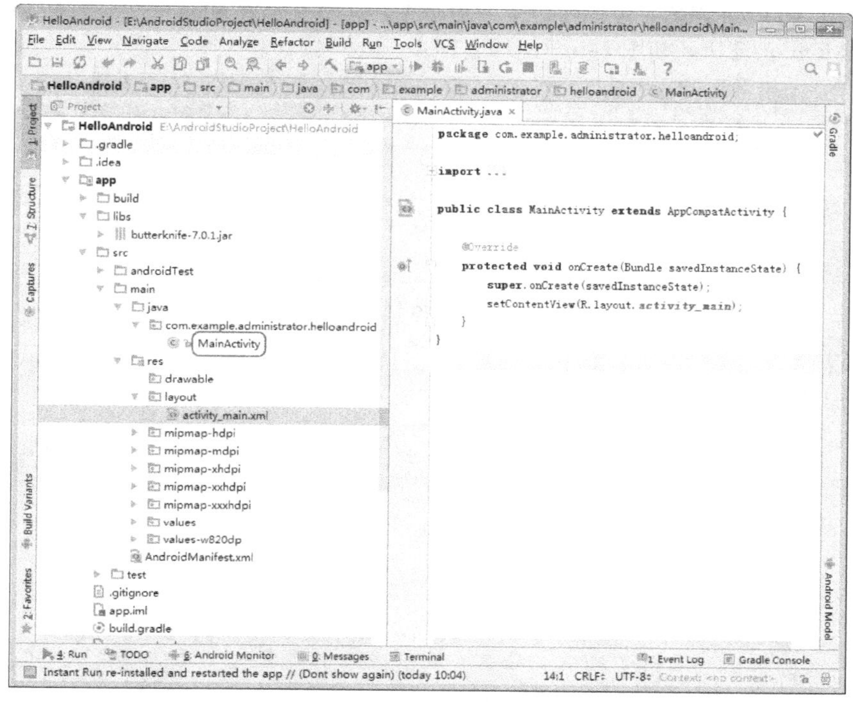

图 1-14　Java 源代码编辑区

（1）编写代码前，进行自动导入包设置，操作步骤如下。

- 执行菜单命令 File→Settings，打开 Settings 对话框，如图 1-15 所示。
- 在左侧的目录中选择 Editor→General→Auto Import，在右侧勾选 Add unambiguous imports on the fly 复选框。

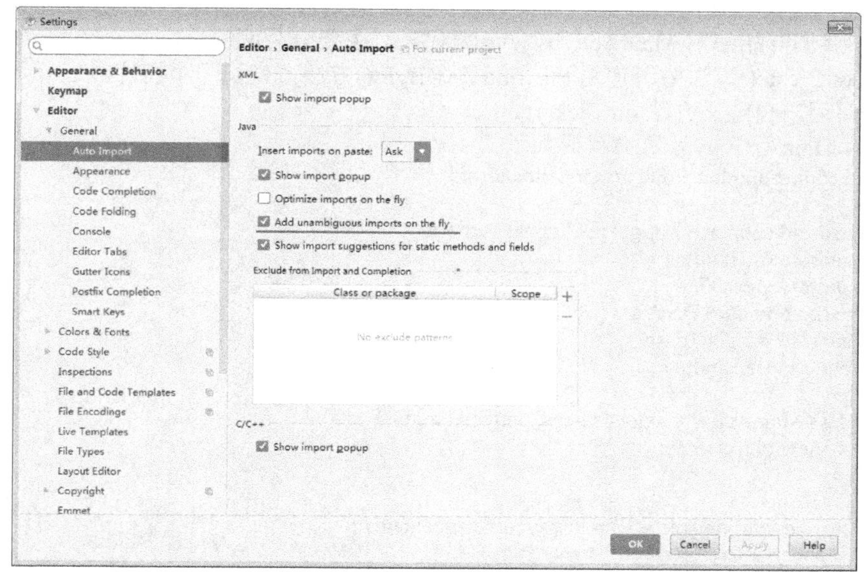

图 1-15　Settings 对话框

（2）代码格式化（Reformat Code）的快捷方式为 Ctrl+Alt+L 组合键。

（3）在 Android Studio 中使用 LogCat 方法。在开发过程中，若出现问题，就需要使用 LogCat 方法快速抓取 Android 崩溃日志。执行菜单命令 View→Tools Windows→Android Monitor，打开 Android Monitor 窗口，如图 1-16 所示。

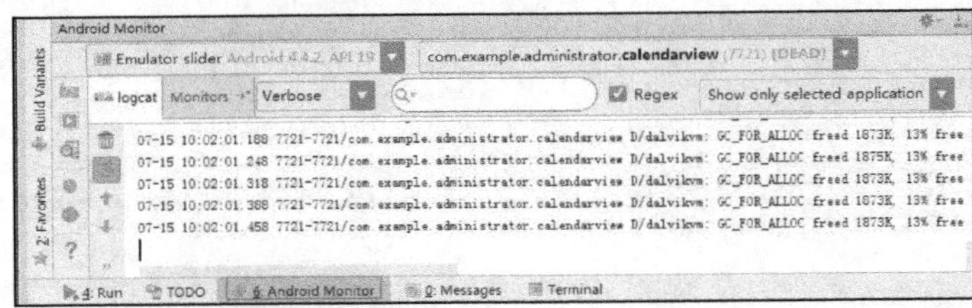

图 1-16　Android Monitor 窗口

或者执行菜单命令 Tools→Android→Android Device Monitor，打开 Android Device Monitor 窗口，如图 1-17 所示。

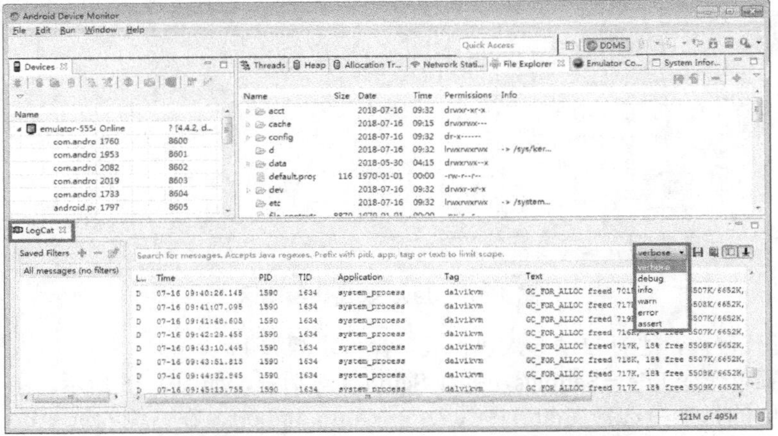

图 1-17　Android Device Monitor 窗口

下面，我们开始编写 MainActivity.java 源代码，主要分为两部分。

- 在 onCreate()方法中，使用 this.findViewById()方法，查找一个控件。
- 编写用户自定义方法 onClick()。

本例的 MainActivity.java 源代码如下：

```
package com.example.administrator.helloandroid;

import android.support.v7.app.AppCompatActivity;
import android.os.Bundle;
import android.view.View;
import android.widget.TextView;
import android.widget.Toast;
import butterknife.Bind;

public class MainActivity extends AppCompatActivity {
    TextView tvWelcome;

    @Override
    protected void onCreate(Bundle savedInstanceState) {
        super.onCreate(savedInstanceState);
        setContentView(R.layout.activity_main);
```

```
        tvWelcome=(TextView)this.findViewById(R.id.tvWelcome);
        tvWelcome.setText("欢迎来到 Android 世界！");
    }

    public void onClick(View v){
        switch (v.getId()){
            case R.id.btnStart:
                Toast.makeText(this,"单击了开始",Toast.LENGTH_LONG).show();
                break;
            case R.id.btnExit:
                System.exit(0);
                break;
        }
    }
}
```

4．项目运行

可以在真实手机或者虚拟手机上运行项目，其中，虚拟手机是指 Android 虚拟仿真器（AVD）。

（1）创建和运行 AVD。单击工具栏上的“AVD Manager”按钮，如图 1-18 所示，弹出 Android Virtual Device Manager 对话框，如图 1-19 所示，开始创建 AVD。

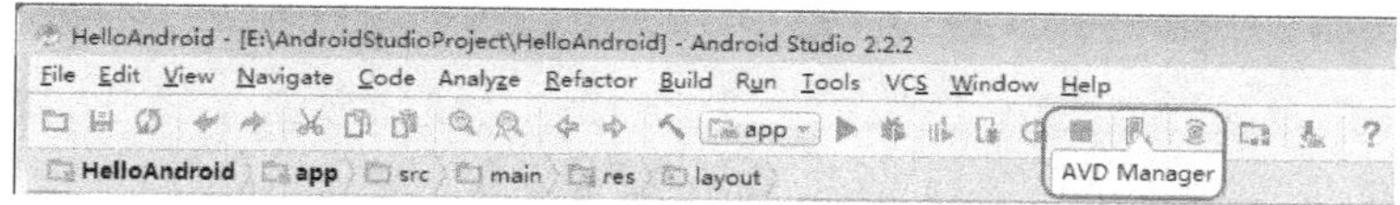

图 1-18　单击工具栏上的“AVD Manager”按钮

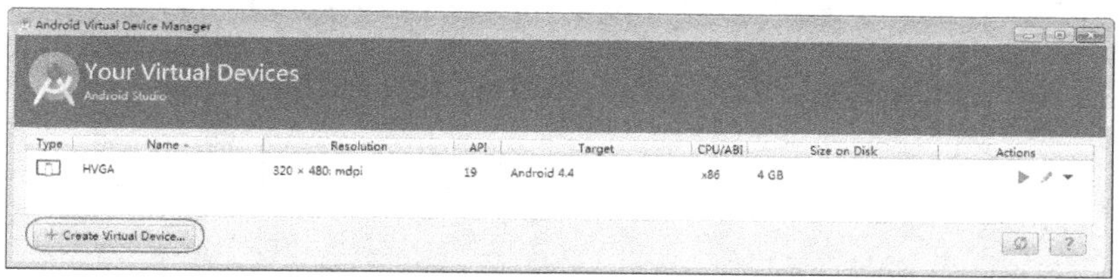

图 1-19　Android Virtual Device Manager 对话框

单击“Create Virtual Device”按钮，弹出如图 1-20 所示的 Virtual Device Configuration 对话框，按照如图 1-20 所示的内容对模拟器进行配置。

图 1-20　Virtual Device Configuration 对话框

单击“Next”按钮，进行模拟器参数设置，如图 1-21 所示。若要对模拟器进行高级设置，可以单击“Show Advanced Settings”按钮，对 RAM、VM heap、Internal Storage 和 SD card 进行设置，如图 1-22 所示。

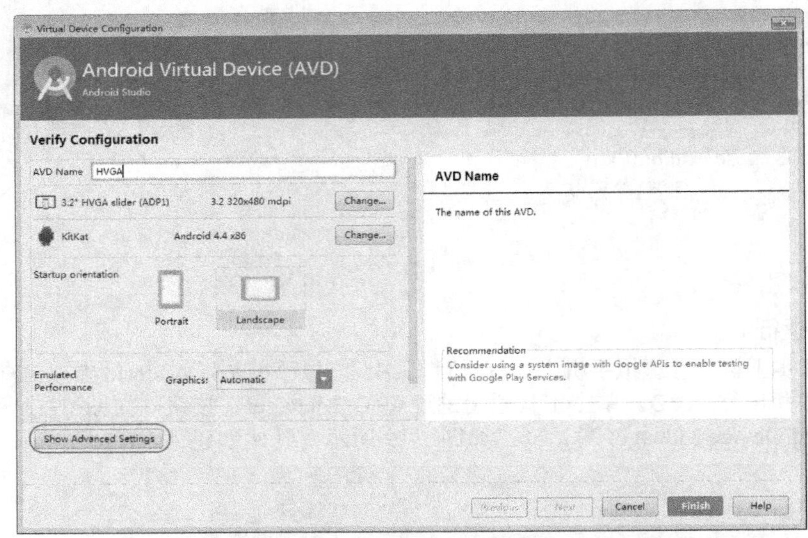

图 1-21 模拟器参数设置

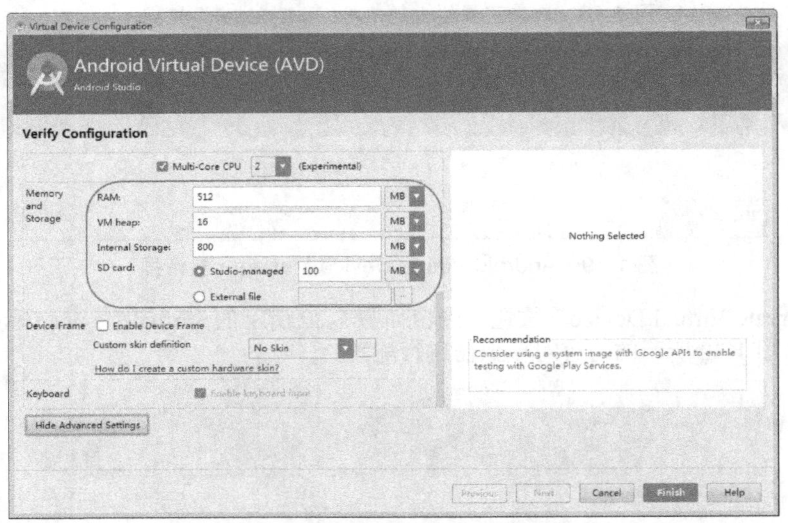

图 1-22 对模拟器进行高级设置

设置完成后，单击“Finish”按钮完成操作。

提示：本书第 5、6 章需要使用 SD 卡，因此要为其开辟空间。但要注意，开辟的空间满足程序测试即可，切不可将空间开辟得太大。

（2）运行项目。单击工具栏上的 ▶（Run）按钮，开始运行项目。单击 ■（Stop）按钮，结束运行项目。如图 1-23 所示。

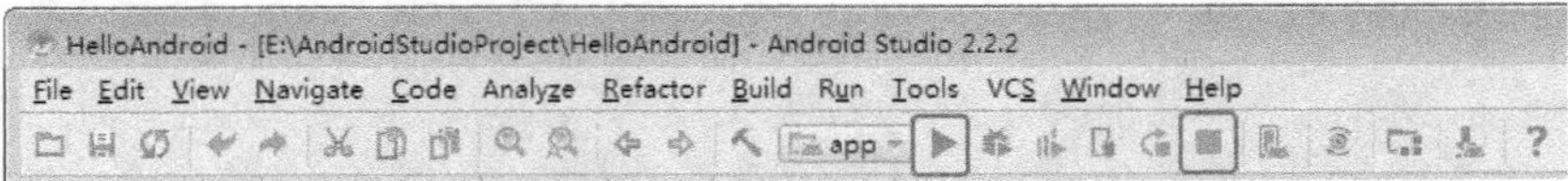

图 1-23 开始运行项目与结束运行项目

1.4　Android 项目结构分析

如图 1-24 所示为 Project 项目结构模式，下面对其中一些要素进行介绍。

1. .gradle 和.idea

.gradle 和.idea 两个文件夹用于存放一些 Android Studio 自动生成的文件，请勿手动编辑。

2. app

app 文件夹的结构如图 1-25 所示，特别介绍三个文件夹：第一，manifests 文件夹用于存放 AndroidManifest.xml 项目的配置文件；第二，java 文件夹中的 com.example.administrator.helloandroid 包用于存放 java 的源文件；第三，res 文件夹包含项目用到的所有图片、布局、字符串等资源，其中，图片存放于 drawable 子文件夹，布局存放于 layout 文件夹，字符串存放于 values 文件夹。

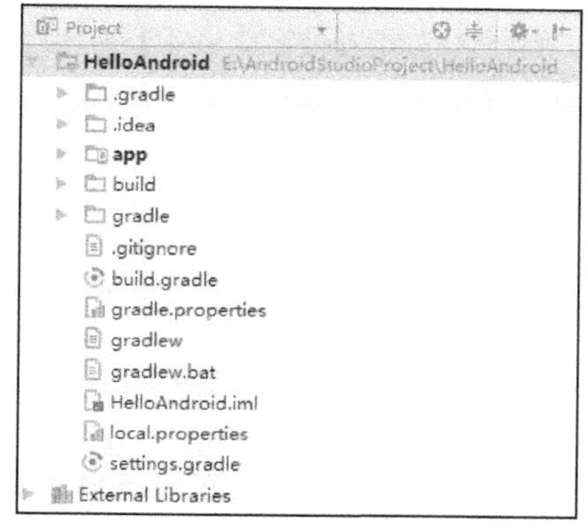

图 1-24　Project 项目结构模式

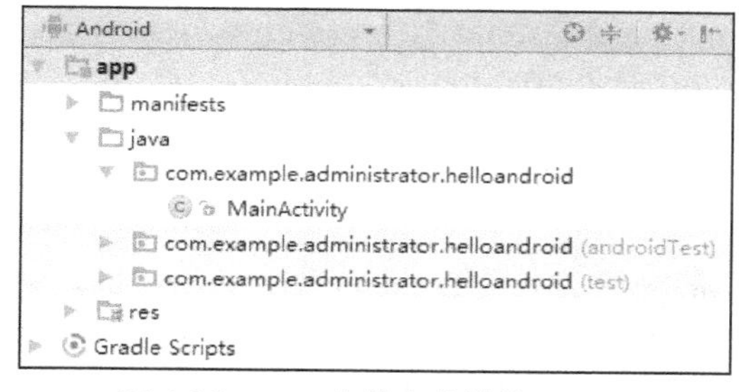

图 1-25　app 文件夹的结构

3. build

build 文件夹主要包含一些在编译时自动生成的文件。

4. gradle（构建）

gradle 文件夹包含 gradle wrapper 的配置文件。

5. 其他文件

.gitignore 文件用于将指定的文件夹或文件排除在版本控制外。

build.gradle 是项目全局的 gradle 构建脚本文件。

gradle.properties 文件是全局的 gradle 配置文件，在该文件中配置的属性将会影响项目中所有的 gradle 编译脚本。

gradlew 和 gradle.bat 两个文件用于在命令行界面中执行 gradle 命令，其中，gradlew 是在 Linux 或 Mac 系统中使用的，gradle.bat 是在 Windows 系统中使用的。

HelloAndroid.iml 文件是所有 IntelliJ IDEA 项目都会自动生成的文件，用来标识当前是一个 IntelliJ IDEA 项目，无须修改该文件中的任何内容。

local.properties 文件用于指定本机中的 Android SDK 路径，通常情况下，其内容都是自动生成的，无须修改。若 SDK 的位置发生变化，将 local.properties 文件中的路径做相应修改，使其指向 SDK 的新位置即可。

settings.gradle 文件用于指定项目中所有引入的模块。由于 HelloAndroid 项目中只有一个 app 模块，因此 settings.gradle 文件只引入了 app 模块。

1.5 Android 学习资料

常用的 Android 学习资料可以通过以下两种途径获得。

第一种，从官方网站下载 Android API 文档，Android API 窗口如图 1-26 所示。

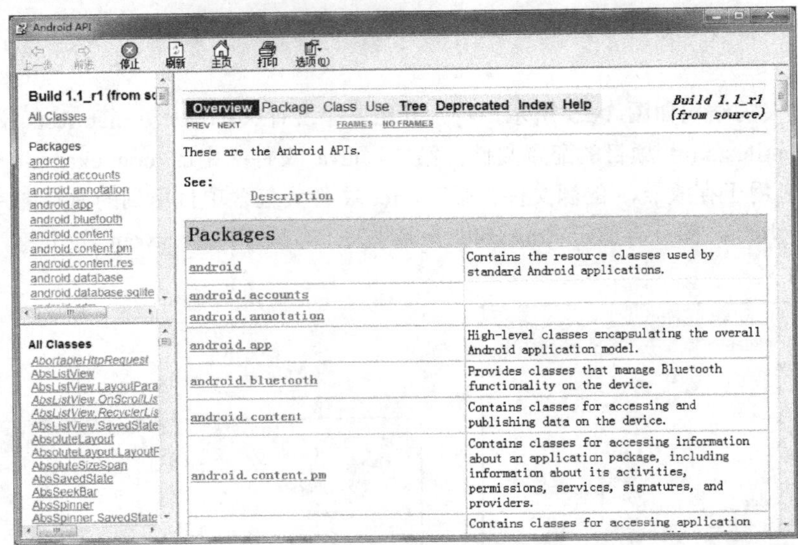

图 1-26 Android API 窗口

第二种，使用 Android Studio 自带的注释文档。执行菜单命令 File→Settings，打开 Settings 对话框，在左侧的目录中选择 Editor→General，在右侧的区域勾选 Show quick documentation on mouse move 复选框，如图 1-27 所示。

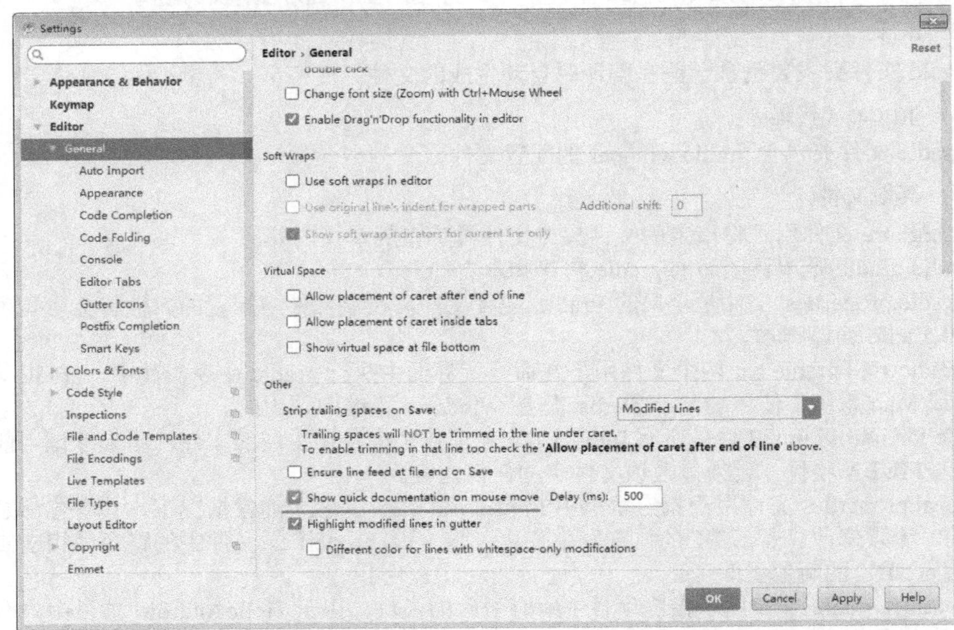

图 1-27 在 Settings 对话框中设置 Android Studio 自带的注释文档

1.6　练习题

1. Android 一词的本义为“机器人”，同时也是（　）于 2007 年 11 月 5 日宣布的基于 Linux 平台的开源手机操作系统的名称。

A．微软公司　　B．英特尔公司　　C．谷歌公司　　D．摩托罗拉公司

2．（　）系统不能作为手机操作系统。

A．Android　　B．Windows Mobile　　C．Windows Vista　　D．iOS

3．Android 是以（　）系统为基础的操作系统。

A．Java　　B．Unix　　C．Linux　　D．Windows

4．Android 平台的 Java 虚拟机是（　）。

A．Dalvik　　B．JVM　　C．KVM　　D．Framework

5．Android 平台的 Java 虚拟机中运行的文件其后缀名是（　）。

A．.apk　　B．.class　　C．.dex　　D．.framework

6．Android 系统安装的软件是（　）格式。

A．.jar　　B．.java　　C．.apk　　D．.sisx

7．（　）不属于 Android 智能手机的缺点。

A．高耗电　　B．易死机

C．容易感染手机病毒　　D．有丰富的应用程序可供选择

8．“Ice Cream Sandwich”是（　）版本的代号。

A．Android 2.3　　B．Android 3.0　　C．Android 4.0.1　　D．Android 5.1

9．通过安装更多的应用程序来丰富手机的功能，这是指智能手机的（　）特点。

A．开放性　　B．扩展性　　C．可再生性　　D．多任务处理

10．在 AndroidManifest.xml 文件中，<activity>元素的（　）属性用于指定活动的类名。

A．android:class_name　　B．android:class.name

C．android:name　　D．activity:name

1.7　作业

1．常见的移动平台有哪些？它们的开发语言各是什么？

2．什么是 Android API 级别（level）？它有什么作用？

3．如何评价 Android 系统的优点与不足？

4．Android 架构分为四层，它们分别是什么？每层的主要作用是什么？

5．Google 公司开发的 Android 平台的 Java 虚拟机（即 Dalvik 虚拟机）与 Sun 公司开发的 Java 虚拟机有什么区别？

6．Android 应用程序的四大组件是什么？请分别简述。

7．AndroidManifest.xml 文件主要包括哪些信息？

第2章 图形界面——计算器项目

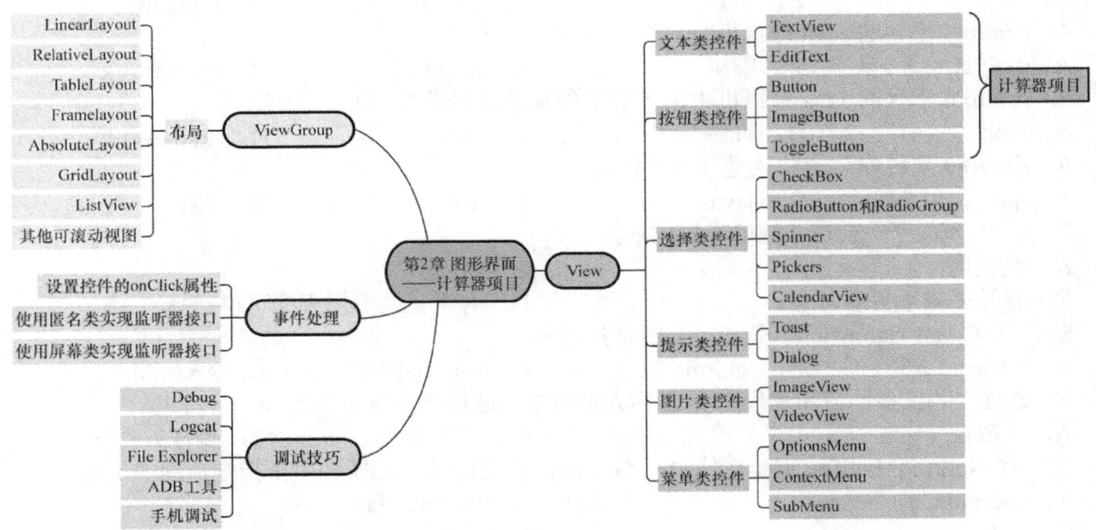

本章知识要点思维导图

在第 1 章中，我们学习了 Android 架构、Android 应用程序组件。本章将通过引入计算器项目，介绍基本的图形用户界面（GUI）的设计方法及添加功能性元素的方法。

2.1 需求分析

开发一个简单的计算器项目，该程序只能进行加、减、乘、除运算。要求界面美观、使用方便。为降低编程难度，本计算器不支持连续计算和混合运算，例如，只支持 2+3=5，不支持 2+3+4=9。

2.2 界面设计

计算器项目的界面如图 2-1 所示，具体内容包括 1 个文本显示框，用于显示用户的按键输入值及计算结果；18 个按钮，即 0～9 数字键，加（+）、减（−）、乘（*）、除（/）、小数点（.）、等于（=），以及清除按钮（CLEAR）和退格按钮（BACKSPACE）。

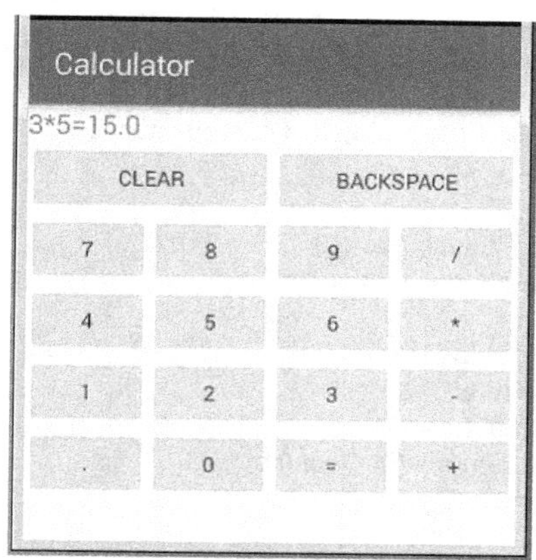

图 2-1　计算器项目的界面

2.3　实施

2.3.1　创建项目

创建一个名为 Calculator 的项目，为简单起见，在开发过程中只使用其默认的布局文件 activity_main.xml 和 MainActivity 类。

2.3.2　界面实现

计算器项目的界面设计思想：外层采用垂直线性布局，内层嵌套水平线性布局。本项目中 activity_main.xml 的图形控件及其 Text、ID 属性见表 2-1。

表 2-1　图形控件及其 Text、ID 属性

控件	Text 属性	ID 属性	控件	Text 属性	ID 属性	控件	Text 属性	ID 属性	控件	Text 属性	ID 属性
TextView		tvResult									
Button	CLEAR	btnClear				Button	BACKSPASE	btnBackSpace			
Button	7	btn7	Button	8	btn8	Button	9	btn9	Button	/	btnDivide
Button	4	btn4	Button	5	btn5	Button	6	btn6	Button	*	btnMultiply
Button	1	btn1	Button	2	btn2	Button	3	btn3	Button	-	btnMinus
Button	.	btnDot	Button	0	btn0	Button	=	btnEqual	Button	+	btnPlus

在本项目中，为所有按钮指定相同的 onClick 属性，其事件处理的方法名全部为 onClick。上述操作完成后的 activity_main.xml 代码如下：

```
<?xml version="1.0" encoding="utf-8"?>
<LinearLayout xmlns:android="http://schemas.android.com/apk/res/android"
    xmlns:app="http://schemas.android.com/apk/res-auto"
    xmlns:tools="http://schemas.android.com/tools"
    android:layout_width="match_parent"
    android:layout_height="match_parent"
    android:orientation="vertical"
    tools:context=".MainActivity">

    <TextView
        android:id="@+id/tvResult"
```

```
        android:layout_width="wrap_content"
        android:layout_height="wrap_content"
        android:text="Medium Text"
        android:textAppearance="?android:attr/textAppearanceMedium" />

    <LinearLayout
        android:layout_width="match_parent"
        android:layout_height="wrap_content">
        <Button
            android:id="@+id/btnClear"
            android:layout_width="0dp"
            android:layout_weight="1"
            android:layout_height="wrap_content"
            android:onClick=" onClick"
            android:text="CLEAR" />
        <Button
            android:id="@+id/btnBackSpace"
            android:layout_width="0dp"
            android:layout_weight="1"
            android:layout_height="wrap_content"
            android:onClick=" onClick"
            android:text="BACKSPACE" />
    </LinearLayout>

    <!-- 第 1 行 -->

    <LinearLayout
        android:layout_width="match_parent"
        android:layout_height="wrap_content" >

        <Button
            android:id="@+id/btn7"
            android:layout_width="wrap_content"
            android:layout_weight="1"
            android:layout_height="wrap_content"
            android:onClick=" onClick"
            android:text="7" />

        <Button
            android:layout_weight="1"
            android:id="@+id/btn8"
            android:layout_width="wrap_content"
            android:layout_height="wrap_content"
            android:onClick=" onClick"
            android:text="8" />

        <Button
            android:layout_weight="1"
            android:id="@+id/btn9"
            android:layout_width="wrap_content"
            android:layout_height="wrap_content"
            android:onClick=" onClick"
            android:text="9" />

        <Button
            android:layout_weight="1"
            android:id="@+id/btnDevide"
            android:layout_width="wrap_content"
```

```
            android:layout_height="wrap_content"
            android:onClick=" onClick"
            android:text="/" />
    </LinearLayout>

    <!-- 第 2 行 -->

    <LinearLayout
        android:layout_width="match_parent"
        android:layout_height="wrap_content" >

        <Button
            android:layout_weight="1"
            android:id="@+id/btn4"
            android:layout_width="wrap_content"
            android:layout_height="wrap_content"
            android:onClick=" onClick"
            android:text="4" />

        <Button
            android:layout_weight="1"
            android:id="@+id/btn5"
            android:layout_width="wrap_content"
            android:layout_height="wrap_content"
            android:onClick=" onClick"
            android:text="5" />

        <Button
            android:id="@+id/btn6"
            android:layout_weight="1"
            android:layout_width="wrap_content"
            android:layout_height="wrap_content"
            android:onClick=" onClick"
            android:text="6" />

        <Button
            android:layout_weight="1"
            android:id="@+id/btnMultiply"
            android:layout_width="wrap_content"
            android:layout_height="wrap_content"
            android:onClick=" onClick"
            android:text="*" />
    </LinearLayout>

    <!-- 第 3 行 -->

    <LinearLayout
        android:layout_width="match_parent"
        android:layout_height="wrap_content" >

        <Button
            android:layout_weight="1"
            android:id="@+id/btn1"
            android:layout_width="wrap_content"
            android:layout_height="wrap_content"
            android:onClick=" onClick"
            android:text="1" />
```

```
    <Button
        android:layout_weight="1"
        android:id="@+id/btn2"
        android:layout_width="wrap_content"
        android:layout_height="wrap_content"
        android:onClick=" onClick"
        android:text="2" />

    <Button
        android:id="@+id/btn3"
        android:layout_width="wrap_content"
        android:layout_height="wrap_content"
        android:layout_weight="1"
        android:onClick=" onClick"
        android:text="3" />

    <Button
        android:layout_weight="1"
        android:id="@+id/btnMinus"
        android:layout_width="wrap_content"
        android:layout_height="wrap_content"
        android:onClick=" onClick"
        android:text="-" />
</LinearLayout>

<!-- 第4行 -->

<LinearLayout
    android:layout_width="match_parent"
    android:layout_height="wrap_content" >

    <Button
        android:layout_weight="1"
        android:id="@+id/btnDot"
        android:layout_width="wrap_content"
        android:layout_height="wrap_content"
        android:onClick=" onClick"
        android:text="." />

    <Button
        android:layout_weight="1"
        android:id="@+id/btn0"
        android:layout_width="wrap_content"
        android:layout_height="wrap_content"
        android:onClick=" onClick"
        android:text="0" />

    <Button
        android:layout_weight="1"
        android:id="@+id/btnEqual"
        android:layout_width="wrap_content"
        android:layout_height="wrap_content"
        android:onClick=" onClick"
        android:text="=" />

    <Button
        android:layout_weight="1"
        android:id="@+id/btnPlus"
```

```
            android:layout_width="wrap_content"
            android:layout_height="wrap_content"
            android:onClick=" onClick"
            android:text="+" />
    </LinearLayout>

</LinearLayout>
```

2.3.3 Java 代码

Activity 类用于实现项目的功能，包括对按钮的响应及计算数值。

计算器项目的 MainActivity.java 代码如下：

```
package com.example.administrator. calculator;

import android.support.v7.app.AppCompatActivity;
import android.os.Bundle;
import android.view.View;
import android.widget.Button;
import android.widget.TextView;

import java.util.regex.Matcher;
import java.util.regex.Pattern;

public class MainActivity extends AppCompatActivity {

    TextView tvResult;
    @Override
    protected void onCreate(Bundle savedInstanceState) {
        super.onCreate(savedInstanceState);
        setContentView(R.layout.activity_main);

        tvResult = (TextView) this.findViewById(R.id.tvResult);
        tvResult.setText("");
    }
    public void onClick(View v) {
        Button b = (Button) v;
        String btnText = b.getText().toString();
        String tvText = tvResult.getText().toString();
        switch (v.getId()) {
            case R.id.btnClear:
                tvResult.setText("");
                break;
            case R.id.btn0:
            case R.id.btn1:
            case R.id.btn2:
            case R.id.btn3:
            case R.id.btn4:
            case R.id.btn5:
            case R.id.btn6:
            case R.id.btn7:
            case R.id.btn8:
            case R.id.btn9:
            case R.id.btnDot:
            case R.id.btnPlus:
            case R.id.btnMinus:
            case R.id.btnMultiply:
            case R.id.btnDevide:
                tvResult.setText(tvText + btnText);
```

```
                break;
            case R.id.btnEqual:
                // 计算结果
                Pattern p = Pattern.compile("(\\d+)([\\+\\-\\*\\/])(\\d+)");
                Matcher m = p.matcher(tvText);
                if (m.find()) {
                    double d1 = Double.parseDouble(m.group(1));
                    double d2 = Double.parseDouble(m.group(3));
                    double d3 = 0;
                    if ("+".equals(m.group(2))) {
                        d3 = d1 + d2;
                    }
                    if ("-".equals(m.group(2))) {
                        d3 = d1 - d2;
                    }
                    if ("*".equals(m.group(2))) {
                        d3 = d1 * d2;
                    }
                    if ("/".equals(m.group(2))) {
                        d3 = d1 / d2;
                    }
                    tvResult.setText(tvText + btnText + d3);
                }
                break;
            case R.id.btnBackSpace:
                if(tvResult.getText().toString().length()!=0) {
                    tvResult.setText(tvResult.getText().toString().substring(0,
                    tvResult.getText().toString().length() - 1));
                }
                break;
        }
    }
}
```

2.3.4 运行测试

将项目在 AVD 上运行，测试其是否符合需求分析中的要求。

2.4 界面设计基础

图形界面的设计方法有两种：第一，直接使用 Java 语言创建和操作图形控件，图形界面的定义和功能的实现都通过 Java 语言完成；第二，通过 xml 定义图形界面，图形界面的定义保存在单独的 xml 文件中，功能的实现则使用 Java 语言完成。两者相比，前者对程序员的要求较高，后者比较适合初学者，因此本书主要讨论第二种方法。

2.4.1 View 和 ViewGroup

Android 应用程序的图形界面由以下两种对象构成，如图 2-2 所示。

- View 是一种可以在屏幕上绘制、显示，并与用户交互的对象，如按钮、文本框、单选钮等。
- ViewGroup 是一种用于定义布局接口并可以容纳其他 View 和 ViewGroup 的对象，ViewGroup 相当于其他 View 和 ViewGroup 的容器，如 LinearLayout。

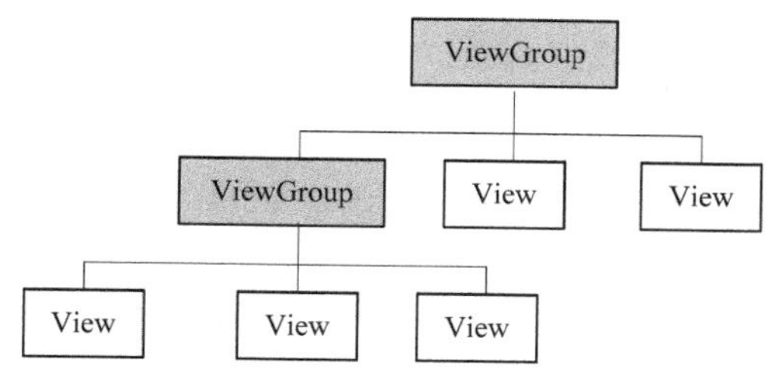

图 2-2　由 View 和 ViewGroup 构成的树状结构

在计算器项目的图形界面中，使用了 1 个总的垂直线性布局，在该布局中嵌套了 1 个 TextView 控件和 5 个水平线性布局，布局及控件关系如图 2-3 所示。

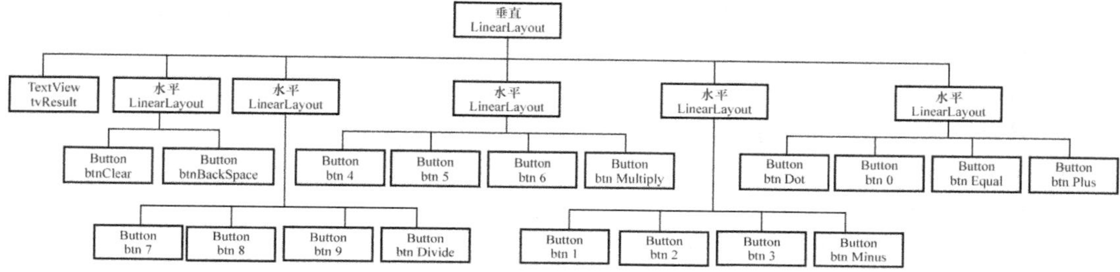

图 2-3　计算器项目中的布局及控件关系

通过对照计算器的界面（见图 2-1）、计算器项目中的布局及控件关系（见图 2-3）以及图形界面的 xml 文件，读者可以加深对界面设计中控件关系的理解。

在计算机应用软件的图形界面设计中，通常采用绝对坐标来放置图形界面元素（控件）；而在移动应用软件的图形界面设计中，由于手机的屏幕尺寸和分辨率的多样性，无法使用绝对坐标来安排界面元素（控件）的位置，而是使用各种布局，以相对坐标的方式来安排元素（控件）的位置。

2.4.2　基本概念

我们先来介绍一些基本概念，以便进行移动应用软件的图形界面设计。

1．像素单位

Android 系统支持多种像素单位，分别如下。

- px（像素）：屏幕上的点。
- in（英寸）：长度单位。
- mm（毫米）：长度单位。
- pt（磅）：1/72 英寸。
- dp（与密度无关的像素）：一种基于屏幕密度的抽象单位。在每英寸 160 点的显示器上，1dp = 1px。
- dip：与 dp 相同，多用于 Google 公司的产品示例中。
- sp（与刻度无关的像素）：与 dp 类似，但可以根据用户的字体大小首选项进行缩放。

注意：开发 Android 程序时，建议读者使用 dp 和 sp 这两种像素单位，前者用于一般场合，后者专用于字体大小。

2．分辨率

不同的手机屏幕有不同的分辨率，常见的分辨率有以下几种。

- VGA（Video Graphics Array）：早期的计算机屏幕分辨率规格，分辨率为 640×480 像

素，长宽比为 4:3。

- HVGA（Half-size VGA）：HVGA 的分辨率是 VGA 的一半，即 480×320 像素，长宽比为 4:3。
- QVGA（Quarter VGA）：QVGA 的分辨率是 VGA 的 1/4，即 320×240 像素，长宽比为 4:3。
- WVGA（Wide VGA）：从字面理解，WVGA 是扩大的 VGA，其分辨率为 800×480 像素，长宽比为 16:10。
- WQVGA（Wide Quarter VGA）：从字面理解，WQVGA 是扩大的 QVGA，其分辨率比 QVGA 高、比 VGA 低，一般为 400×240 像素或 480×272 像素，长宽比为 16:9。

3．像素密度

像素密度（density）表示每英寸有多少个显示点，与分辨率是不同的概念。

- WQVGA 规格中，density=120。
- QVGA 规格中，density=120。
- HVGA 规格中，density=160。
- WVGA 规格中，density=240。

像素密度还与标签资源有关，关系如下。

- 当屏幕 density=240 时，使用 hdpi 标签的资源。
- 当屏幕 density=160 时，使用 mdpi 标签的资源。
- 当屏幕 density=120 时，使用 ldpi 标签的资源。

4．颜色的表示形式

颜色的表示形式有两种，如“#rrggbb”和“#aarrggbb”，前一种形式表示颜色的 RGB 分量，例如，“#FF0000”表示红色，“#0000FF”表示蓝色，“#888888”表示灰色；后一种形式还增加了颜色的 Alpha 值（透明度），例如，“#000000”表示黑色，而“#FF000000”表示不透明的黑色。

2.4.3 共有属性

有些属性是多数 View 和 ViewGroup 共有的，用户可以在 xml 文件中设置，也可以通过 Java 代码进行赋值。

1．ID

通常情况下，每个 View 和 ViewGroup 都关联了一个整数类型的 ID 值，用于唯一地标识这个 View 或 ViewGroup。在应用程序编译后，该 ID 值是一个整数值，但是在布局文件中，该 ID 值却表现为 View 或 ViewGroup 中 android:id 属性的一个字符串，比如：

```
android:id="@+id/my_button"
```

其中，@符号表示需要将其后的字符作为一个 ID 资源进行处理，后接的+号表示这是一个新的 ID，需要将该 ID 加入资源类（R.java）中。Android 框架也提供了一些 ID 资源，当引用这类资源时，则要用 Android 命名空间来代替+号，比如：

```
android:id="@android:id/empty"
```

其中，@符号之后的 android 表示引用的 ID 资源是在 android.R.java 资源类中的，而不是项目本地的 R.java 资源类。

当需要从 Activity 类引用布局文件中的 View 和 ViewGroup 时，标准做法如下。

（1）先在布局文件中定义 View 和 ViewGroup，并指定其 ID，代码如下：

```
<Button android:id="@+id/my_button"
    android:layout_width="wrap_content"
```

```
        android:layout_height="wrap_content"
        android:text="@string/my_button_text"/>
```

这时系统自动在资源文件 R.java 中加入 ID 资源，代码如下：

```
public final class R {
        // ......
        public static final class id {
                public static final int my_button =0x7f090002;
                // .......
```

（2）然后在 Activity 类中，通过以下代码引用该 ID 所代表的 View 和 ViewGroup：

```
Button myButton = (Button) findViewById(R.id.my_button);
```

这段代码通常在 Activity 类的 onCreate()方法中。

2．宽度和高度

为控件设置宽度和高度是一件很烦琐的事情，但正是这些细节决定了界面的美观程度。与宽度相关的属性有以下几项。

- android:width：指定宽度，单位建议使用 dp。
- android:minWidth：最小宽度，单位建议使用 dp。
- android:maxWidth：最大宽度，单位建议使用 dp。
- android:layout_width：相对于父视图的宽度（值为 match_parent 或 wrap_content）。

在多数场合，采用 android:layout_width 方式设置宽度比较方便，并且通常其值取 match_parent。android:layout_width 的三个可用值的含义分别如下。

- fill_parent：与父控件等宽，由于用词不贴切，在 Android 2.2 版本后改为 match_parent。
- match_parent：与父控件等宽（与 fill_parent 同义），建议使用该名称。
- wrap_content：根据内容改变大小。

与高度相关的属性和上述与宽度相关的属性类似。其中，android:layout_height 的取值一般为 wrap_content。

3．权重

权重与宽度和高度有关，是指多个控件占用空间的相对比重。例如，在水平布局中，有三个水平排列的控件 A、B 和 C，其中 A 未指定权重，B 和 C 的权重分别是 1 和 2。那么，在保证 A 占用其指定空间的前提下（A 未指定权重，是弱势群体，得到优先保证），B 占用剩余空间中的 1/3，C 占用剩余空间中的 2/3，显示结果如图 2-4 所示，布局文件的代码如下：

```
<?xml version="1.0" encoding="utf-8"?>
<LinearLayout xmlns:android="http://schemas.android.com/apk/res/android"
    android:layout_width="match_parent"
    android:layout_height="match_parent"
    android:orientation="vertical" >

    <LinearLayout
        android:layout_width="match_parent"
        android:layout_height="wrap_content" >

        <Button
            android:id="@+id/button1"
            android:layout_width="wrap_content"
            android:layout_height="wrap_content"
            android:text="A" />

        <Button
            android:id="@+id/button2"
            android:layout_height="wrap_content"
```

```
            android:layout_weight="1"
            android:text="B" />

        <Button
            android:id="@+id/button3"
            android:layout_height="wrap_content"
            android:layout_weight="2"
            android:text="C" />

    </LinearLayout>

</LinearLayout>
```

注意：这里 B 和 C 不需要设置宽度，因为此时的宽度是在运行时根据剩余的空间自动安排的。但是，A 仍需要设置宽度，并且 A 的宽度能够得到优先保证。

4．对齐方式

Html 的对齐方式标记为 align；而 Android 的对齐方式在 gravity 属性和 layout_gravity 属性中设置，其取值范围见表 2-2。

表 2-2 gravity 属性和 layout_gravity 属性取值范围

值	说　明	值	说　明
top	上对齐	fill_vertical	垂直撑满
bottom	下对齐	center_horizontal	水平居中
left	左对齐	fill_horizontal	水平撑满
right	右对齐	center	水平垂直居中
center_vertical	垂直居中	fill	水平垂直撑满

根据需要，对不同的实体设置对齐，可以分为两种对齐方式。

- android:gravity：用于设置控件上的文字在控件内的位置。例如，将 Button1 的 android:gravity 属性设置为“left | bottom”，即 Button1 上的文字其对齐方式为左下。
- android:layout_gravity：用于设置控件在布局中的位置。例如，将 Button2 的 android:layout_gravity 属性设置为“center”，即 Button2 的对齐方式为水平居中。

显示结果如图 2-5 所示，布局文件的代码如下：

```
<?xml version="1.0" encoding="utf-8"?>
<LinearLayout xmlns:android="http://schemas.android.com/apk/res/android"
    android:layout_width="match_parent"
    android:layout_height="match_parent"
    android:orientation="vertical" >

    <Button
        android:id="@+id/button1"
        android:layout_width="150dp"
        android:layout_height="60dp"
        android:layout_gravity="center"
        android:gravity="left|bottom"
        android:text="Button1" />

    <Button
        android:id="@+id/button2"
        android:layout_width="150dp"
        android:layout_height="80dp"
        android:layout_gravity="center"
        android:gravity="right|top"
        android:text="Button2" />
</LinearLayout>
```

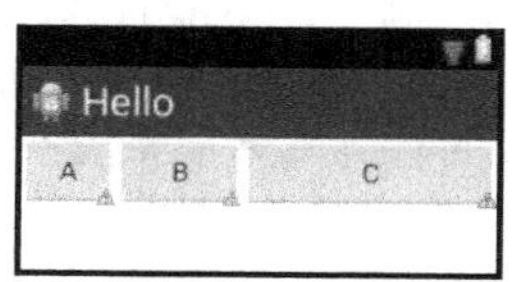

图 2-4　权重

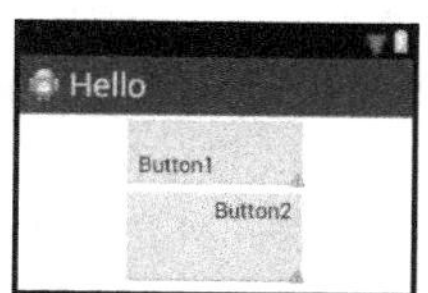

图 2-5　对齐方式

5．颜色

根据视图的不同，颜色有前景颜色、背景颜色（android:background）、阴影颜色（android:shadowColor）、正常文本色（android:textColor）、高亮文本色（android: textColorHighlight）、提示文本色（android:textColorHint）、链接文本色（android: textColorLink）等，取值形式为"#rrggbb"或"#aarrggbb"。

6．文本

文本包括用于正常显示的文本（android:text），也包括用于提示的文本（android:hint）。此外，还可以设置文本的大小（android:textSize，以 sp 为单位）、字体（android:typeface，如 normal、sans、serif、monospace）和样式（android:textStyle，如 bold、italic、bolditalic）。

7．边距

边距有两种概念，即 Margin 和 Padding，每种都可以单独设置，并且形式有多种，其含义如图 2-6 所示。

- android:padding：设置四周的边距为同一值。
- android:paddingBottom：单独设置底边距。
- android:paddingLeft：单独设置左边距。
- android:paddingRight：单独设置右边距。
- android:paddingTop：单独设置上边距。

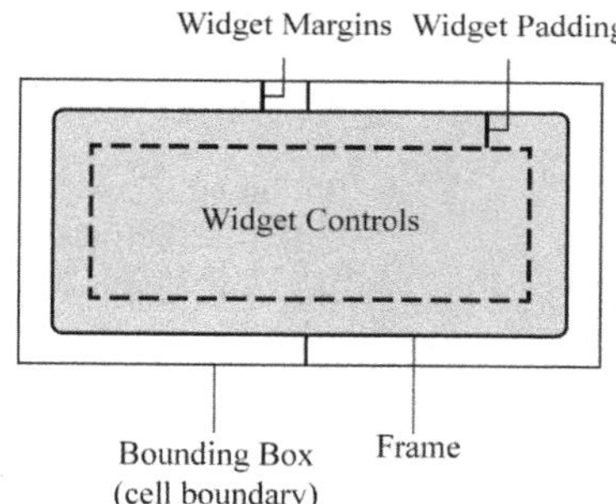

图 2-6　Margin 和 Padding 的含义

2.5　事件处理

对屏幕中控件的事件处理方式有以下三种。

第一种，设置控件的 onClick 属性，绑定一个用户自定义方法，该方式适合屏幕中具有多个同类控件需要监听的情况；第二种，在控件上设置监听器，并通过匿名类的方式完善对应方法的代码，该方式适合屏幕中单个或少数几个控件需要监听的情况；第三种，让控件所在的屏幕类直接实现监听器接口，该方式适合屏幕中多个同类或不同类控件需要监听的情况。

2.5.1　设置控件的 onClick 属性

在布局文件中指定控件的 android:onClick 属性，代码如下：

```
<Button
    android:id="@+id/button1"
    android:layout_width="wrap_content"
    android:layout_height="wrap_content"
    android:onClick="onClick"
    android:text="Button" />
```

然后在 Activity 类中编写 onClick 方法，代码如下：

```
public void onClick(View v) {
    // do something
}
```

该方法的签名有三个要求：第一，必须是 public（公共）；第二，不能有返回值，即使用 void；第三，带一个 View 类型的参数。

该方法的优点是实现的代码直观、方便；缺点是将控件的行为写入布局文件中，不利于布局与行为的分离。此外，如果需要处理的控件数量较多时，对应的方法也很多，会使代码显得比较混乱，因此，建议使用匿名类实现监听器接口。

2.5.2 使用匿名类实现监听器接口

使用匿名类实现监听器接口时，在布局文件中不需要指定 android:onClick 属性，代码如下：

```
<Button
    android:id="@+id/button1"
    android:layout_width="wrap_content"
    android:layout_height="wrap_content"
    android:text="Button" />
```

在 Activity 类中实现监听器接口，代码如下：

```
public class MainActivity extends AppCompatActivity {

    @Override
    protected void onCreate(Bundle savedInstanceState) {
        super.onCreate(savedInstanceState);
        setContentView(R.layout.activity_main);

        Button button = (Button) this.findViewById(R.id.button1);
        button.setOnClickListener(new View.OnClickListener() {
            //双击 OnClickListener，然后按 Alt+Enter 组合键

            @Override
            public void onClick(View arg0) {
                // do something
            }
        });
    }
}
```

双击 View 后面的 OnClickListener，然后按 Alt+Enter 组合键，在弹出的菜单中选择 Implement methods 提示项，如图 2-7 所示。

这种方法适合需要处理的控件数量较少的情况。

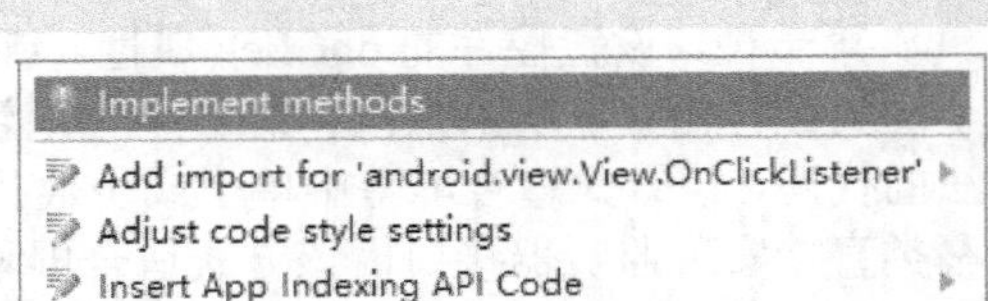

图 2-7 选择 Implement methods 提示项

2.5.3 使用屏幕类实现监听器接口

使用屏幕类实现监听器接口，关键代码如下：

```
public class MainActivity extends AppCompatActivity implements OnClickListener{

    @Override
    protected void onCreate(Bundle savedInstanceState) {
        super.onCreate(savedInstanceState);
        setContentView(R.layout.activity_main);

        Button button = (Button) this.findViewById(R.id.button1);
        button.setOnClickListener(this) ;     //双击 this，然后按 Alt+Enter 组合键
```

```
    }

    @Override
    public void onClick(View v) {
        switch(v.getId()){
        case R.id.button1:
            // do something
            break;
        // 更多控件
        }
    }
```

双击上述代码中 implements 后面的 OnClickListener，然后按 Alt+Enter 组合键，在弹出的菜单中选择 Make 'MainActivity' implement 'android.view.View.OnClickListener'提示项，如图 2-8 所示。

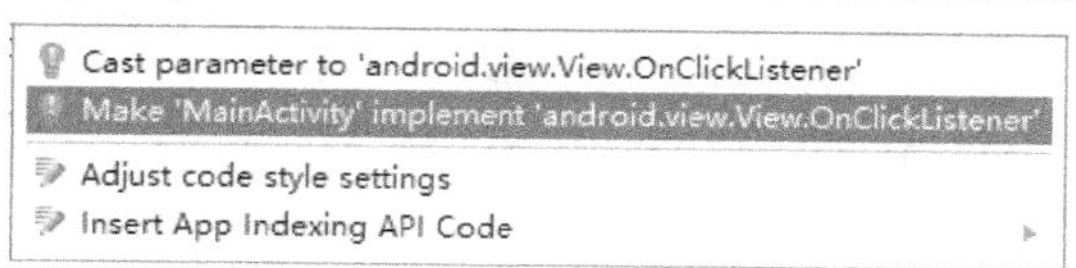

图 2-8　选择 Make 'MainActivity' implement 'android.view.View.OnClickListener'提示项

这种方法适合用一个事件处理程序中有多个控件的情况。

2.6　布局

Android 布局是界面开发的重要一环，Android 有多种布局，常用的布局方式有线性布局（LinearLayout）、相对布局（RelativeLayout）和其他布局。

2.6.1　线性布局

线性布局是指多个控件在布局内按顺序排列成一行或一列。根据控件的排列方向，线性布局分为两种：水平线性布局（Horizontal LinearLayout）和垂直线性布局（Vertical LinearLayout）。一般情况下，通过线性布局的属性 android:orientation 设置排列方向，默认值为水平排列（horizontal）。例如，android:orientation="vertical" 表示设置为垂直排列。设置方法为 setOrientation()，可以通过 Java 代码来设置。

布局内的控件可以定义控件自身的属性。

- 线性布局内控件的填充方式。所有线性布局内的控件都必须指定自己的填充方式，方法为设置控件的 android:layout_width 和 android: layout_height 属性，取值范围有三种：第一种，具体的像素值，如 20px；第二种，wrap_content，表示按控件文本实际长度显示；第三种，match_parent（旧称为 fill_parent），表示填充剩余的所有可用空间。
- 线性布局内控件的权重。如果想把线性布局内的控件按比例显示，就需要设置控件的权重。例如，一行有两个控件，想把其中一个控件所占的空间设置为另一个控件的两倍，可以把前一个控件的 android:layout_weight 设置为 1，后者设置为 2。

线性布局内控件的权重问题可参照 2.4.3 节的“权重”部分。

下面举例说明嵌套的线性布局，代码如下：

```
<?xml version="1.0" encoding="utf-8"?>
<LinearLayout xmlns:android="http://schemas.android.com/apk/res/android"
    android:id="@+id/LinearLayout2"
    android:layout_width="match_parent"
    android:layout_height="match_parent"
    android:orientation="vertical" >
```

```
<LinearLayout
    android:layout_width="match_parent"
    android:layout_height="wrap_content" >

    <TextView
        android:id="@+id/textView1"
        android:layout_width="wrap_content"
        android:layout_height="wrap_content"
        android:paddingLeft="@dimen/activity_horizontal_margin"
        android:text="账号："
        android:textAppearance="?android:attr/textAppearanceLarge" />

    <EditText
        android:id="@+id/editText1"
        android:layout_width="wrap_content"
        android:layout_height="wrap_content"
        android:layout_weight="1"
        android:ems="10" >

        <requestFocus />
    </EditText>

</LinearLayout>

<LinearLayout
    android:layout_width="match_parent"
    android:layout_height="wrap_content" >

    <TextView
        android:id="@+id/textView2"
        android:layout_width="wrap_content"
        android:layout_height="wrap_content"
        android:paddingLeft="@dimen/activity_vertical_margin"
        android:text="密码："
        android:textAppearance="?android:attr/textAppearanceLarge" />

    <EditText
        android:id="@+id/editText2"
        android:layout_width="wrap_content"
        android:layout_height="wrap_content"
        android:layout_weight="1"
        android:ems="10"
        android:inputType="textPassword" />

</LinearLayout>

<LinearLayout
    android:layout_width="match_parent"
    android:layout_height="wrap_content"
    android:gravity="right" >

    <Button
        android:id="@+id/button1"
        android:layout_width="wrap_content"
        android:layout_height="wrap_content"
        android:text="登录" />

    <Button
```

```
                android:id="@+id/button2"
                android:layout_width="wrap_content"
                android:layout_height="wrap_content"
                android:text="取消" />

        </LinearLayout>

    </LinearLayout>
```

本例中，根元素采用垂直线性布局，此外，还有三个水平线性布局，在每个水平线性布局中再放置对应的控件，这样构成了嵌套的线性布局，显示结果如图 2-9 所示。

读者应掌握线性布局的嵌套方法，并结合项目实际情况进行合理嵌套，从而为后续开发各种复杂布局奠定基础。

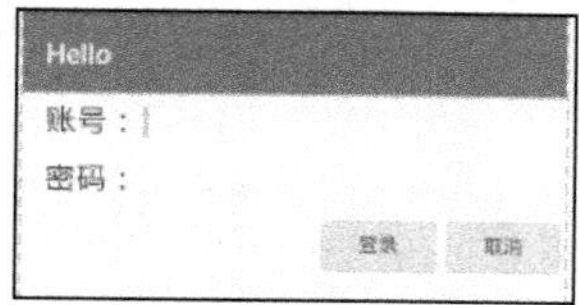

图 2-9　嵌套的线性布局

2.6.2　相对布局

相对布局是指布局内的控件通过确定与父容器或同一容器中其他控件的相对位置进行定位，布局内的控件可以设置自己的属性，常用属性见表 2-3。

表 2-3　相对布局内控件的常用属性

属 性 设 置	对应的 xml 代码	含　义
属性值为 true 或 false	android:layout_centerHrizontal	水平居中
	android:layout_centerVertical	垂直居中
	android:layout_centerInparent	相对父元素完全居中
	android:layout_alignParentBottom	贴紧父元素的下边缘
	android:layout_alignParentLeft	贴紧父元素的左边缘
	android:layout_alignParentRight	贴紧父元素的右边缘
	android:layout_alignParentTop	贴紧父元素的上边缘
	android:layout_alignWithParentIfMissing	如果找不到对应的兄弟元素，则以父元素为参照物
属性值必须为 id 的引用名，如“@id/id-name”	android:layout_below	在某元素的下方
	android:layout_above	在某元素的上方
	android:layout_toLeftOf	在某元素的左边
	android:layout_toRightOf	在某元素的右边
	android:layout_alignTop	本元素的上边缘和某元素的上边缘对齐
	android:layout_alignLeft	本元素的左边缘和某元素的左边缘对齐
	android:layout_alignBottom	本元素的下边缘和某元素的下边缘对齐
	android:layout_alignRight	本元素的右边缘和某元素的右边缘对齐
属性值为具体的像素值，如 30dp	android:layout_marginBottom	到某元素下边缘的距离
	android:layout_marginLeft	到某元素左边缘的距离
	android:layout_marginRight	到某元素右边缘的距离
	android:layout_marginTop	到某元素上边缘的距离

- 相对父容器（布局）的对齐方式：例如，android:layout_alignParentTop 表示控件的顶部与父容器的上边缘对齐，类似的属性还有 android:layout_alignParentBottom、android:layout_alignParentLeft 和 android:layout_alignParentRight。
- 相对同一容器中其他控件的位置：例如，android:layout_above 表示控件在另一控件的上面，类似的还有 android:layout_below、android:layout_toLeftOf 和 android:layout_toRightOf。
- 相对同一容器中其他控件的对齐方式：例如，android:layout_alignTop 表示控件与另一控件上边缘对齐，类似的还有 android:layout_alignBottom、android:layout_alignLeft 和 android:layout_alignRight。

下面举例说明，如图 2-10 所示的相对布局，其布局文件的代码如下：

```
<?xml version="1.0" encoding="utf-8"?>
<RelativeLayout xmlns:android="http://schemas.android.com/apk/res/android"
    android:layout_width="match_parent"
    android:layout_height="match_parent" >

    <Button
        android:id="@+id/Button1"
        android:layout_width="wrap_content"
        android:layout_height="wrap_content"
        android:layout_alignParentLeft="true"
        android:layout_alignParentTop="true"
        android:layout_marginLeft="38dp"
        android:layout_marginTop="45dp"
        android:text="Button1" />

    <Button
        android:id="@+id/Button2"
        android:layout_width="wrap_content"
        android:layout_height="wrap_content"
        android:layout_alignLeft="@+id/Button1"
        android:layout_below="@+id/Button1"
        android:layout_marginLeft="57dp"
        android:layout_marginTop="32dp"
        android:text="Button2" />

</RelativeLayout>
```

在本例中，Button1 相对父元素（RelativeLayout），与其左对齐并留出间距 38dp，与其上对齐并留出间距 45dp。

Button2 相对控件 Button1，在其下方并留出间距 32dp，与其左对齐并留出间距 57dp，图 2-10 中的辅助线可以帮助我们理解这一点。

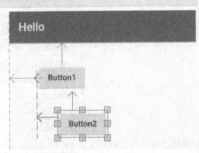

图 2-10　相对布局

2.6.3　其他布局

1．表格布局

表格布局（TableLayout）是线性布局的一个子类。表格布局类似 HTML 中的 Table。每个表格布局里有表格行（TableRow），表格行里可以具体定义每个元素。

2．帧布局

帧布局（FrameLayout）通常只能放一个视图，因为第二个视图会覆盖前一个视图。

3．绝对布局

绝对布局（AbsoluteLayout）采用 *X* 和 *Y* 坐标值定位控件的绝对位置。这种布局方式也比较简单，左上角的坐标位置是（0,0）。但是，当水平坐标与垂直坐标切换时，往往会出问题，而且存在多个元素时，计算比较麻烦。

4．网格布局

网格布局（GridLayout）的效果与表格布局相似。

5．列表视图

列表视图（ListView）用于显示可滚动的单列列表。由于是可滚动的，其作用和写法都与前述的布局不同，并且名称也改为视图。

列表视图适用于显示动态内容，比如在运行时，从数组或数据库中取得显示的数据，这时需要将列表视图绑定到一个 Adapter 接口的实例，这个实例用来从外部（数组或数据库）取得用于显示的数据。

Adapter 接口的实现方式有多种，其中常用的两种方式如下。

- ArrayAdapter：用于从数组中获得数据。
- SimpleCursorAdapter：用于从访问数据库时生成的游标中获得数据。

这里以数组为例说明使用列表视图的方法。先在布局文件中添加一个 ListView，代码如下：

```
<RelativeLayout xmlns:android="http://schemas.android.com/apk/res/android"
    xmlns:tools="http://schemas.android.com/tools"
    android:layout_width="match_parent"
    android:layout_height="match_parent"
    android:paddingBottom="@dimen/activity_vertical_margin"
    android:paddingLeft="@dimen/activity_horizontal_margin"
    android:paddingRight="@dimen/activity_horizontal_margin"
    android:paddingTop="@dimen/activity_vertical_margin"
    tools:context=".MainActivity" >

    <ListView
        android:id="@+id/myListView"
        android:layout_width="match_parent"
        android:layout_height="wrap_content"
        android:layout_alignParentLeft="true"
        android:layout_alignParentTop="true" >
    </ListView>

</RelativeLayout>
```

然后在 Activity 类中加入如下代码：

```
package com.example.administrator. hello;

import android.support.v7.app.AppCompatActivity;
import android.os.Bundle;
import android.widget.ArrayAdapter;
import android.widget.ListView;

public class MainActivity extends AppCompatActivity {
    String[] myStringArray = { "one", "two", "three" };

    @Override
    protected void onCreate(Bundle savedInstanceState) {
        super.onCreate(savedInstanceState);
        setContentView(R.layout.activity_main);

        ArrayAdapter adapter = new ArrayAdapter<String>(this,
                android.R.layout.simple_list_item_1, myStringArray);

        ListView listView = (ListView) findViewById(R.id.myListView);
        listView.setAdapter(adapter);
    }
}
```

如果底层的数据发生动态变化，则要调用 adapter.notifyDataSetChanged()方法通知 ListView 重新刷新显示。

对于 ListView，可能还需要对单击事件进行处理，即设置 ListView 的 OnItemClick Listener 监听器，Activity 类的完整代码如下：

```
package com.example.administrator. hello;

import android.support.v7.app.AppCompatActivity;
import android.os.Bundle;
import android.view.View;
import android.widget.AdapterView;
import android.widget.AdapterView.OnItemClickListener;
import android.widget.ArrayAdapter;
import android.widget.ListView;
import android.widget.Toast;

public class MainActivity extends AppCompatActivity {
	String[] myStringArray = { "one", "two", "three" };

	@Override
	protected void onCreate(Bundle savedInstanceState) {
		super.onCreate(savedInstanceState);
		setContentView(R.layout.activity_main);

		ArrayAdapter adapter = new ArrayAdapter<String>(this,
				android.R.layout.simple_list_item_1, myStringArray);

		ListView listView = (ListView) findViewById(R.id.myListView);
		listView.setAdapter(adapter);
		listView.setOnItemClickListener(new OnItemClickListener() {
		public void onItemClick(AdapterView parent, View v, int position,
				long id) {
			String item = parent.getItemAtPosition(position).toString();
			Toast.makeText(MainActivity.this, "You clicked " + position + ": " + item,
					Toast.LENGTH_LONG).show();
		});
	}
}
```

ListView 的执行效果如图 2-11 所示。

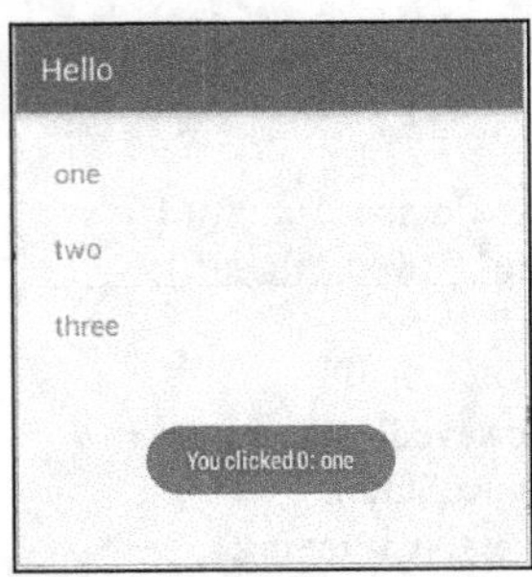

图 2-11　ListView 的执行效果（单击“one”时）

注意：当数组中含有更多数据且超出屏幕范围时，才能看到滚动的效果。

6．其他可滚动视图

其他可滚动视图包括 GridView、ScrollView、HorizontalScrollView 等。GridView 用于显示可滚动的由行和列组成的网格列表。ScrollView 用于显示垂直滚动列表，HorizontalScrollView 用于显示水平滚动列表。

每种布局均有自己适用的范围，另外，这些布局元素可以相互嵌套应用，从而制作出美观的界面。

2.7　常用控件

在工具箱中可以看到所有控件，这些控件按不同的标准进行分类，其中，最常用的为 Widgets 控件，如图 2-12 所示。

图 2-12　常用的 Widgets 控件

2.7.1　文本类控件

1．TextView

TextView 控件是不可编辑的文本标签，它有超过 100 种属性，通过设置属性可以非常细致地调整其外观和行为。部分属性见表 2-4。

表 2-4　TextView 控件的部分属性

属　性　名	说　　明
android:id	控件的唯一标识（ID）
android:layout_width	相对于布局的宽度，值为 wrap_content 时根据内容调整，值为 match_parent 时充满布局宽度
android:layout_height	相对于布局的高度，值为 wrap_content 时根据内容调整，值为 match_parent 时充满布局高度
android:layout_alignLeft	左侧与某个控件对齐，需指定控件的 ID
android:layout_below	位于某个控件之下，需指定控件的 ID
android:layout_marginTop	与上方（layout_below）控件的相对距离
android:layout_marginLeft	与左侧（layout_alignLeft）控件的相对距离
android:layout_gravity	相对于线性布局的对齐方式，取值有 top、bottom、left、right、center、center_vertical 等。可以设置一个或多个，用“|”隔开，比如 right | bottom 表示右下方
android:ems	设置 TextView 的宽度为 *n* 个字符的宽度
android:maxems	设置 TextView 的宽度为最长为 *n* 个字符的宽度，与 ems 同时使用时覆盖 ems 属性
android:minems	设置 TextView 的宽度为最短为 *n* 个字符的宽度，与 ems 同时使用时覆盖 ems 属性
android:maxLength	限制输入的字符数。例如，设置参数值为 5，那么仅可以输入 5 个汉字/数字/英文字母
android:lines	设置文本的行数。例如，设置行数为 2，则显示 2 行，即使第 2 行没有数据
android:maxLines	设置文本的最大显示行数，与 width 或 layout_width 结合使用时，超出行宽的部分自动换行，超出最大行数的部分不显示
android:minLines	设置文本的最小行数，与 lines 属性类似

续表

属 性 名	说 明
android:textAppearance	设置文字外观。例如，"?android:attr/textAppearanceLarge" 是大号字体
android:textColor	设置文本颜色
android:textColorHighlight	设置被选中的文字的底色，默认为蓝色
android:textColorHint	设置提示信息文字的颜色，默认为灰色。与 hint 属性配合使用
android:textColorLink	文字链接的颜色
android:textSize	设置文字大小，推荐度量单位使用 sp，如"15sp"
android:textStyle	设置字形。例如，bold 表示粗体、italic 表示斜体。可以设置一个或多个，用"|"隔开，比如"bold | italic"表示加粗倾斜
android:typeface	设置文本字体，必须是以下常量值之一：normal、sans、serif、monospace
android:height	设置文本区域的高度，支持的度量单位：px/dp/sp/in/mm
android:maxHeight	设置文本区域的最大高度
android:minHeight	设置文本区域的最小高度
android:width	设置文本区域的宽度，支持的度量单位：px/dp/sp/in/mm
android:maxWidth	设置文本区域的最大宽度
android:minWidth	设置文本区域的最小宽度
android:gravity	设置文本的对齐方式，取值为 top、bottom、left、right、center、center_vertical 等。可以设置一个或多个，用"|"隔开，如"right | bottom"表示右下

其中，android:textAppearance 属性有三个值可供选择：textAppearanceLarge、textAppearanceMedium 和 textAppearanceSmall。

为了使用 Java 语言访问 TextView 控件，上述多数属性可以通过相应的方法设置，如表 2-5 所示。另外，TextView 控件有些方法是没有对应属性的。

表 2-5 TextView 控件部分属性对应的方法及说明

方 法 名	说 明
void setText(CharSequence text)	设置 TextView 控件的文本
CharSequence getText()	获取 TextView 控件的内容，返回 Editable 对象，若要取得其文本，需要写为 getText().toString()

注意：这里只列出了部分方法，更多方法的含义可参考 Android API 文档。

2. EditText

EditText 控件是可编辑的文本输入框，它是 TextView 的子类，所以它拥有 TextView 的所有属性，其中与文本编辑有关的属性见表 2-6。

表 2-6 EditText 控件的文本编辑属性

属 性 名	说 明
android:inputType	设定输入类型，取值为 text、textEmailAddress、number、numberSigned、numberDecimal 等
android:password	密码框属性，取值为 true/false。当设置为 true 时，可以让 EditText 显示的内容自动变为*号，即输入的内容会在输入后 1 秒内变成*号
android:cursorVisible	是否显示光标，取值为 true/false
android:editable	是否可编辑，取值为 true/false
android:singleLine	是否强制单行输入，取值为 true/false

注意：EditText 控件的大部分属性是在 TextView 中定义的。

其中，属性 android:inputType 有许多可选项，为此，工具箱中提供了 EditText 控件的各种变体供开发者选择，如图 2-13 所示，用户可以根据需要，直接选择对应的 EditText 控件，提高开发速度。

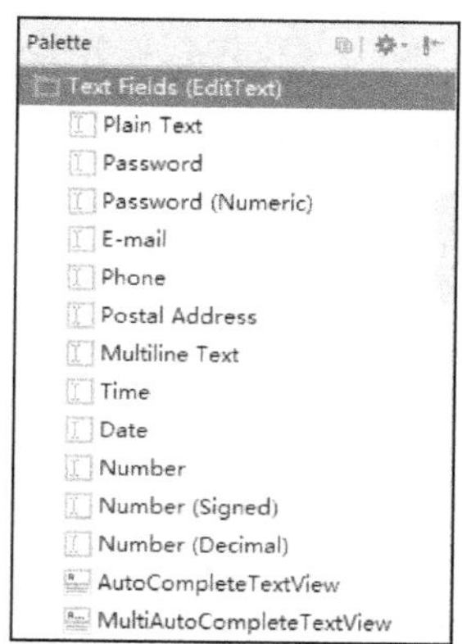

图 2-13　EditText 控件的各种变体

为了使用 Java 语言访问 EditText 控件，EditText 控件的属性可以通过相应的方法设置，如表 2-7 所示。

表 2-7　EditText 控件部分属性对应的方法及说明

方 法 名	说　明
void setText(CharSequence text)	设置 EditText 控件的文本
Editable getText()	获取 EditText 控件的内容，返回 Editable 对象，要取得其文本，需要写为 getText().toString()

本章的计算器项目涉及相关的 EditText 控件。下面再举一个有关 EditText 控件的例子，布局文件的代码如下：

```
<?xml version="1.0" encoding="utf-8"?>
<LinearLayout xmlns:android="http://schemas.android.com/apk/res/android"
    android:layout_width="match_parent"
    android:layout_height="match_parent"
    android:orientation="vertical" >

    <LinearLayout
        android:layout_width="match_parent"
        android:layout_height="wrap_content" >

        <TextView
            android:id="@+id/textView1"
            android:layout_width="wrap_content"
            android:layout_height="wrap_content"
            android:text="请输入"
            android:textAppearance="?android:attr/textAppearanceLarge" />

        <EditText
            android:id="@+id/input"
            android:layout_width="wrap_content"
            android:layout_height="wrap_content"
            android:layout_weight="1"
            android:ems="10"
            android:text="Android" >

            <requestFocus />
        </EditText>

    </LinearLayout>
```

```
    <LinearLayout
        android:layout_width="match_parent"
        android:layout_height="wrap_content"
        android:gravity="right" >

        <Button
            android:id="@+id/ok"
            android:layout_width="wrap_content"
            android:layout_height="wrap_content"
            android:text="确认" />

        <Button
            android:id="@+id/reset"
            android:layout_width="wrap_content"
            android:layout_height="wrap_content"
            android:text="重置" />

    </LinearLayout>

</LinearLayout>
```

Activity 类的代码如下：

```
package com.example.administrator. hello;

import android.support.v7.app.AppCompatActivity;
import android.os.Bundle;
import android.view.View;
import android.view.View.OnClickListener;
import android.widget.Button;
import android.widget.EditText;
import android.widget.Toast;

public class MainActivity extends AppCompatActivity implements OnClickListener {
    EditText etInput;

    @Override
    protected void onCreate(Bundle savedInstanceState) {
        super.onCreate(savedInstanceState);
        setContentView(R.layout.activity_main);

        etInput = (EditText) this.findViewById(R.id.input);

        Button btn = (Button) this.findViewById(R.id.ok);
        btn.setOnClickListener(this);
        btn = (Button) this.findViewById(R.id.reset);
        btn.setOnClickListener(this);
    }

    @Override
    public void onClick(View v) {
        switch (v.getId()) {
        case R.id.ok:
            Toast.makeText(this,
                    "Your input is: " + etInput.getText().toString(),
                    Toast.LENGTH_LONG).show();
            break;
        case R.id.reset:
            etInput.setText("Android");
```

```
                break;
            }
        }
}
```

运行上述代码，界面效果如图 2-14 所示

图 2-14　EditText 控件举例——界面效果

2.7.2　按钮类控件

按钮的外观包含文字（如 Alarm ）、图标（如 ⏰ ）或兼有文字和图标（如 ⏰ Alarm ）。因此，上述三种类别对应三种定义按钮的方法。

1．Button

使用 Button 控件，生成的按钮只包含文字，代码为：

```
<Button
    android:layout_width="wrap_content"
    android:layout_height="wrap_content"
    android:text="@string/button_text"
    ... />
```

2．ImageButton

使用 ImageButton 控件，生成的按钮只包含图标，代码为：

```
<ImageButton
    android:layout_width="wrap_content"
    android:layout_height="wrap_content"
    android:src="@drawable/button_icon"
    ... />
```

其中，图标文件的名称为 button_icon，文件格式包括 png、jpg、gif 等，该文件应该复制到项目的 drawable 目录中，并且每个目录都应该有一份内容相同但分辨率不同的图片文件，Android 系统在程序运行时会根据屏幕分辨率自动选择大小合适的图片用于显示。

3．ToggleButton

在 Button 按钮中，若同时指定文字和图标，则使用如下代码：

```
<ToggleButton
    android:layout_width="wrap_content"
    android:layout_height="wrap_content"
    android:text="ToggleButton"
    android:drawableLeft="@drawable/button_icon"
    ... />
```

2.7.3 选择类控件

1. CheckBox

CheckBox 控件是复选框，代码为：

```
<CheckBox
    android:id="@+id/checkBox1"
    android:layout_width="wrap_content"
    android:layout_height="wrap_content"
    android:text="CheckBox1" />
```

2. RadioButton 和 RadioGroup

RadioButton 为单选钮，RadioGroup 为单选钮组。使用单选钮时需要注意，应该将多个互斥的单选钮放在同一个单选钮组中，代码为：

```
<RadioGroup
    android:id="@+id/radioGroup1"
    android:layout_width="wrap_content"
    android:layout_height="wrap_content" >

    <RadioButton
        android:id="@+id/radio1"
        android:layout_width="wrap_content"
        android:layout_height="wrap_content"
        android:checked="true"
        android:text="RadioButton1" />

    <RadioButton
        android:id="@+id/radio2"
        android:layout_width="wrap_content"
        android:layout_height="wrap_content"
        android:text="RadioButton2" />
</RadioGroup>
```

下面举例说明 Activity 类与复选框和单选钮的交互功能，布局文件的代码如下：

```
<?xml version="1.0" encoding="utf-8"?>
<LinearLayout xmlns:android="http://schemas.android.com/apk/res/android"
    android:layout_width="match_parent"
    android:layout_height="match_parent"
    android:orientation="vertical" >

    <CheckBox
        android:id="@+id/checkBox1"
        android:layout_width="wrap_content"
        android:layout_height="wrap_content"
        android:text="CheckBox1" />

    <CheckBox
        android:id="@+id/checkBox2"
        android:layout_width="wrap_content"
        android:layout_height="wrap_content"
        android:text="CheckBox2" />

    <CheckBox
        android:id="@+id/checkBox3"
        android:layout_width="wrap_content"
        android:layout_height="wrap_content"
        android:text="CheckBox3" />
```

```
    <RadioGroup
        android:id="@+id/radioGroup1"
        android:layout_width="wrap_content"
        android:layout_height="wrap_content" >

        <RadioButton
            android:id="@+id/radio1"
            android:layout_width="wrap_content"
            android:layout_height="wrap_content"
            android:checked="true"
            android:text="RadioButton1" />

        <RadioButton
            android:id="@+id/radio2"
            android:layout_width="wrap_content"
            android:layout_height="wrap_content"
            android:text="RadioButton2" />
    </RadioGroup>

    <Button
        android:id="@+id/ok"
        android:layout_width="wrap_content"
        android:layout_height="wrap_content"
        android:text="确认" />

</LinearLayout>
```

Activity 类的代码如下：

```
package com.example.administrator. hello;

import android.support.v7.app.AppCompatActivity;
import android.os.Bundle;
import android.util.Log;
import android.view.View;
import android.view.View.OnClickListener;
import android.widget.Button;
import android.widget.CheckBox;
import android.widget.RadioButton;

public class MainActivity extends AppCompatActivity {
    CheckBox cb1;
    CheckBox cb2;
    CheckBox cb3;

    RadioButton rb1;
    RadioButton rb2;

    @Override
    protected void onCreate(Bundle savedInstanceState) {
        super.onCreate(savedInstanceState);
        setContentView(R.layout.activity_main);

        cb1 = (CheckBox) this.findViewById(R.id.checkBox1);
        cb2 = (CheckBox) this.findViewById(R.id.checkBox2);
        cb3 = (CheckBox) this.findViewById(R.id.checkBox3);

        rb1 = (RadioButton) this.findViewById(R.id.radio1);
        rb2 = (RadioButton) this.findViewById(R.id.radio2);
```

```
        Button btn = (Button) this.findViewById(R.id.ok);
        btn.setOnClickListener(new OnClickListener(){

            @Override
            public void onClick(View v) {
                Log.i("CheckBox","用户选择的结果:");
                if(cb1.isChecked()){
                    Log.i("CheckBox","选择了 CheckBox1");
                }
                if(cb2.isChecked()){
                    Log.i("CheckBox","选择了 CheckBox2");
                }
                if(cb3.isChecked()){
                    Log.i("CheckBox","选择了 CheckBox3");
                }
                if(rb1.isChecked()){
                    Log.i("CheckBox","选择了 RadioButton1");
                }
                if(rb2.isChecked()){
                    Log.i("CheckBox","选择了 RadioButton2");
                }

            }});
    }
}
```

上述代码的输出结果在 LogCat 中显示。

3. Spinner

Spinner 的作用为生成下拉菜单，编写步骤如下。

（1）编写布局文件。在布局文件中添加一个 Spinner 控件，代码如下：

```
<?xml version="1.0" encoding="utf-8"?>
<LinearLayout xmlns:android="http://schemas.android.com/apk/res/android"
    android:layout_width="match_parent"
    android:layout_height="match_parent"
    android:orientation="vertical" >

    <Spinner
        android:id="@+id/spinner1"
        android:layout_width="match_parent"
        android:layout_height="wrap_content" />

</LinearLayout>
```

（2）创建资源文件 array.xml。在 values 目录中创建一个名为 array.xml 的资源文件，在该文件中保存了下拉菜单的选择项，其中，name 属性就是数据的 ID，将被 R.java 文件引用，代码如下：

```
<?xml version="1.0" encoding="utf-8"?>
<resources>

    <string-array name="choice">
        <item>One</item>
        <item>two</item>
        <item>three</item>
        <item>four</item>
    </string-array>

</resources>
```

（3）对 Activity 类进行设置，应当注意以下三项。

①引用 Spinner 控件。

②为 Spinner 控件设置选择项。通过 ArrayAdapter 将 Spinner 控件与资源文件 array.xml 绑定起来。

③为 Spinner 控件设置事件监听器。Spinner 控件的事件监听器有多种，通常使用 OnItemSelectedListener，即选择某项后触发事件。

Activity 类的代码如下：

```
package com.example.administrator. hello;

import android.support.v7.app.AppCompatActivity;
import android.os.Bundle;
import android.view.View;
import android.widget.AdapterView;
import android.widget.AdapterView.OnItemSelectedListener;
import android.widget.ArrayAdapter;
import android.widget.Spinner;
import android.widget.Toast;

public class MainActivity extends AppCompatActivity {

    @Override
    protected void onCreate(Bundle savedInstanceState) {
        super.onCreate(savedInstanceState);
        setContentView(R.layout.activity_main);

        //引用 Spinner 控件
        Spinner spinner = (Spinner) findViewById(R.id.spinner1);
        //为 Spinner 控件设置选择项
        ArrayAdapter<CharSequence> adapter = ArrayAdapter.createFromResource(
                this, R.array.choice, android.R.layout.simple_spinner_item);

        adapter.setDropDownViewResource(android.R.layout.simple_spinner_dropdown_item);
        spinner.setAdapter(adapter);

        //为 Spinner 控件设置事件监听器
        spinner.setOnItemSelectedListener(new OnItemSelectedListener() {
            @Override
            public void onItemSelected(AdapterView<?> av, View v, int i, long l) {
                Toast.makeText(MainActivity.this, av.getItemAtPosition(i).toString(),
                        Toast.LENGTH_LONG).show();
            }

            @Override
            public void onNothingSelected(AdapterView<?> av) {
                Toast.makeText(me, "没有选择", Toast.LENGTH_LONG).show();
            }
        });
    }

}
```

4. Pickers

Pickers 控件可以提供日期选择器和时间选择器，下面举例说明，代码如下：

```
<?xml version="1.0" encoding="utf-8"?>
<LinearLayout xmlns:android="http://schemas.android.com/apk/res/android"
    android:layout_width="match_parent"
```

```
    android:layout_height="match_parent"
    android:orientation="vertical" >

    <DatePicker
        android:id="@+id/datePicker1"
        android:layout_width="wrap_content"
        android:layout_height="wrap_content" />

    <LinearLayout
        android:layout_width="match_parent"
        android:layout_height="wrap_content" >

        <TimePicker
            android:id="@+id/timePicker1"
            android:layout_width="wrap_content"
            android:layout_height="wrap_content" />

        <Button
            android:id="@+id/ok"
            android:layout_width="wrap_content"
            android:layout_height="wrap_content"
            android:layout_gravity="bottom|right"
            android:text="确认" />

    </LinearLayout>

</LinearLayout>
```

Activity 类代码如下：

```
package com.example.administrator. hello;

import android.support.v7.app.AppCompatActivity;
import android.os.Bundle;
import android.view.View;
import android.view.View.OnClickListener;
import android.widget.Button;
import android.widget.DatePicker;
import android.widget.DatePicker.OnDateChangedListener;
import android.widget.TimePicker;
import android.widget.TimePicker.OnTimeChangedListener;
import android.widget.Toast;

public class MainActivity extends AppCompatActivity implements OnClickListener {
    static Activity me;

    private int year;
    private int month;
    private int day;
    private int hour;
    private int minute;

    @Override
    protected void onCreate(Bundle savedInstanceState) {
        super.onCreate(savedInstanceState);
        setContentView(R.layout.activity_main);

        me = this;

        DatePicker dp = (DatePicker) me.findViewById(R.id.datePicker1);
```

```
            TimePicker tp = (TimePicker) me.findViewById(R.id.timePicker1);

            dp.init(year, month, day, new OnDateChangedListener() {
                @Override
                public void onDateChanged(DatePicker view, int year, int month,
                        int day) {
                    MainActivity.this.year = year;
                    MainActivity.this.month = month;
                    MainActivity.this.day = day;
                }
            });

            tp.setOnTimeChangedListener(new OnTimeChangedListener() {
                @Override
                public void onTimeChanged(TimePicker view, int hour, int minute) {
                    MainActivity.this.hour = hour;
                    MainActivity.this.minute = minute;
                }
            });

            Button btn = (Button) this.findViewById(R.id.ok);
            btn.setOnClickListener(this);
        }

        @Override
        public void onClick(View v) {
            // 注意 month 从 0 开始
            String date = year + "年" + (month+1) + "月" + day + "日" + hour + "时"
                    + minute + "分";
            Toast.makeText(me, "你的选择是: " + date, Toast.LENGTH_LONG).show();
        }
}
```

运行代码后，界面效果如图 2-15 所示。

5. CalendarView

CalendarView（日历视图）控件是一个高度定制的日期选择器，可以满足多选日期的需求。例如，车票预订和酒店入住时，需要选择起止日期，如图 2-16 所示。

图 2-15　Pickers 控件举例——界面效果

图 2-16　日历视图

下面举例说明，先编写 activity_main.xml，代码如下：

```
<?xml version="1.0" encoding="utf-8"?>
<RelativeLayout xmlns:android="http://schemas.android.com/apk/res/android"
    xmlns:tools="http://schemas.android.com/tools"
```

```
    android:id="@+id/activity_main"
    android:layout_width="match_parent"
    android:layout_height="match_parent"
    android:paddingBottom="@dimen/activity_vertical_margin"
    android:paddingLeft="@dimen/activity_horizontal_margin"
    android:paddingRight="@dimen/activity_horizontal_margin"
    android:paddingTop="@dimen/activity_vertical_margin"
    tools:context="com.example.administrator.calendarview.MainActivity">

    <TextView
        android:id="@+id/textView"
        android:layout_width="wrap_content"
        android:layout_height="50dp"
        android:layout_alignParentTop="true"
        android:layout_centerHorizontal="true"
        android:gravity="center"
        android:text="Large Text"
        android:textColor="#FF0000"
        android:textAppearance="@style/Base.TextAppearance.AppCompat.Large"
        />

    <!—所选日期所在月的日期显示为蓝色字体，所选日期两旁的竖线为红色，所选日期所在的那
周背景色为灰色，所选日期所在月之外的其他日期显示为黑色字体-->

    <CalendarView
        android:id="@+id/calendarView"
        android:layout_width="match_parent"
        android:layout_height="match_parent"
        android:layout_below="@id/textView"
        android:focusedMonthDateColor="#12c8cf"
        android:selectedDateVerticalBar="@color/red_price"
        android:selectedWeekBackgroundColor="#c0c0c0"
        android:unfocusedMonthDateColor="#4c4948"
        />

</RelativeLayout>
```

在日历视图中选择一个日期后，结果会在最上方的 TextView 控件中显示。对应的源代码如下：

```
package com.example.administrator.calendarview;

import android.support.v7.app.AppCompatActivity;
import android.os.Bundle;
import android.widget.CalendarView;
import android.widget.TextView;

import java.util.Calendar;

public class MainActivity extends AppCompatActivity {
    TextView textView;
    CalendarView calendarView;

    @Override
    protected void onCreate(Bundle savedInstanceState) {
        super.onCreate(savedInstanceState);
        setContentView(R.layout.activity_main);

        textView = (TextView) this.findViewById(R.id.textView);
        calendarView = (CalendarView) this.findViewById(R.id.calendarView);
```

```
        //初始化日期
        Calendar calender = Calendar.getInstance();
        String datas = calender.get(Calendar.YEAR) + "年" + (calender.get(Calendar.MONTH) + 1) + "
月" + calender.get(Calendar.DAY_OF_MONTH) + "日";
        textView.setText(datas);

        chooseCalender();
    }

    /**
     * 得到选择的日期
     */
    private void chooseCalender() {
        calendarView.setOnDateChangeListener(new CalendarView.OnDateChangeListener() {
            @Override
            public void onSelectedDayChange(CalendarView view, int year, int month, int
dayOfMonth) {
                String dates = year + "-" + ((month + 1) < 10 ? "0" + (month + 1) : (month + 1)) + "-"
+ (dayOfMonth < 10 ? "0" + dayOfMonth : dayOfMonth);
                textView.setText(dates);
            }
        });
    }
}
```

本例的 CalendarView 控件只能实现单选功能，如果要实现多选功能，还需要自定义 CalendarView 控件。感兴趣的读者可以参考本书的配套资源，自学相关内容，以便更好地完成 2.10 节的实训内容。

2.7.4 提示类控件

提示类控件都是在程序运行过程中产生的，因此它不必也不可能在布局文件中被静态指定，它只能在 Activity 类中由 Java 代码动态生成。

1. Toast

Toast 是一种简单的信息提示方式，例如：

```
Toast.makeText(this, "你的选择是: " + date, Toast.LENGTH_LONG).show();
```

注意：第一个参数是 Toast 所在的 Activity 类的实例。

2. Dialog

在 Android 中，Dialog（对话框）的使用方法稍显复杂。虽然 Dialog 不是抽象类，但它的构造方法为 protected，所以不能直接将其实例化，而应该将它的子类实例化，其中，最常用的子类是 AlertDialog。例如，弹出简单对话框，代码如下：

```
public void onClick(View v) {
    AlertDialog.Builder builder = new AlertDialog.Builder(me);
    builder.setMessage("这是对话的提示信息").setTitle("对话框");
    AlertDialog dialog = builder.create();
    dialog.show();
}
```

上述几行还可以合并，所有对象都是匿名的，代码如下：

```
public void onClick(View v) {
    new AlertDialog.Builder(me).setMessage("这是对话的提示信息").setTitle("对话框")
            .create().show();
}
```

通常，Dialog 的下方会有几个按钮供用户选择，代码如下：

```
public void onClick(View v) {
        AlertDialog.Builder builder = new AlertDialog.Builder(me);
        builder.setMessage("这是对话的提示信息").setTitle("对话框");

        builder.setPositiveButton("确认", new DialogInterface.OnClickListener() {
                public void onClick(DialogInterface dialog, int id) {
                        Toast.makeText(me, "你选择了确认", Toast.LENGTH_SHORT).show();
                }
        });

        builder.setNeutralButton("中性选择", new DialogInterface.OnClickListener() {
                public void onClick(DialogInterface dialog, int id) {
                        Toast.makeText(me, "你选择了中性选择", Toast.LENGTH_SHORT).show();
                }
        });

        builder.setNegativeButton("放弃", new DialogInterface.OnClickListener() {
                public void onClick(DialogInterface dialog, int id) {
                        Toast.makeText(me, "你选择了放弃", Toast.LENGTH_SHORT).show();
                }
        });

        AlertDialog dialog = builder.create();
        dialog.show();
}
```

注意：有时需要连续出现两次对话框，让用户依次进行操作。但是，应当注意代码的正确性，比如下面的代码就无法实现目的：

```
public void onClick(View v) {
        new AlertDialog.Builder(me).setMessage("请确认，这是第 1 个对话框……").setTitle("对话框 1")
                        .setPositiveButton("确认", null).create().show();

        new AlertDialog.Builder(me).setMessage("再次确认").setTitle("对话框 2")
        .setPositiveButton("确认", null).create().show();
}
```

运行上述代码可以发现，对话框的显示顺序与预期相反，这是因为显示对话框 1 后，系统并不等待用户的选择操作，而是立即显示对话框 2，并把对话框 1 覆盖；当用户在对话框 2 中进行选择后，对话框 2 隐去，重新显示对话框 1，所以导致对话框的显示顺序是相反的。正确的写法是将第 2 个对话框放在第 1 个对话框的事件处理代码中，具体代码如下：

```
public void onClick(View v) {
        new AlertDialog.Builder(me).setMessage("请确认，这是第 1 个对话框……")
                        .setTitle("对话框 1")
                        .setPositiveButton("确认", new DialogInterface.OnClickListener() {
                                public void onClick(DialogInterface dialog, int id) {
                                        new AlertDialog.Builder(me).setMessage("再次确认")
                                                        .setTitle("对话框 2").setPositiveButton("确认", null)
                                                        .create().show();
                                }
                        }).create().show();

}
```

2.7.5　图片类控件

1．ImageView

ImageView 控件用于显示图片，下面举例说明。

（1）准备两张用于切换的图片，即 android1.png 和 android2.png。编写布局文件，该文件只有一个 ImageView 控件，代码如下：

```
<?xml version="1.0" encoding="utf-8"?>
<LinearLayout xmlns:android="http://schemas.android.com/apk/res/android"
    android:layout_width="match_parent"
    android:layout_height="match_parent"
    android:orientation="vertical" >

    <ImageView
        android:id="@+id/imageView1"
        android:layout_width="wrap_content"
        android:layout_height="wrap_content"
        android:src="@drawable/android1" />

</LinearLayout>
```

Activity 类代码如下：

```
package com.example.administrator. imageview;

import android.support.v7.app.AppCompatActivity;
import android.os.Bundle;
import android.view.View;
import android.view.View.OnClickListener;
import android.widget.ImageView;

public class MainActivity extends AppCompatActivity {
    boolean flag;
    ImageView iv;

    @Override
    protected void onCreate(Bundle savedInstanceState) {
        super.onCreate(savedInstanceState);
        setContentView(R.layout.activity_main);

        flag=true;
        iv = (ImageView) this.findViewById(R.id.imageView1);
        iv.setClickable(true);                          // 设置图片为可单击的
        iv.setOnClickListener(new OnClickListener(){

            @Override
            public void onClick(View v) {               // 单击图片
                if(flag){
                    iv.setImageResource(R.drawable.android2);  // 在两张图片之间切换
                }else{
                    iv.setImageResource(R.drawable.android1);
                }
                flag = !flag;

            }});
    }
}
```

运行上述代码，得到的图片效果如图 2-17 所示。

为达到最佳的显示效果，根据 Android 系统的特质，应该为不同屏幕分辨率的移动设备准备不同的图片。在 Android 系统中，屏幕规格分为四种，即 small、normal、large 和 xlarge；像素密度也分为四种，即 small、normal、large 和 xlarge。因此，每张图片都应该按照四种分辨率生成四张图片，这四种分辨率之比为 3:4:6:8，四张图片分别保存在项目资源目录下的四个子目录中。

例如，有一张图片，其常规分辨率为 48×48 像素，则在不同的目录下保存为不同分辨率的图片，其图片比例如图 2-18 所示，四种分辨率分别如下。

drawable-ldpi（低分辨率），图片分辨率为 36×36 像素。

drawable-mdpi（中分辨率），图片分辨率为 48×48 像素。

drawable-hdpi（高分辨率），图片分辨率为 72×72 像素。

drawable-xhdpi（超高分辨率），图片分辨率为 96×96 像素。

图 2-17　图片效果

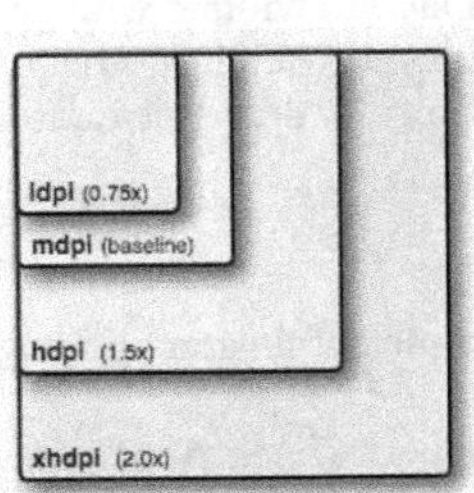

图 2-18　不同分辨率时的图片比例

2．VideoView

VideoView 控件用于播放视频，相关内容将在第 6 章进行介绍。

2.7.6　菜单类控件

Android 的菜单有三种：选项菜单（OptionsMenu）、上下文菜单（ContextMenu）和子菜单（SubMenu）。

1．选项菜单

选项菜单在 Android 系统中发挥着重要作用，创建项目时默认生成的代码中就包含了完整的选项菜单代码，这些代码包括两部分：一部分是菜单文件，它是一个 xml 文件，位于 res/menu 目录下；另一部分是在 Activity 中的 Java 代码，其功能是加载菜单和为菜单提供事件处理。

（1）菜单文件。菜单文件的创建方法是在 res 目录下，新建 menu 文件夹，然后右击 menu 文件夹，在弹出的菜单中，选择 New→Menu resource file 命令，并输入文件名 main，按 Enter 键完成操作，默认生成的菜单文件是 main.xml，内容如下：

```
<menu xmlns:android="http://schemas.android.com/apk/res/android" >

    <item
        android:id="@+id/action_settings"
        android:orderInCategory="100"
        android:title="@string/action_settings"/>

</menu>
```

每个菜单文件对应一个菜单，而菜单文件中的每个 item 均是一个菜单项。有了菜单文件后，就能在 Activity 类中对菜单进行操作了。

（2）Activity 类。在 Activity 类中对菜单进行操作的内容主要包括两部分：一部分是加载菜单（onCreateOptionsMenu），另一部分是事件处理（onOptionsItemSelected）。具体添加过程：在 MainActivity 类代码的空白处右击，在弹出的菜单中，选择 Generate→Override Methods→android.app.Activity 命令，然后选择对应的加载菜单或事件处理方法即可。

①加载菜单的代码在默认生成的 Activity 类中已经存在，代码如下：

```
@Override
public boolean onCreateOptionsMenu(Menu menu) {
        getMenuInflater().inflate(R.menu.main, menu);
        return true;
}
```

注意：onCreateOptionsMenu()是一种回调方法，Activity 类启动时会自动调用这个方法，不需要程序员编写调用代码。如果应用程序不需要选项菜单，可以删除这段代码，否则它将加载 R.menu.main 中的菜单项。

②事件处理的代码则需要程序员编写，代码如下：

```
@Override
public boolean onOptionsItemSelected(MenuItem item) {
        Log.v("Menu","select menu"+item.getItemId());
        switch (item.getItemId()) {
        case R.id.item1:
            Toast.makeText(this, "菜单 1 被单击了", Toast.LENGTH_LONG).show();
            Log.v("Menu","menu1");
            break;
        case R.id.item2:
            Toast.makeText(this, "菜单 2 被单击了", Toast.LENGTH_LONG).show();
            Log.v("Menu","menu2");
            break;
        }
        return false;
}
```

onOptionsItemSelected()也是一种回调方法，程序员只需编写其中的处理代码即可。Android 系统还为菜单提供了一些其他的回调方法，如 onOptionsMenuClosed()和 onPrepareOptionsMenu()。

2．上下文菜单

在 Android Studio 环境中使用上下文菜单时，一般需要注册（绑定）到界面上的某个控件，如 TextView、ListView 等。下面举例介绍上下文菜单。

activity_main.xml 代码如下：

```
<?xml version="1.0" encoding="utf-8"?>
<RelativeLayout
    xmlns:android="http://schemas.android.com/apk/res/android"
    xmlns:tools="http://schemas.android.com/tools"
    android:id="@+id/activity_main"
    android:layout_width="match_parent"
    android:layout_height="match_parent"
    android:paddingLeft="@dimen/activity_horizontal_margin"
    android:paddingRight="@dimen/activity_horizontal_margin"
    android:paddingTop="@dimen/activity_vertical_margin"
    android:paddingBottom="@dimen/activity_vertical_margin"
    tools:context="com.example.administrator.contextmenu.MainActivity">

    <TextView
```

```
            android:text="LargeText"
            android:textAppearance="@style/Base.TextAppearance.AppCompat.Large"
            android:layout_width="wrap_content"
            android:layout_height="wrap_content"
            android:layout_alignParentTop="true"
            android:layout_centerHorizontal="true"
            android:layout_marginTop="50dp"
            android:id="@+id/textView1" />
</RelativeLayout>
```

MainActivity.java 代码如下：

```
package com.example.administrator.contextmenu;

import android.animation.ArgbEvaluator;
import android.content.ComponentName;
import android.graphics.Color;
import android.provider.CalendarContract;
import android.support.v7.app.AppCompatActivity;
import android.os.Bundle;
import android.view.ContextMenu;
import android.view.Menu;
import android.view.MenuInflater;
import android.view.MenuItem;
import android.view.View;
import android.widget.TextView;

public class MainActivity extends AppCompatActivity {
    TextView textView1;

    @Override
    protected void onCreate(Bundle savedInstanceState) {
        super.onCreate(savedInstanceState);
        setContentView(R.layout.activity_main);

        textView1 = (TextView) this.findViewById(R.id.textView1);
        registerForContextMenu(textView1);
    }

    @Override
    public void onCreateContextMenu(ContextMenu menu, View v, ContextMenu.ContextMenuInfo
menuInfo) {
        super.onCreateContextMenu(menu, v, menuInfo);
        menu.setHeaderTitle("颜色设置");
        menu.setHeaderIcon(R.drawable.setup);
        menu.add(0, 0, 0, "红色");
        menu.add(0, 1, 0, "绿色");
        menu.add(0, 2, 0, "蓝色");
    }

    @Override
    public boolean onContextItemSelected(MenuItem item) {
        switch (item.getItemId()) {
            case 0:
                textView1.setTextColor(Color.rgb(255,0,0));
                break;
            case 1:
                textView1.setTextColor(Color.rgb(0,255,0));
                break;
            case 2:
```

```
                    textView1.setTextColor(Color.rgb(0,0,255));
                    break;
            }
            return super.onContextItemSelected(item);
        }

        @Override
        public void onContextMenuClosed(Menu menu) {
            super.onContextMenuClosed(menu);
        }
}
```

插入 Override Methods 时，在 MainActivity 类代码的空白处右击，在弹出的菜单中，选择 Generate→Override Methods 命令，如图 2-19 所示。本例需要的三个方法，需要在 android.app.Activity 包中寻找，如图 2-20 所示。

最终，本例的运行效果如图 2-21 所示。

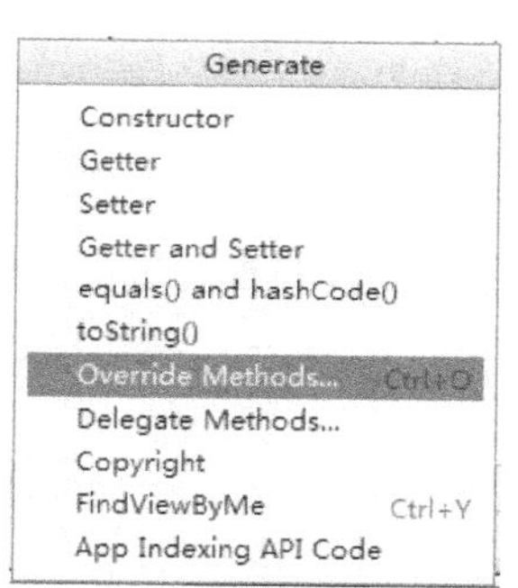

图 2-19　选择 Generate→Override Methods 命令

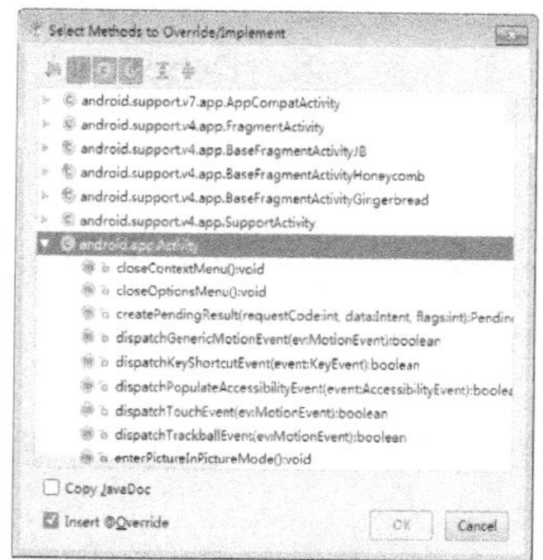

图 2-20　寻找三个方法

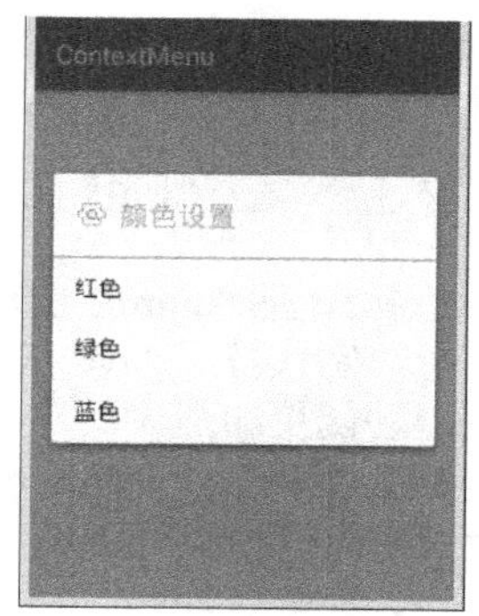

图 2-21　上下文菜单运行效果

3．子菜单

子菜单是一种将功能相同或相似的分组进行多级显示的菜单。

创建子菜单的步骤如下。

（1）覆盖 Activity 类的 onCreateOptionsMenu()方法，调用 Menu 的 addSubMenu()方法添加子菜单。

（2）调用 SubMenu 的 add()方法，添加子菜单。

（3）覆盖 onContextItemSelected()方法，响应子菜单的单击事件。

关于子菜单的更多详细内容，请参考本书的配套资源。

2.8　调试技巧

Android Studio 可以通过 Debug 方式进行程序断点调试，观察程序运行情况；通过 LogCat 生成的日志查看运行错误提示，进行程序纠错；通过 File Explorer 可以查看模拟手机的内置卡和 SD 卡中文件的生成情况；通过 ADB 通用调试工具发出 ADB 命令，实现与模拟手机的交互；此外，也可以直接使用真实手机进行 App 调试。

2.8.1　Debug

与 Java SE 相似，Android Studio 支持对 Android 项目执行断点调试的 Debug 方式，如图 2-22 所示；也可以通过暂停应用程序后观察代码并访问变量，调试运行的应用程序。

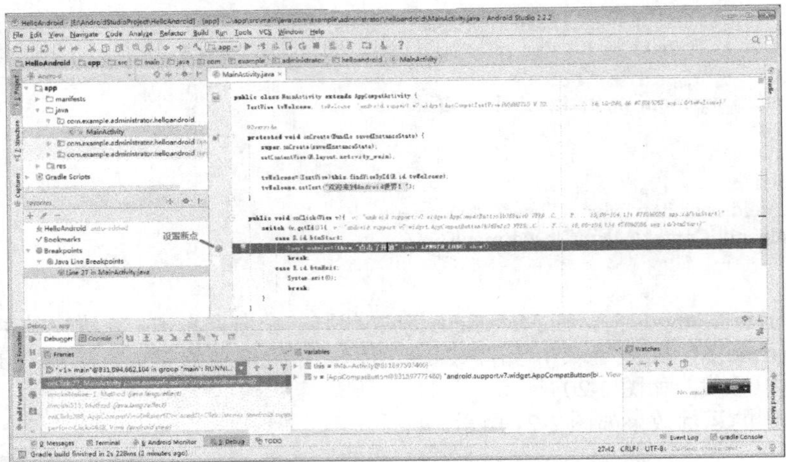

图 2-22　Debug 方式

2.8.2　LogCat

LogCat 是 Android SDK 支持的一款实用程序，可供真实手机或 AVD 在运行过程中输出调试信息，它代替了 Java 在 Console 上的标准输出（System.out.print），LogCat 提供的信息更加丰富，功能更加强大。

执行菜单命令 Tools→Android→Android Device Monitor，打开如图 2-23 所示的界面，然后选择 LogCat 选项卡。

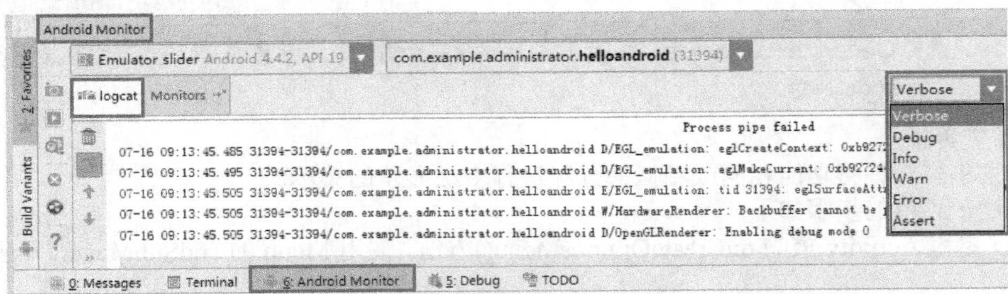

图 2-23　选择 LogCat 选项卡

1．调试信息的内容

LogCat 程序收集了系统和各种应用程序的日志，包括以下内容。

- 仿真器引发错误时的堆栈跟踪。
- 使用 android.util 包的 Log 类的方法从应用程序打印的消息。

2．调试信息的种类

Android 的调试信息分为五类，其严重程度依次增加，分别如下。

- Verbose 调试信息：表示详细信息。
- Debug 调试信息：表示调试信息（Debug）。
- Info 调试信息：表示一般提示性信息（Information）。
- Warn 调试信息：表示警告性信息（Warning）。
- Error 调试信息：表示错误信息。

3．调试信息的显示

调试信息通常在 LogCat 中显示。Android Studio 开发环境被广泛应用前，用户一般使用的 Eclipse 环境中没有 LogCat 窗口。但是，可以执行菜单命令 Window→Show View→Others，在打开的窗口中选择 Android→LogCat，并单击“OK”按钮即可。

LogCat 调试信息显示窗口的右上方有五个图标，代表各种过滤器，可过滤输出相应的调试信息，分别如下。

- V：不过滤，输出所有调试信息，包括 Verbose、Debug、Info、Warn、Error。
- D：Debug 过滤器，输出 Debug、Info、Warn、Error 调试信息。
- I：Info 过滤器，输出 Info、Warn、Error 调试信息。
- W：Warn 过滤器，输出 Warn 和 Error 调试信息。
- E：Error 过滤器，只输出 Error 调试信息。

4．项目中输出调试信息

Android Log 提供添加五种调试信息对应的方法，分别如下。

- Log.v(String tag, String msg)：添加 Verbose 调试信息的方法。
- Log.d(String tag, String msg)：添加 Debug 调试信息的方法。
- Log.i(String tag, String msg)：添加 Info 调试信息的方法。
- Log.w(String tag, String msg)：添加 Warn 调试信息的方法。
- Log.e(String tag, String msg)：添加 Error 调试信息的方法。

其中 tag 为调试信息标签名称，msg 为添加的调试信息。

例如，执行如下代码：

```
public void startClick(View v) {
    Log.i("MainActivity", "用户单击了“开始”按钮，激发相应的事件处理代码");
    Toast.makeText(this, "单击了开始", Toast.LENGTH_SHORT).show();
}
```

运行这段代码，当用户单击“开始”按钮后，会在 LogCat 窗口中显示所有的调试信息。

2.8.3　File Explorer

File Explorer（文件浏览器）列出了模拟手机中所有的文件和文件夹。

执行菜单命令 Tools→Android→Android Device Monitor，打开如图 2-24 所示的界面，然后选择 File Explorer 选项卡。

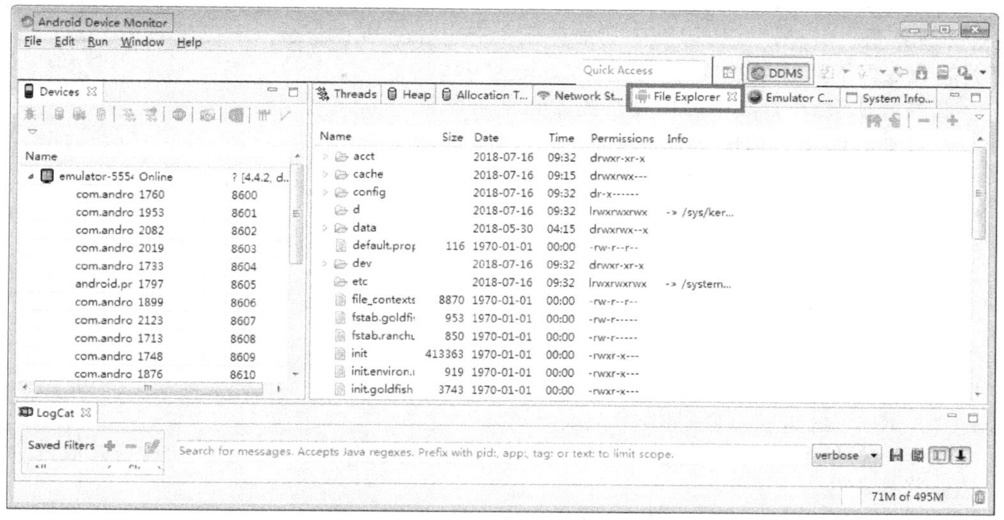

图 2-24　选择 File Explorer 选项卡

File Explorer 可以查看 Android 系统中的文件，并且可以复制（包括从计算机中复制到 Android 系统，以及从 Android 系统复制到计算机中）文件、删除文件。

通过 File Explorer 可以浏览 Android 的整个文件系统，可以看出，Android 系统就是基于 Linux 系统开发的，其中有 dev、etc、data、storage 等目录。通过访问目录“data/data/项目包名”，可找到保存在内置卡中的文件；通过访问目录“storage/emulated”，可找到保存在 SD 卡中的文件。

2.8.4 ADB 工具

ADB（Android Debug Bridge）是一个可以与 AVD（虚拟仿真器）进行通信的多功能命令行工具。在 Android Studio 环境下使用 ADB 工具与在 Eclipse 环境下使用 ADB 工具的路径一致，也位于 SDK 目录的 platform→tools 子目录下，如图 2-25 所示。

1. 登录 AVD

ADB 最常用的用途就是直接登录 AVD，对 AVD 进行交互式操作，如图 2-26 所示。

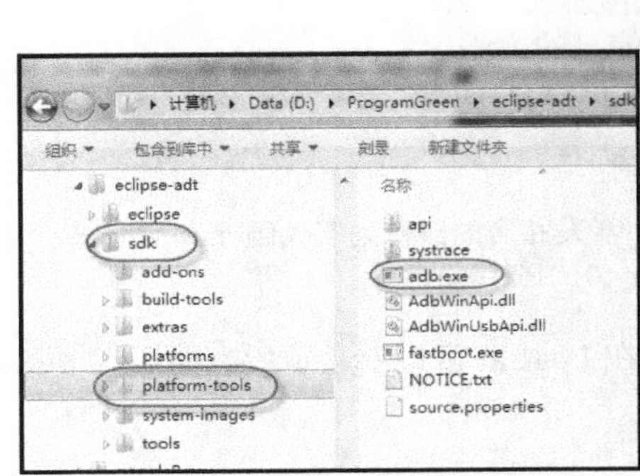

图 2-25 ADB 工具所在的路径

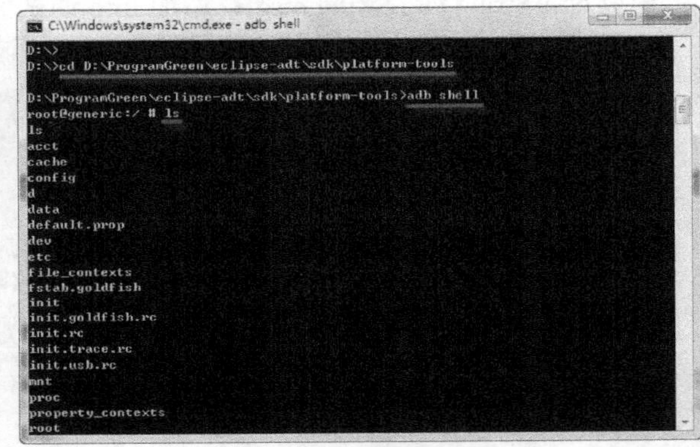

图 2-26 登录 AVD

打开命令行窗口，使用 CD 命令进入 ADB 所在的目录，然后输入以下命令：

```
adb shell
```

这时就进入 Android 的 shell 界面了，相当于以管理员的身份进入了 Android 的底层核心 Linux 管理界面。但是，在该界面中只能使用部分 Linux 核心命令，不仅没有与 Linux 图形界面有关的命令，而且还缺少 vi 等常见的 Linux 工具。

下面列举在 shell 界面中可以使用的 Linux 命令。

（1）目录的创建、切换与查看命令。

- mkdir <目录名>：创建指定的目录。
- cd <目录名>：切换到指定的目录。
- ls：列出目录的内容。
- pwd：列出当前所在目录的名字。

（2）目录与文件的复制、删除与移动命令。

- cp <源目录或文件> <目标目录或文件>：复制目录或文件。
- rm <文件>：删除指定的文件。
- rm -R -f <目录>：强制删除指定的整个目录。
- mv <旧目录或文件> <新目录或文件>：目录或文件改名，也可以改名到其他目录中。

（3）其他命令。

- cat /proc/cpuinfo：查看 CPU 的信息。

2. ADB 常用命令

（1）查看设备，代码如下：

```
adb devices
```

该命令用于查看当前连接的设备，连接到计算机的 Android 设备或模拟器将会以列表的方式显示。

（2）安装软件，代码如下：

```
adb install <apk 文件路径>
```

该命令将指定的 apk 文件安装到设备上。

（3）卸载软件，代码如下：

```
adb uninstall <软件名>
```

其中，软件名使用软件的包名作为唯一标识。

如果在 uninstall 后增加-k，表示卸载软件但保留配置及缓存文件。代码如下：

```
adb uninstall -k <软件名>
```

（4）登录设备的 shell 界面，代码如下：

```
adb shell
```

该命令用于登录虚拟仿真器的 shell 界面，进入交互状态。

登录 AVD 的 shell 界面，然后执行命令，最后退出界面，即直接运行设备命令，相当于执行远程命令，代码如下：

```
adb shell <command 命令>
```

（5）从计算机中发送文件到设备，代码如下：

```
adb push <本地路径> <远程路径>
```

该命令可以把计算机中的文件或文件夹复制到 AVD 中。

（6）从设备中将文件下载到计算机中，代码如下：

```
adb pull <远程路径> <本地路径>
```

该命令可以把 AVD 中的文件或文件夹复制到计算机中。

（7）显示帮助信息，代码如下：

```
adb help
```

（8）退出命令，代码如下：

```
exit
```

2.8.5　手机调试

如果使用手机进行调试，则需要下载 USB 驱动程序，操作方法如下。

单击“SDK Manager”按钮，弹出如图 2-27 所示的 Default Settings 对话框，在对话框的左侧选择 Android SDK 选项，在右侧选择 SDK Tools 选项卡，选中“Google USB Driver,rev 11”复选框，单击“OK”按钮，即可安装 USB 驱动程序。

使用 USB 数据线连接手机，需要在计算机中安装相关的 USB 驱动程序，而在手机中需要设置“USB 调试”模式。当在计算机和手机中设置成功后，使用 adb devices 命令可以查看手机设置，这样就能在手机上调试程序了。

注意：若通过 USB 数据线连接手机，并在手机上对应用程序进行调试，则必须安装相应的驱动程序。

若想在手机中进行调试，需要先将项目输出为 apk 文件，文件路径为 app→build→outputs→apk，如图 2-28 所示。再通过 USB 数据线将 apk 文件直接复制到手机中，或者通过 QQ 等传输方式，将 apk 文件发送到手机中，然后进行调试。apk 文件压缩包中的内容如图 2-29 所示。

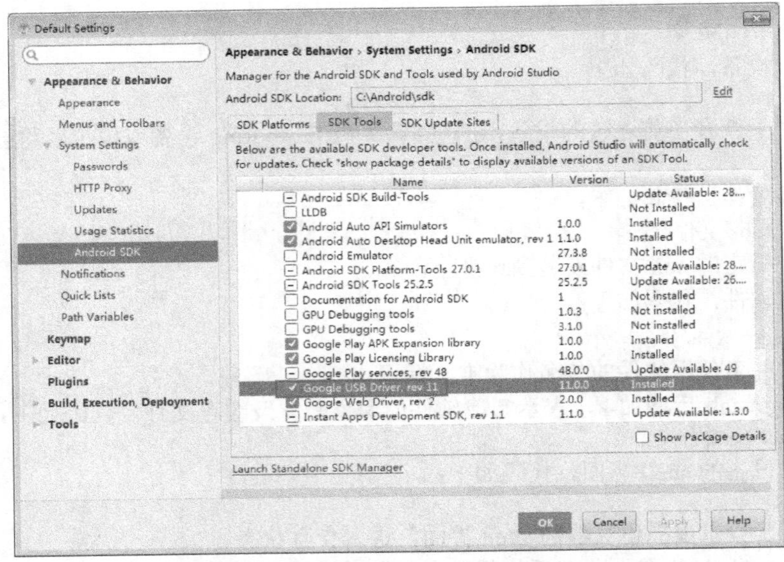

图 2-27　Default Settings 对话框

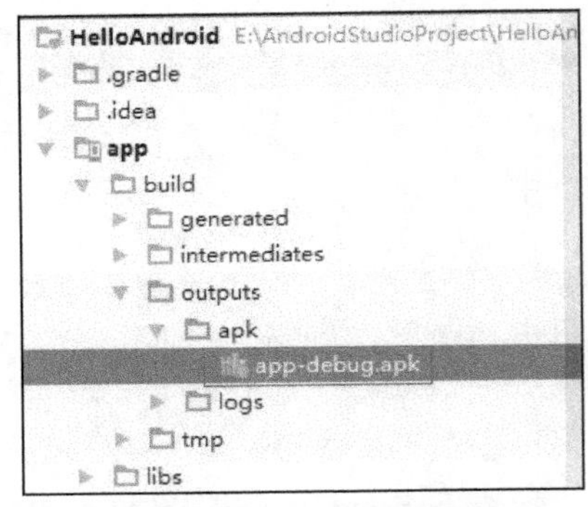

图 2-28　apk 文件所在的路径

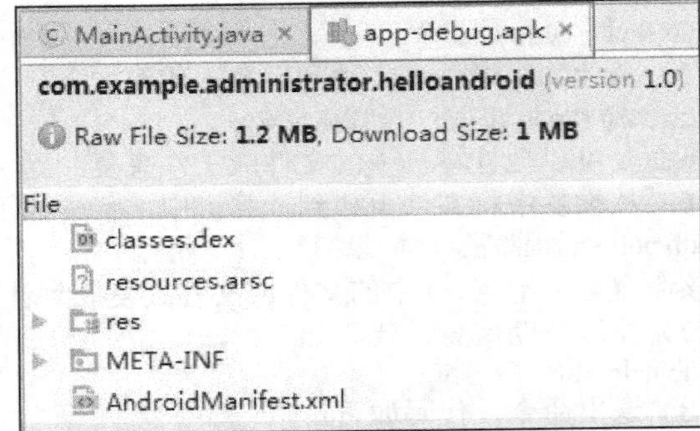

图 2-29　apk 文件压缩包中的内容

2.9　实训：完善计算器项目

围绕本章的计算器项目，增加以下需求。

- 添加连续计算功能，比如能够实现 2+3+4=9，(2+3)×4=20。
- 改进界面，使其美观一些，比如排齐按钮、改变底色或文字颜色。

实训要求： 建议读者手动输入所有代码，不从配套资源中复制、粘贴，适当增加实训次数，提高熟练程度。

2.10　实训：实现日期多选功能

本节的实训项目要求读者完成自定义 CalendarView 控件，实现日期多选功能，界面效果如图 2-30 所示，功能要求如下。

- 自定义控件上所有的文字均采用 TextView 控件进行显示。
- 创建 DateItem 日期项类。
- 创建 CalendarGridViewAdapter 适配器类。

- 把某月的日期列表适配到 GridView 中。
- 让用户通过单击动作选中日期，并修改用户所选日期的背景色（setBackgroundColor）。

日	一	二	三	四	五	六
25	26	27	28	29	30	1
2	3	4	5	6	7	8
9	10	11	12	13	14	15
16	17	18	19	20	21	22
23	24	25	26	27	28	29
30	31	1	2	3	4	5

图 2-30　日期多选界面效果

2.11　实训：设计用户注册的 Activity

本节的实训项目为设计用户注册的 Activity。activity_main.xml 布局文件如图 2-31 所示，界面效果如图 2-32 所示。在设计本项目的界面时，activity_main.xml 采用垂直线性布局，内部嵌套水平线性布局。

图 2-31　activity_main.xml 布局文件

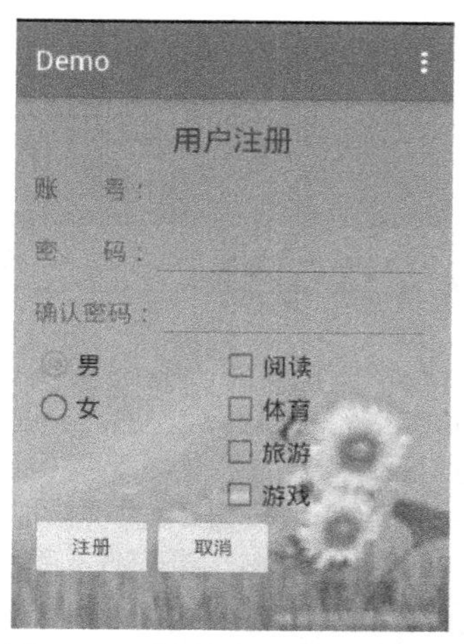

图 2-32　用户注册的界面效果

本项目的主要功能如下。

（1）如果“密码”和“确认密码”不一致，提示“两次密码不一致”，并将“确认密码”栏中的 EditText 内容清空。

（2）如果用户要进行注册，则当用户在界面中输入、勾选相关信息后，单击“注册”按钮，先出现提示信息“注册成功”，如图 2-33 所示；再出现“您的信息时：Nancy，123，女，阅读体育”提示信息。

（3）如果用户取消注册，则单击“取消”按钮，界面中自动清除“账号”“密码”“确认密码”栏中的 EditText 内容。

（4）用户单击虚拟键盘上的 Menu 键，在屏幕的下方出现定制的菜单，如图 2-34 所示。

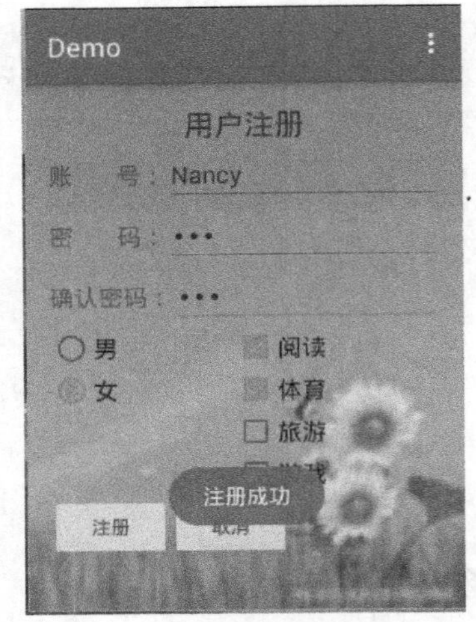

图 2-33 “注册成功”提示信息

图 2-34 用户定制的菜单

（5）用户单击“退出”按钮，出现如图 2-35 所示的普通对话框，当用户单击“放弃”按钮时，界面无反应；当用户单击“退出”按钮时，则程序退出运行。

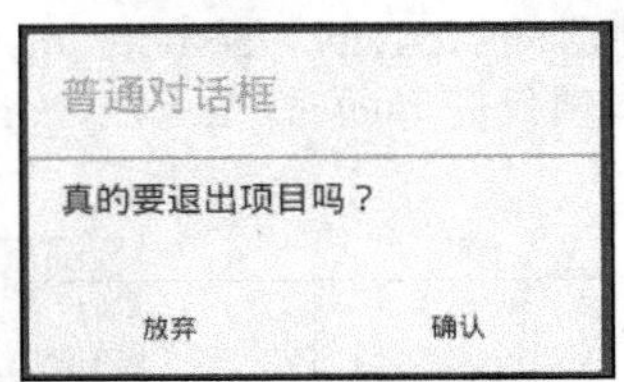

图 2-35 普通对话框

2.12 练习题

1．（ ）可作为 EditText 的提示信息。

A．android:inputType B．android:text C．android:digits D．android:hint

2．下列说法错误的是（ ）。

A．Button 是普通按钮控件，除此之外，还有其他按钮控件

B．TextView 是显示文本控件，TextView 是 EditText 的父类

C．EditText 是编辑文本控件，可以使用它编辑文本

D．ImageView 是显示图片控件，可以设置其属性显示局部图片

3．关于 Android 布局文件常用的长度/大小单位的描述，说法正确的是（ ）。

A．px 单位最精确，通常使用它

B．dp 单位与密度无关，通常使用它

C．pt 单位是最常用的，通常使用它

D．sp 单位与刻度相关，它专用于字体大小

4．关于适配器（Adapter，如 ArrayAdapter）的描述，说法正确的是（ ）。

A．它主要用来存储数据 B．它主要用来把数据绑定到控件上

C．它主要用来解析数据 D．它主要用来存储 xml 数据

5．关于 RelativeLayout 的描述，说法正确的是（ ）。

A．该布局为绝对布局，可以自定义控件的坐标位置

B．该布局为切换卡布局，可实现标签切换的功能

C．该布局为相对布局，其中，控件的位置都是相对位置

D．该布局为表格布局，需要配合 TableRow 一起使用

6．对于 xml 布局文件中的视图控件，layout_width 属性的值不可以是（　　）。

A．match_parent　　B．fill_parent　　C．wrap_content　　D．match_content

7．关于 ListView 使用方法的描述，说法错误的是（　　）。

A．要使用 ListView，必须为该 ListView 使用 Adpater 方式传递数据

B．要使用 ListView，该布局文件对应的 Activity 必须继承 ListActivity

C．在 ListView 中，项的视图布局既可以使用内置的布局方式，也可以使用自定义的布局方式

D．ListView 中的每项被选中时，将会触发 ListView 对象的 ItemClick 事件

8．在 Eclipse 环境中进行 Android 程序断点调试时，需要进入（　　）视图

A．Android 视图　　B．DDMS 视图　　C．Java 视图　　D．Debug 视图

9．Android 中的 Log 信息分为（　　）个级别。

A．3　B．4　C．5　D．7

10．如果将手机与计算机连接，但计算机中却无法显示手机，该现象可能和（　　）选项有关。

A．未知源　　B．无线 AP　　C．关于手机　　D．USB 调试

2.13 作业

1．屏幕像素单位有哪些？它们的含义是什么？

2．gravity 和 layout_gravity 的属性有什么区别？

3．在视图属性中，与长、宽有关的属性有哪些？它们有什么区别？

4．属性 id 有什么作用？如何在 Java 代码中使用 id 属性？

5．Android 有哪些常用的控件？它们的作用是什么？

6．EditText 的属性“android:inputType”有什么作用？该属性常用的取值有哪些？这些值的功能是什么？

7．Android 有哪些视图组（布局）？布局特点是什么？

8．Adapter 有什么作用？常用的 Adapter 有哪些？

第3章

Activity与Intent——运动会报名项目

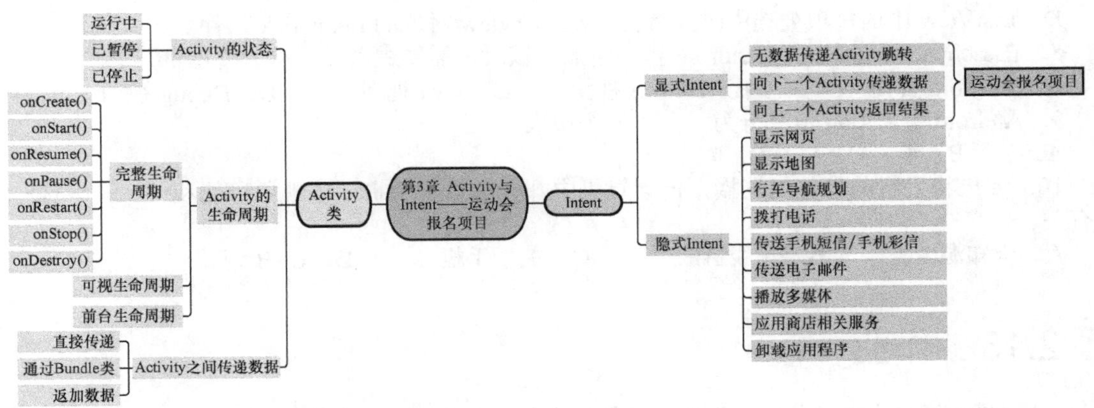

本章知识要点思维导图

在第2章中，我们主要讨论了单个Activity类和布局的设计，各种布局、控件的使用方法及其与Java代码之间的交互操作。本章将重点介绍多个Activity类和布局的设计，Activity类之间的跳转，数据的传递，以及Activity类的运行过程。

3.1 需求分析

开发一个简易的项目程序，可在移动端报名参加运动会赛项，项目要求如下。

（1）本项目包含两个界面，界面1为登录界面。其中，用户名固定为admin，密码固定为123。

（2）界面2为报名界面，每人最多只能选择两个项目，报名数据提交后，会以两个文件的形式保存到SD卡中。其中，一个文件只保存学生的学号，用于检测是否存在重复报名的学生。另一个文件用于保存完整的报名信息。

建议：感兴趣的读者，可以在完成第5章的学习后，对本项目进行修改，将登录与报名数据都存储在数据库中。

3.2 界面设计

本项目的登录界面和报名界面各对应一个Activity类和一个布局文件，界面效果如图3-1所示。

(a)

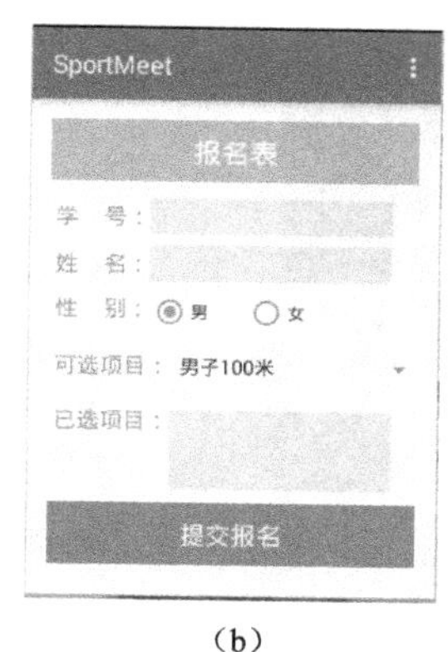

(b)

图 3-1 运动会报名项目界面效果

3.3 系统设计

本项目的系统设计主要包括功能设计、数据保存以及 SD 卡的访问权限。

3.3.1 功能设计

运动会报名项目，其功能设计主要考虑以下四点。

（1）登录界面：请用户输入正确的用户名和密码，如果用户输入正确，则提示“登录成功”，跳转到界面 2；如果用户输入错误，则提示“用户名或密码错误”。

注意：在界面 1 中，无选项菜单。

（2）报名界面：男、女生报名的运动项目被分别存放在两个不同的数组中。初次进入此界面时，控件 Spinner 用于填写男生报名的项目信息。用户可以单击 RadioButton 控件切换男、女选项，从而改变 Spinner 的项目信息。需要注意，填写 Spinner 时采用 ArrayAdapter。

用户输入学号后，在 etSno 中增加 addTextChangedListener 监听器，通过封装的 afterTextChanged()方法，采用 BufferedReader()方法按行读取 SD 卡上的 Sno.txt 文件。系统通过学号判定是否存在学生重复报名。如果检测到某学生重复报名，则提示“同一学号只允许报名一次”，并清空 etSno。

用户输入姓名，选择性别后，再通过 Spinner(etEvent)下拉菜单选择要报名的项目。每选择一个运动项目，该运动项目会被自动添加到下面的 TextView(tvEvent)中。用户可以只选择一个运动项目后就提交信息。

或者，选择完一个运动项目后，在 tvEvent 上长按屏幕，弹出上下文菜单，清空刚才所选信息，重新选择运动项目。每选择一个运动项目，该运动项目会被自动添加到 set 容器类中，通过 set.size()方法，判断所选运动项目是否超过两项。如果超过两项，就提示“报名项目最多为两项”，同时从 set 容器中删除最后一次选择的运动项目。如果没超过两项，则用户单击“提交”按钮后，其学号信息将写入 SD 卡的 Sno.txt 文件中；报名信息将写入 SD 卡的 Submit.txt 文件中。同时，屏幕提示“写文件完成！”。

清空 etSno、etName 和 tvEvent，以及用于保存报名信息的 set 容器，此外，还要清空编程过程中用到的几个内存变量（如 s、select、Event）。

注意：在界面 2 中，有一个选项菜单。

（3）选项菜单：为界面 2 添加菜单，菜单项分别为“退出系统”和“取消”，如图 3-2 所示。

（4）对话框：当用户单击“退出系统”选项菜单时，弹出“退出确认”对话框，如图 3-3 所示。该对话框包含“放弃”和“确定”两个按钮。单击“放弃”按钮，返回报名界面；单击“确定”按钮，则退出项目，结束程序运行。

图 3-2 选项菜单

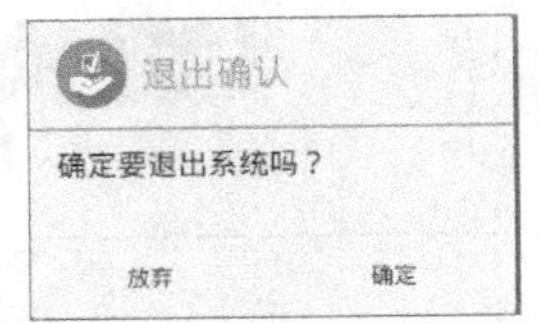

图 3-3 “退出确认”对话框

3.3.2 数据保存

将报名信息写入 SD 卡的文件，同一学号只允许报名一次。Sno.txt 和 Submit.txt 两个文件是可以追加的。

关键代码如下：

```
FileWriter fw = new FileWriter(Environment.getExternalStorageDirectory().getPath()+ "/Sno.txt", true);
FileWriter fw = new FileWriter(Environment.getExternalStorageDirectory().getPath()+ "/Submit.txt", true);
```

3.3.3 给 SD 卡开启访问权限

界面 2 中的操作需要读写 SD 卡，所以要给 SD 卡开启访问权限。把以下代码加入 AndroidManifest.xml 文件中：

```
<uses-permission android:name="android.permission.MOUNT_UNMOUNT_FILESYSTEMS" />
    <uses-permission android:name="android.permission.READ_EXTERNAL_STORAGE"/>
    <uses-permission android:name="android.permission.WRITE_EXTERNAL_STORAGE"/>
```

3.4 实施

实施项目时，要按照需求分析的要求规范设计，实现四项功能：用户信息的登录与验证、用户报名信息的填写、选项菜单功能及对话框功能。

3.4.1 创建项目

创建一个名为 SportMeet 的项目，默认的布局文件 activity_main.xml 和 MainActivity 类作为程序中的界面 1（登录界面）。

创建一个 Activity 类，命名为 EntryActivity，作为用户填写报名信息的界面。

操作步骤：在 app 目录中选择 app→java，找到 com.example.administrator.sportmeet 并右击，在弹出的菜单中选择 New→Activity→Empty Activity 命令，弹出如图 3-4 所示的对话框，填入类名称 EntryActivity。注意，这时系统会自动给出建议的布局文件名，不要修改此文件名。

注意：标题可以使用默认名，也可以改写，改写后的标题文字会保存到 string.xml 文件中。程序运行时，在标题栏可以看到改后的标题文字。

请留意文件名命名的特点：布局文件名只能用小写字母和下画线；类名采用驼峰命名法，因为创建的是 Activity 类，所以类名的末尾要加上“Activity”以示区别。

每次创建一个 Activity 类，同时系统内部将修改两个文件，即 AndroidManifest.xml 和 string.xml，并增加三个文件，即 Activity 类文件、布局文件和菜单文件。

提醒：请留意布局文件和菜单文件的命名的规则。

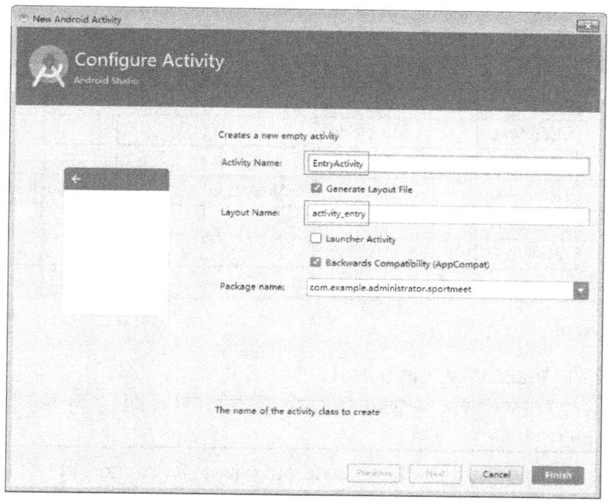

图 3-4　创建 Activity 类

创建 Activity 类之后，AndroidManifest.xml 中增加了以下代码：

```
<activity
        android:name="com.example.sportmeet.MainActivity"
        android:label="@string/app_name" >
        <intent-filter>
                <action android:name="android.intent.action.MAIN" />

                <category android:name="android.intent.category.LAUNCHER" />
        </intent-filter>
</activity>
<activity
        android:name="com.example.sportmeet.EntryActivity"
        android:label="@string/title_activity_entry" >
</activity>
```

string.xml 的内容如下：

```
<?xml version="1.0" encoding="utf-8"?>
<resources>
    <string name="app_name">SportMeet</string>
    <string name="action_settings">Settings</string>
    <string name="hello_world">Hello world!</string>
    <string name="title_activity_entry">EntryActivity</string>
</resources>
```

3.4.2　界面实现

登录界面是垂直线性布局，嵌套水平线性布局；报名界面是垂直线性布局，内嵌一个 ScrollView 控件。

登录界面 activity_main 图形控件的 Text 及 ID 属性见表 3-1，报名界面 activity_entry 图形控件的 Text 及 ID 属性见表 3-2。

表 3-1　activity_main 图形控件的 Text 及 ID 属性

控　件	Text 属性	ID 属性	控　件	Text 属性	ID 属性
TextView	用 户 登 录				
TextView	用户名：		EditText		etPersonName
TextView	密　码：		EditText		etPassword
Button	登　录	btnLogin			

表 3-2　activity_entry 图形控件的 Text 及 ID 属性

控件	Text 属性	ID 属性	控件	Text 属性	ID 属性	控件	Text 属性	ID 属性	控件	Text 属性	ID 属性
TextView	报名表										
TextView	学　号：		EditText		etSno						
TextView	姓　名：		EditText		etName						
TextView	性　别：		RadioGroup			RadioButton		rbtnNan	RadioButton		rbtnNv
TextView	可选项目：		Spinner		spEvent						
TextView	已选项目：		TextView		tvEvent						
Button	提交报名	btnSubmit									

1．activity_main.xml

登录界面的布局文件为 activity_main.xml。

登录界面的背景图片存放于 res→drawable 目录下，图片名称为 bg_login.png。

activity_main.xml 的代码如下：

```
<LinearLayout xmlns:android="http://schemas.android.com/apk/res/android"
    xmlns:tools="http://schemas.android.com/tools"
    android:id="@+id/LinearLayout1"
    android:layout_width="match_parent"
    android:layout_height="match_parent"
    android:orientation="vertical"
    android:paddingBottom="@dimen/activity_vertical_margin"
    android:paddingLeft="@dimen/activity_horizontal_margin"
    android:paddingRight="@dimen/activity_horizontal_margin"
    android:paddingTop="@dimen/activity_vertical_margin"
     android:background="@drawable/bg_login"
    tools:context=".LoginActivity" >

    <TextView
        android:id="@+id/textView1"
        android:layout_width="wrap_content"
        android:layout_height="wrap_content"
        android:text="用 户 登 录"
         android:layout_gravity="center"
        android:textAppearance="?android:attr/textAppearanceLarge" />

    <LinearLayout
        android:layout_width="match_parent"
        android:layout_height="wrap_content"
        android:layout_marginTop="20dp" >

        <TextView
            android:id="@+id/textView2"
            android:layout_width="wrap_content"
            android:layout_height="wrap_content"
            android:text="用户名："
            android:textAppearance="?android:attr/textAppearanceMedium" />

        <EditText
            android:id="@+id/etPersonName"
            android:layout_width="match_parent"
            android:layout_height="wrap_content"
            android:layout_weight="1"
            android:ems="10"
            android:inputType="textPersonName" >
```

```
            <requestFocus />
        </EditText>

    </LinearLayout>

    <LinearLayout
        android:layout_width="match_parent"
        android:layout_height="wrap_content"
        android:layout_marginTop="20dp" >

        <TextView
            android:id="@+id/textView2"
            android:layout_width="wrap_content"
            android:layout_height="wrap_content"
            android:text="密　码："
            android:textAppearance="?android:attr/textAppearanceMedium" />

        <EditText
            android:id="@+id/etPassword"
            android:layout_width="wrap_content"
            android:layout_height="wrap_content"
            android:layout_weight="1"
            android:ems="10"
            android:inputType="textPassword" />

    </LinearLayout>

    <Button
        android:id="@+id/btnLogin"
        android:layout_width="wrap_content"
        android:layout_height="wrap_content"
        android:layout_marginTop="20dp"
            android:layout_gravity="center"
        android:text="登　录" />

</LinearLayout>
```

2．activity_entry.xml

报名界面的布局文件为 activity_entry.xml。

设计报名界面时，应注意以下几项内容。

- 为了使内容能完全显示出来，报名界面的主体部分采用了一个 ScrollView 容器控件。
- round_et.xml 与 etSno、etName、tvEvent 搭配，作为背景的自定义 xml 文件。
- 在 tvEvent 的属性中，maxLines =4。

activity_entry.xml 的代码如下：

```
<LinearLayout xmlns:tools="http://schemas.android.com/tools"
    xmlns:android="http://schemas.android.com/apk/res/android"
    android:id="@+id/LinearLayout1"
    android:layout_width="match_parent"
    android:layout_height="match_parent"
    android:orientation="vertical"
    android:paddingBottom="@dimen/activity_vertical_margin"
    android:paddingLeft="@dimen/activity_horizontal_margin"
    android:paddingRight="@dimen/activity_horizontal_margin"
    android:paddingTop="@dimen/activity_vertical_margin"
    tools:context=".RegActivity" >

    <TextView
```

```
            android:id="@+id/textView1"
            android:layout_width="match_parent"
            android:layout_height="50dp"
            android:background="#FF9912"
            android:gravity="center"
            android:text="报名表"
            android:textAppearance="?android:attr/textAppearanceLarge"
            android:textColor="#FFFFFF" />

        <include
            android:id="@+id/body"
            android:layout_height="wrap_content"
            layout="@layout/scrollview" />

</LinearLayout>
```

3．round_et.xml

报名界面 EditText 背景的自定义文件为 round_et.xml，保存在 res→drawable 目录下，需要单独创建。创建方法：右击 drawable，在弹出的菜单中选择 New→Drawable resource file 命令，弹出如图 3-5 所示的对话框，输入文件名 round_et，路径名称为 drawable，最后单击“OK”按钮。

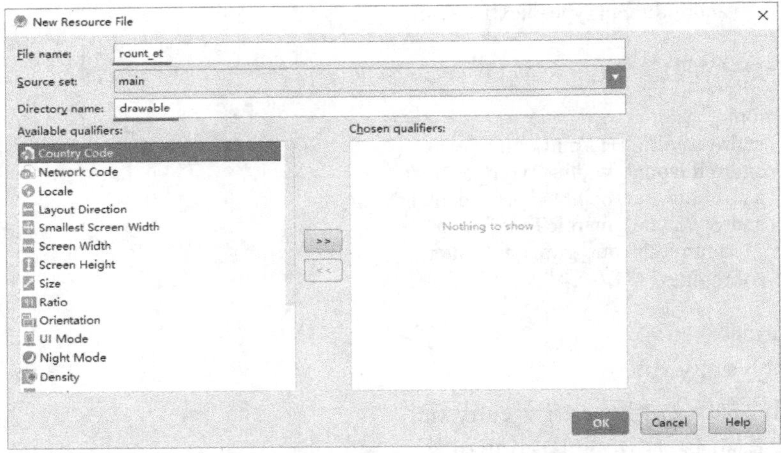

图 3-5　创建 round_et.xml 文件

round_et.xml 的代码如下：

```
<?xml version="1.0" encoding="utf-8"?>
<shape xmlns:android="http://schemas.android.com/apk/res/android"
    android:shape="rectangle" >"
    <solid android:color="#DEDEDE"></solid>

    <corners
        android:topLeftRadius="0dp"
        android:bottomLeftRadius="0dp"
        android:topRightRadius="0dp"
        android:bottomRightRadius="0dp"    />
</shape>
```

4．scroll_view.xml

报名界面主体部分的布局文件为 scroll_view.xml，代码如下：

```
<?xml version="1.0" encoding="utf-8"?>
```

```
<ScrollView xmlns:android="http://schemas.android.com/apk/res/android"
    android:layout_width="match_parent"
    android:layout_height="wrap_content">

  <LinearLayout
        android:layout_width="match_parent"
        android:layout_height="wrap_content"
        android:layout_marginTop="5dp"
        android:orientation="vertical" >

      <LinearLayout
            android:layout_width="match_parent"
            android:layout_height="wrap_content"
            android:layout_marginTop="5dp"
            android:orientation="horizontal" >

            <TextView
                android:id="@+id/textView2"
                android:layout_width="wrap_content"
                android:layout_height="wrap_content"
                android:text=" 学   号: "
                android:textAppearance="?android:attr/textAppearanceMedium" />

            <EditText
                android:id="@+id/etSno"
                android:layout_width="190dp"
                android:layout_height="30dp"
                android:background="@drawable/round_et"
                android:ems="10" />
        </LinearLayout>

        <LinearLayout
            android:layout_width="match_parent"
            android:layout_height="wrap_content"
            android:layout_marginTop="20dp"
            android:orientation="horizontal" >

            <TextView
                android:id="@+id/textView3"
                android:layout_width="wrap_content"
                android:layout_height="wrap_content"
                android:text=" 姓   名: "
                android:textAppearance="?android:attr/textAppearanceMedium" />

            <EditText
                android:id="@+id/etName"
                android:layout_width="190dp"
                android:layout_height="30dp"
                android:background="@drawable/round_et"
                android:ems="10" />
        </LinearLayout>

        <LinearLayout
            android:layout_width="match_parent"
            android:layout_height="wrap_content"
            android:layout_marginTop="20dp"
            android:orientation="horizontal" >
```

```
        <TextView
            android:id="@+id/textView4"
            android:layout_width="wrap_content"
            android:layout_height="wrap_content"
            android:text=" 性  别： "
            android:textAppearance="?android:attr/textAppearanceMedium" />

        <RadioGroup
            android:id="@+id/radioGroup1"
            android:layout_width="wrap_content"
            android:layout_height="wrap_content"
             android:orientation="horizontal" >

            <RadioButton
                android:id="@+id/rbtnNan"
                android:layout_width="wrap_content"
                android:layout_height="wrap_content"
                android:layout_marginRight="30dp"
                android:layout_weight="1"
                android:checked="true"
                android:text="男" />

            <RadioButton
                android:id="@+id/rbtnNv"
                android:layout_width="wrap_content"
                android:layout_height="wrap_content"
                android:layout_weight="1"
                android:text="女" />
        </RadioGroup>
    </LinearLayout>

    <LinearLayout
        android:layout_width="match_parent"
        android:layout_height="wrap_content"
        android:layout_marginTop="20dp"
        android:orientation="horizontal" >

        <TextView
            android:id="@+id/textView5"
            android:layout_width="wrap_content"
            android:layout_height="wrap_content"
            android:text=" 可选项目： "
            android:textAppearance="?android:attr/textAppearanceMedium" />

        <Spinner
            android:id="@+id/spEvent"
            android:layout_width="190dp"
            android:layout_height="wrap_content" />
    </LinearLayout>

    <LinearLayout
        android:layout_width="match_parent"
        android:layout_height="wrap_content"
        android:layout_marginTop="20dp"
        android:orientation="horizontal" >

        <TextView
            android:id="@+id/textView6"
```

```
                android:layout_width="wrap_content"
                android:layout_height="wrap_content"
                android:text=" 已选项目："
                android:textAppearance="?android:attr/textAppearanceMedium" />

            <TextView
                android:id="@+id/tvEvent"
                android:layout_width="175dp"
                android:layout_height="70dp"
                android:background="@drawable/round_et"
                android:maxLines="4"
                android:text="Small Text"
                android:textAppearance="?android:attr/textAppearanceSmall"
                android:textColor="#FF0000" />

        </LinearLayout>

            <Button
android:id="@+id/btnSubmit"
            android:layout_width="match_parent"
            android:layout_height="50dp"
            android:layout_marginTop="10dp"
            android:background="#00BF00"
            android:text="提交报名"
            android:textColor="#FFFFFF"
            android:textSize="20sp" />

    </LinearLayout>

</ScrollView>
```

3.4.3　Java 代码

1．MainActivity.java

登录界面的 MainActivity.java 代码如下：

```
package com.example.administrator.sportmeet;

import android.support.v7.app.AppCompatActivity;
import android.content.Intent;
import android.os.Bundle;
import android.view.View;
import android.view.View.OnClickListener;
import android.widget.Button;
import android.widget.EditText;
import android.widget.Toast;

public class MainActivity extends AppCompatActivity{
    EditText etPersonName,etPassword;
    Button btnLogin;

    @Override
    protected void onCreate(Bundle savedInstanceState) {
        super.onCreate(savedInstanceState);
        setContentView(R.layout.activity_main);

        etPersonName=(EditText) this.findViewById(R.id.etPersonName);
        etPassword=(EditText) this.findViewById(R.id.etPassword);
```

```
            btnLogin=(Button)this.findViewById(R.id.btnLogin);
            btnLogin.setOnClickListener(new OnClickListener(){

            @Override
            public void onClick(View arg0) {
                if("admin".equals(etPersonName.getText().toString()) &&
                        "123".equals(etPassword.getText().toString()))
                {
                    //提示“登录成功”，然后再跳转到界面 2
                    Toast.makeText(MainActivity.this, "登录成功",1).show();
                    Intent i=new Intent(MainActivity.this,EntryActivity.class);
                    finish();
                    startActivity(i);
                }
                else
                {
                    //账号或密码错误，并清除 EditText
                    Toast.makeText(MainActivity.this, "用户名或密码错误",1).show();
                    etPersonName.setText("");
                    etPassword.setText("");
                }
            }});
        }
}
```

2. EntryActicity.java

报名界面的 EntryActivity.java 代码如下：

```
package com.example.administrator.sportmeet;

import java.io.BufferedReader;
import java.io.FileNotFoundException;
import java.io.FileReader;
import java.io.FileWriter;
import java.io.IOException;
import java.lang.reflect.Method;
import java.util.HashSet;
import java.util.LinkedList;
import java.util.List;
import java.util.Set;

import android.support.v7.app.AppCompatActivity;
import android.app.AlertDialog;
import android.content.DialogInterface;
import android.os.Bundle;
import android.os.Environment;
import android.text.Editable;
import android.text.TextWatcher;
import android.view.ContextMenu;
import android.view.Menu;
import android.view.MenuItem;
import android.view.View;
import android.view.View.OnClickListener;
import android.widget.AdapterView;
import android.widget.AdapterView.OnItemSelectedListener;
import android.widget.ArrayAdapter;
import android.widget.Button;
import android.widget.EditText;
```

```
import android.widget.RadioButton;
import android.widget.Spinner;
import android.widget.TextView;
import android.widget.Toast;

public class EntryActivity extends AppCompatActivity implements OnClickListener {
    EditText etSno, etName;
    RadioButton rbtnNan, rbtnNv;
    Button btnSubmit;
    Spinner spEvent;
    String[] eventNan = new String[] { "男子 100 米", "200 米", "400 米", "800 米",
            "1500 米", "5000 米", "4×100 米", "4×400 米", "110 米跨栏", "跳高", "跳远", "三级跳",
                "铁饼", "铅球", "标枪" };
    String[] eventNv = new String[] { "女子 100 米", "200 米", "400 米", "800 米",
            "1500 米", "3000 米", "4×100 米", "4×400 米", "110 米跨栏", "跳高", "跳远", "铁饼",
                "铅球", "标枪" };
    TextView tvEvent;
    String Sno, Name, Gender = "男", Event = "";
    String s="", select = "";
    Set set = new HashSet();
    List list = new LinkedList();
    boolean isSpinnerFirst = true;

    @Override
    protected void onCreate(Bundle savedInstanceState) {
        super.onCreate(savedInstanceState);
        setContentView(R.layout.activity_entry);

        etSno = (EditText) this.findViewById(R.id.etSno);
        etSno.addTextChangedListener(new TextWatcher() {

            @Override
            public void afterTextChanged(Editable arg0) {
                try {
                    BufferedReader br = new BufferedReader(new FileReader(
                            Environment.getExternalStorageDirectory().getPath()
                                    + "/Sno.txt"));
                    String result = "";
                    while ((result = br.readLine()) != null)
                        list.add(result);
                    for (int i = 0; i < list.size(); i++) {
                        String str = (String) list.get(i);
                        if (str.equals(etSno.getText().toString())) {
                            Toast.makeText(EntryActivity.this, "同一学号只允许报名一次", 1)
                                    .show();
                            etSno.setText("");
                        }
                    }
                } catch (FileNotFoundException e) {
                    e.printStackTrace();
                } catch (IOException e) {
                    e.printStackTrace();
                }
            }

            @Override
            public void beforeTextChanged(CharSequence arg0, int arg1,
                    int arg2, int arg3) {
```

```
        }

        @Override
        public void onTextChanged(CharSequence arg0, int arg1, int arg2,
                int arg3) {


        }
    });

    etName = (EditText) this.findViewById(R.id.etName);

    rbtnNan = (RadioButton) this.findViewById(R.id.rbtnNan);
    rbtnNan.setOnClickListener(this);
    rbtnNv = (RadioButton) this.findViewById(R.id.rbtnNv);
    rbtnNv.setOnClickListener(this);

    tvEvent = (TextView) this.findViewById(R.id.tvEvent);
    tvEvent.setText("");

    spEvent = (Spinner) this.findViewById(R.id.spEvent);
    ArrayAdapter adapterNan = new ArrayAdapter(this,
            android.R.layout.simple_spinner_dropdown_item, eventNan);
    spEvent.setAdapter(adapterNan);
    spEvent.setOnItemSelectedListener(new OnItemSelectedListener() {

        @Override
        public void onItemSelected(AdapterView<?> av, View v, int position,
                long arg3) {
            if (isSpinnerFirst) {
                // 使 Spinner 的第一项默认不被选中
                isSpinnerFirst = false;
                return;
            }

            Event = av.getItemAtPosition(position).toString();
            set.add(Event);
            select += Event + "\r\n";
            if (set.size() > 2) {
                Toast.makeText(EntryActivity.this, "报名项目最多为两项", 1).show();
                set.remove(Event);
            } else {
                tvEvent.setText(select);
            }
        }

        @Override
        public void onNothingSelected(AdapterView<?> arg0) {

        }
    });

    registerForContextMenu(tvEvent);

    btnSubmit = (Button) this.findViewById(R.id.btnSubmit);
    btnSubmit.setOnClickListener(this);
```

```
}

@Override
public boolean onCreateOptionsMenu(Menu menu) { // 使用此方法调用 OptionsMenu()
    getMenuInflater().inflate(R.menu.entry, menu);
    return true;
}

@Override
public boolean onMenuOpened(int featureId, Menu menu) { // 菜单打开后发生的动作
    if (menu != null) {
        if (menu.getClass().getSimpleName().equalsIgnoreCase("MenuBuilder")) {
            try {
                Method method = menu.getClass().getDeclaredMethod(
                        "setOptionalIconsVisible", Boolean.TYPE);
                method.setAccessible(true);
                method.invoke(menu, true);
            } catch (Exception e) {
                e.printStackTrace();
            }
        }
    }
    return super.onMenuOpened(featureId, menu);
}

@Override
public boolean onOptionsItemSelected(MenuItem item) { // 选中菜单项后发生的动作
    switch (item.getItemId()) {
    case R.id.menu_exit:
        AlertDialog.Builder builder = new AlertDialog.Builder(this);
        builder.setMessage("确定要退出系统吗？").setTitle("退出确认")
                .setIcon(R.drawable.ok);

        builder.setPositiveButton("确定",
                new DialogInterface.OnClickListener() {

                    @Override
                    public void onClick(DialogInterface arg0, int arg1) {
                        System.exit(0);
                    }
                });

        builder.setNegativeButton("放弃",
                new DialogInterface.OnClickListener() {

                    @Override
                    public void onClick(DialogInterface arg0, int arg1) {

                    }
                });

        AlertDialog dialog = builder.create();
        dialog.show();

        break;
```

```
        case R.id.menu_cancel:
            break;
        }
        return super.onOptionsItemSelected(item);
    }

    @Override
    public void onCreateContextMenu(ContextMenu menu, View v,
            ContextMenu.ContextMenuInfo menuInfo) {
        super.onCreateContextMenu(menu, v, menuInfo);
        menu.add(0, 0, 0, "清空");
    }

    @Override
    public boolean onContextItemSelected(MenuItem item) {
        switch (item.getItemId()) {
        case 0:
            select="";
            tvEvent.setText("");
            set.clear();
            break;
        }
        return super.onContextItemSelected(item);
    }

    @Override
    public void onContextMenuClosed(Menu menu) {
        super.onContextMenuClosed(menu);
    }

    @Override
    public void onClick(View v) {
        switch (v.getId()) {
        case R.id.rbtnNan:
            isSpinnerFirst=true;
            if (rbtnNan.isChecked()) {
                Gender = "男";
                ArrayAdapter adapterNan = new ArrayAdapter(this,
                        android.R.layout.simple_spinner_dropdown_item, eventNan);
                spEvent.setAdapter(adapterNan);
            }
            break;
        case R.id.rbtnNv:
            isSpinnerFirst=true;
            if (rbtnNv.isChecked()) {
                Gender = "女";
                ArrayAdapter adapterNv = new ArrayAdapter(this,
                        android.R.layout.simple_spinner_dropdown_item, eventNv);
                spEvent.setAdapter(adapterNv);
            }
            break;
        case R.id.btnSubmit:
            try {
                FileWriter fw = new FileWriter(Environment
                        .getExternalStorageDirectory().getPath() + "/Sno.txt",
                        true);
                String s = etSno.getText().toString() + "\r\n";
```

```
                fw.write(s);
                fw.flush();
                fw.close();
            } catch (IOException e) {
                e.printStackTrace();
            }

            try {
                FileWriter fw = new FileWriter(Environment
                        .getExternalStorageDirectory().getPath()
                        + "/Submit.txt", true);
                String s = etSno.getText().toString() + ","
                        + etName.getText().toString() + "," + Gender + ","
                        + select + "\r\n";
                fw.write(s);
                fw.flush();
                fw.close();
                Toast.makeText(this, "写文件完成！ ", 0).show();
            } catch (IOException e) {
                e.printStackTrace();
            }

            etSno.setText("");
            etName.setText("");
            tvEvent.setText("");
            s="";
            Event="";
            select="";
            set.clear();
            break;
        }
    }
}
```

3.4.4　运行测试

将项目在 AVD 上运行，测试其是否与需求分析中的要求相符。

如图 3-6 所示为保存后的 Sno.txt 和 Submit.txt，即学号和报名信息，查看路径是为 File Explorer→storage→sdcard。

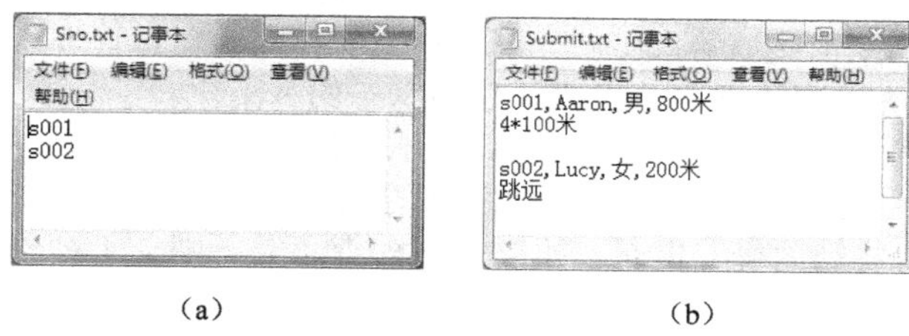

（a）　　　　　　　　　　（b）

图 3-6　保存后的 Sno.txt 和 Submit.txt

我们在开发项目时，一般会涉及多界面问题，如果每次都按顺序从界面 1 开始测试，就会浪费很多测试时间。为了提高效率，我们可以打开 AndroidManifest.xml 文件，切换程序运行的入口程序。学习过 C#的读者可能会想到，在 Program.cs 文件中的 Main()方法里修改 Application.Run(new Form1())，其操作思路和此处的切换程序运行的入口程序是类似的。

运动会报名项目涉及两个界面，界面 1 是登录界面，界面 2 是报名界面。只要在两个不同的 Activity 之间移动<intent-filter>…</intent-filter>代码段即可，如图 3-7 所示。

```
<application
    android:allowBackup="true"
    android:icon="@drawable/icon"
    android:label="用户管理"
    android:theme="@style/AppTheme" >
    <activity
        android:name="com.example.usergl.MainActivity"
        android:label="用户管理" >
        <intent-filter>
            <action android:name="android.intent.action.MAIN" />

            <category android:name="android.intent.category.LAUNCHER" />
        </intent-filter>
    </activity>

    <activity
        android:name="com.example.usergl.WeihuActivity"
        android:label="@string/title_activity_weihu" >

    </activity>
</application>
```

图 3-7 切换程序运行的入口程序

3.5 Activity 的状态与生命周期

一般情况下，新建一个界面时，总会自带一个屏幕类的 onCreate()方法。在后续的学习过程中，还会遇到 onPause()、onResume()等方法。但是，这些方法有什么作用呢？哪些代码应该写入对应的哪种方法呢？若想解决这些问题，就需要读者对 Activity 的生命周期有较深入的了解，从而更好地编写高质量的程序。

Activity 是 Android 程序中最主要的一种组件，它相当于一个用户界面，并且用户能够通过 Activity 与应用程序进行交互。

一个应用程序可以有多个 Activity，应用程序启动时显示的 Activity 被称为主 Activity。

3.5.1 Activity 的状态

Activity 有生命周期，它的状态有以下三种。

- 运行中：当 Activity 在屏幕前台时（位于当前任务堆栈的顶部），它是激活或运行状态。它就是响应用户操作的 Activity。
- 已暂停：当 Activity 上面有另外一个 Activity，且当前的 Activity 失去了焦点但对用户仍可见时，它处于暂停状态。若当前的 Activity 之上的 Activity 没有完全覆盖屏幕（或是透明的），则被暂停的 Activity 对用户仍可见，并且它处于存活状态（即被暂停的 Activity 保留着所有的状态和成员信息，并且保持着与窗口管理器的连接）。但是，如果系统内存不足，便会杀死当前处于暂停状态的 Activity。
- 已停止：当 Activity 完全被另一个 Activity 覆盖时，则当前的 Activity 处于停止状态。处于停止状态的 Activity 仍保留所有的状态和成员信息，但对用户是不可见的，所以它的窗口将被隐藏。但是，如果系统内存不足，则会杀死当前处于停滞状态的 Activity。

例如，当用户使用手机玩游戏时，恰好有电话呼入，这时的手机屏幕会被来电提醒所在的 Activity 覆盖，则游戏所在的 Activity 会进入停止状态，如果系统内存不够时，就会杀死游戏所在的 Activity，即游戏的运行状态会消失。所以，这就要求程序设计师应充分考虑此类情况，即当游戏所在的 Activity 进入停止状态时，要保存游戏的状态信息，以便游戏再次开启时能恢复到停止前的状态。

当 Activity 从一种状态转变为另一种状态时，为了实现状态的保存或恢复，可以使用以下七种调用方法，如表 3-3 所示。

表 3-3　Activity 的七种调用方法

方　法	描　述
void onCreate(Bundle savedInstanceState)	当第一次创建 Activity 时被调用。该方法可用于所有初始化设置：创建视图、绑定数据至列表等
void onStart()	当 Activity 对用户可见时被调用
void onRestart()	当 Activity 停止后，且在再次启动前被调用
void onResume()	在 Activity 与用户进行交互之前被调用。此时的 Activity 位于堆栈顶部，并接受用户输入
void onPause()	当系统将要启动另一个 Activity 时被调用。该方法主要用于保存状态，且该方法的所有动作应在短时间内完成，因为下一个 Activity 必须等到该方法返回后才能继续
void onStop()	当 Activity 不再对用户可见时被调用。该方法可能发生的情形有：当前的 Activity 被销毁，或者另一个 Activity（现存的或是最新的）回到运行状态并覆盖了当前的 Activity
void onDestroy()	在 Activity 销毁前被调用。该方法是 Activity 接收的最后一种调用方法

3.5.2　Activity 的生命周期

Activity 的生命周期可以分为多个阶段，包括完整生命周期、可视生命周期和前台生命周期。

1. 完整生命周期

Activity 的完整生命周期如图 3-8 所示。Activity 的完整生命周期自第一次调用 onCreate()方法开始，直到调用 onDestroy()方法为止。Activity 在 onCreate()方法中设置所有“全局”状态以完成初始化，而在 onDestroy()方法中释放所有系统资源。例如，Activity 的一个线程要在后台运行，准备从网络下载数据，则 Activity 会在 onCreate()方法中创建线程，而在 onDestroy()方法中销毁线程。

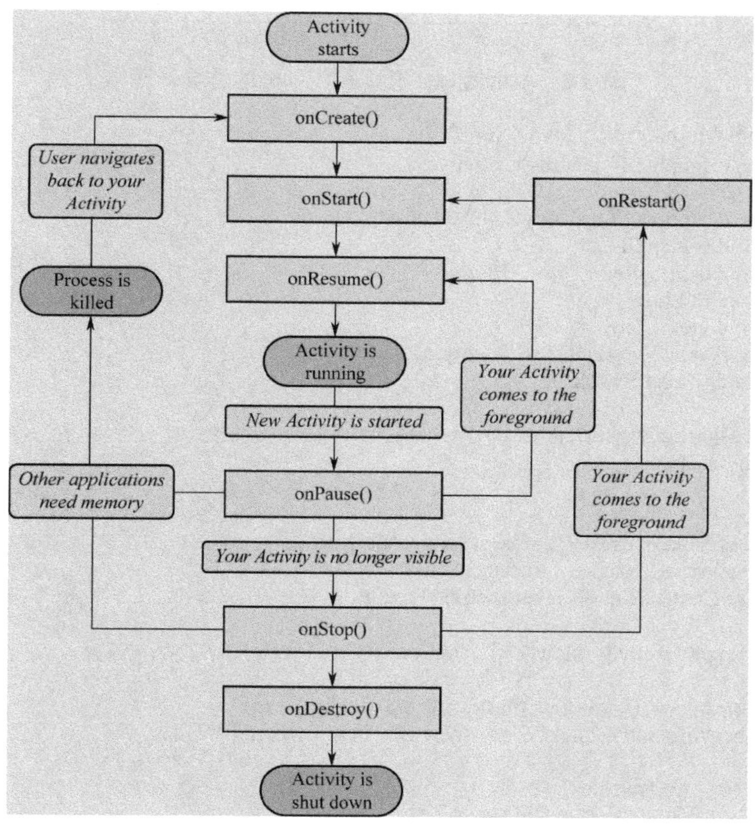

图 3-8　Activity 的生命周期

2．可视生命周期

Activity 的可视生命周期自 onStart()方法调用开始，直到相应的 onStop()方法调用结束。在此期间，即使Activity不在前台或者不与用户进行交互，用户依然可以在屏幕上看到Activity。在 onStart()方法和 onStop()方法之间，可以保留向用户显示当前 Activity 所需的资源。例如，当用户看不见显示的内容时，可以在 onStart()方法中注册一个 BroadcastReceiver 来监控影响 UI 的变化，而在 onStop()方法中进行注销。onStart()方法和 onStop()方法可以根据应用程序是否对用户可见而被多次调用。

3．前台生命周期

Activity 的前台生命周期自 onResume()方法调用开始，直到相应的 onPause()方法调用结束。在此期间，Activity 位于前台的最上面并与用户进行交互。Activity 会经常在暂停和恢复两种状态之间进行转换。例如，当设备转入休眠状态或者有新的Activity启动时，当前的Activity将调用 onPause()方法；当 Activity 获得结果或者接收到新的 Intent 时会调用 onResume()方法。

下面，我们举例说明 Activity 生命周期的各阶段变化，如图 3-9 所示。

（1）在图 3-9 中，界面 1 和界面 2 的界面设计代码分别为 activity_main.xml 和 activity_second. xml，请读者扫描下方的二维码查看。

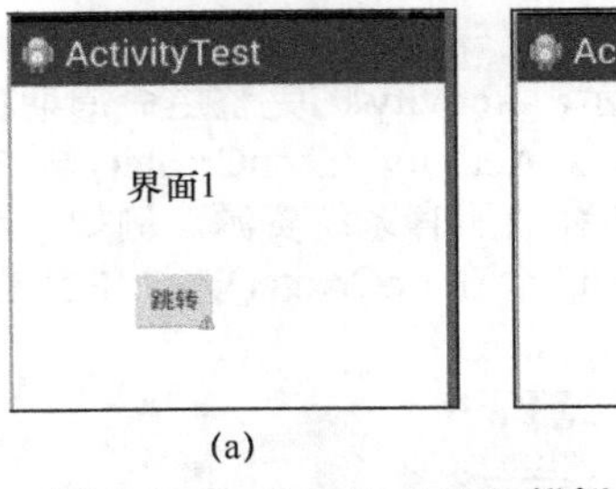

(a)

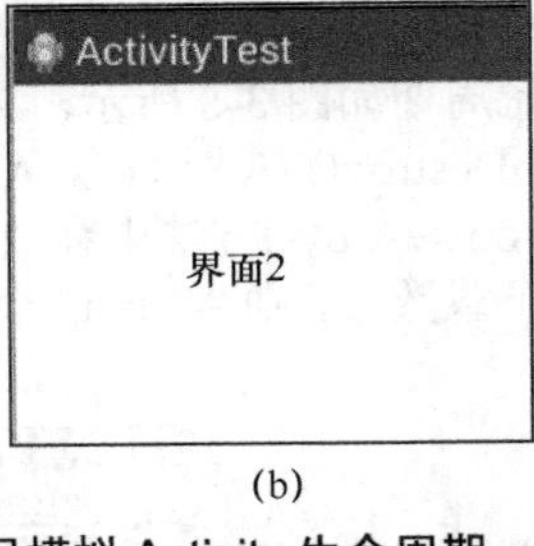

(b)

activity_main.xml

activity_second.xml

图 3-9　ActivityTest 项目模拟 Activity 生命周期

（2）本例的 MainActivity.java 代码如下：

```
package com.example.administrator.activitytest;

import android.support.v7.app.AppCompatActivity;
import android.content.Intent;
import android.os.Bundle;
import android.util.Log;
import android.view.View;
import android.view.View.OnClickListener;
import android.widget.Button;

public class MainActivity extends AppCompatActivity {
    Button btnJump;

    @Override
    protected void onCreate(Bundle savedInstanceState) {
        super.onCreate(savedInstanceState);
        setContentView(R.layout.activity_main);

        Log.v("Activity 生命周期", "onCreate()");

        btnJump=(Button)this.findViewById(R.id.btnJump);
        btnJump.setOnClickListener(new OnClickListener(){

            @Override
            public void onClick(View arg0) {
                Intent i=new Intent(MainActivity.this,SecondActivity.class);
```

```
                    startActivity(i);
                }});
    }

    @Override
    protected void onDestroy() {
        super.onDestroy();
        Log.v("Activity 生命周期", "onDestroy()");
    }

    @Override
    protected void onPause() {
        super.onPause();
        Log.v("Activity 生命周期", "onPause()");
    }

    @Override
    protected void onRestart() {
        super.onRestart();
        Log.v("Activity 生命周期", "onRestart()");
    }

    @Override
    protected void onResume() {
        super.onResume();
        Log.v("Activity 生命周期", "onResume()");
    }

    @Override
    protected void onStart() {
        super.onStart();
        Log.v("Activity 生命周期", "onStart()");
    }

    @Override
    protected void onStop() {
        super.onStop();
        Log.v("Activity 生命周期", "onStop()");
    }
}
```

（3）SecondActivity.java 不需要进行编程，请读者观察界面 1。

（4）为了便于读者观察 Activity 生命周期的各阶段，以及七种调用方法的使用情况，我们运行 ActivityTest 项目，通过模拟器切换两个界面并对项目运行过程中的 LogCat 信息进行过滤，观察各方法出现的顺序。操作步骤如下。

①选择如图 3-10 所示的“Android Device Monitor”窗口左下方的 LogCat 选项卡，单击左侧的绿色“+”号，增加一个过滤器。在弹出的如图 3-11 所示的对话框中设置过滤器，输入 Filter Name（过滤器名称）及 by Log Tag（被过滤标识）。

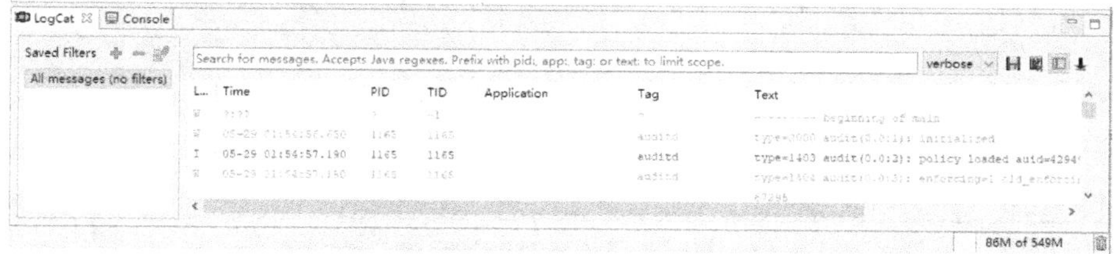

图 3-10　选择 logCat 选项卡

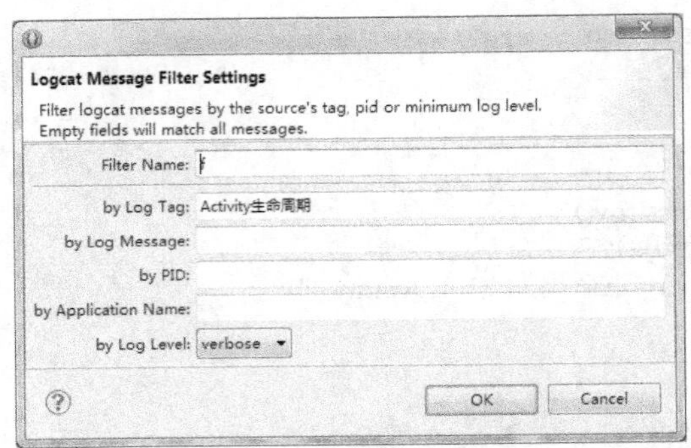

图 3-11 设置过滤器

②项目运行后，会看到如图所示的界面 1 出现在模拟器的前台，此时调用了三种方法，先后顺序分别为 onCreate()、onStart()和 onResume()，如图 3-12 所示。

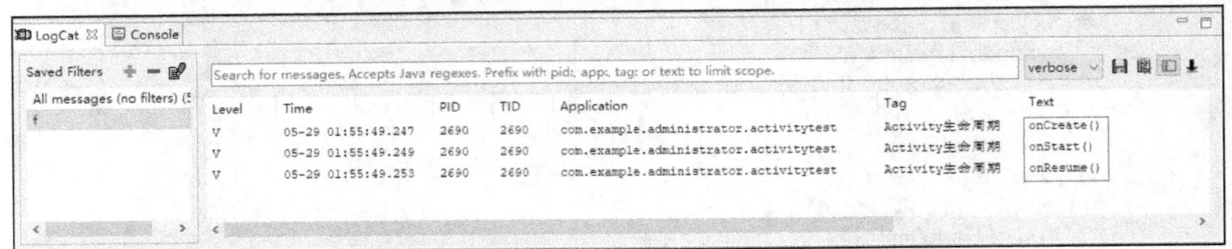

图 3-12 界面 1 出现时调用的三种方法

③单击界面 1 中的“跳转”按钮，跳转到界面 2。请注意，不要对界面 1 使用 finish()方法。此时，会看到界面 1 转入模拟器的后台运行，界面 2 出现在模拟器的前台，这时调用了两种方法，先后顺序分别为 onPause()和 onStop()，如图 3-13 所示。

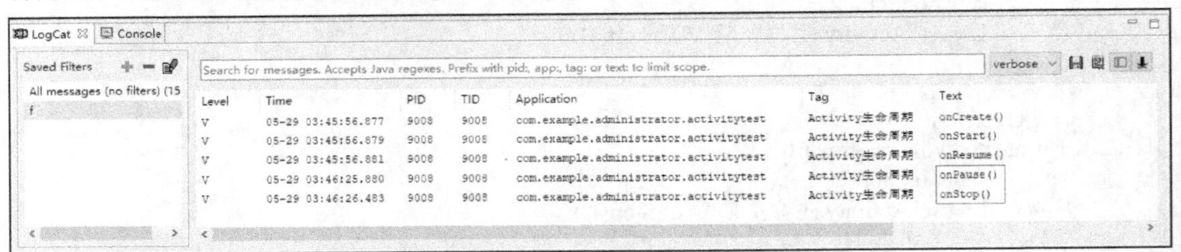

图 3-13 界面 2 出现时调用的两种方法

④单击模拟器右侧的返回按钮◁，返回界面 1。此时，会看到界面 2 转入模拟器的后台运行，界面 1 重新出现在模拟器的前台，这时调用了三种方法，先后顺序分别为 onRestart()、onStart()和 onResume()，如图 3-14 所示。

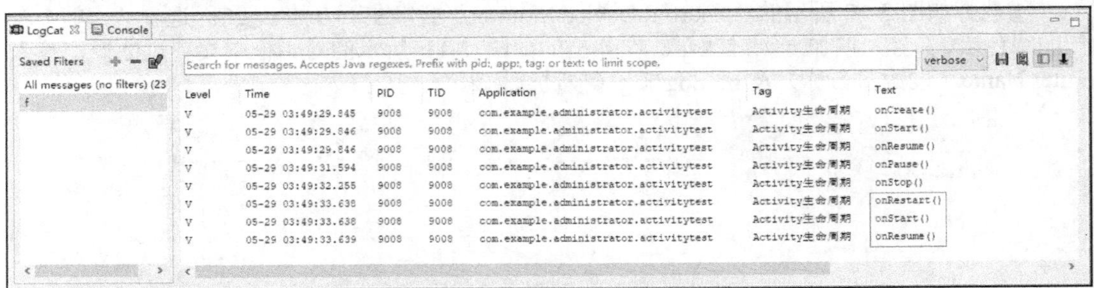

图 3-14 界面 1 再次出现时调用的三种方法

⑤再次单击模拟器右侧的返回按钮◁，退出项目。此时，界面返回模拟器桌面，这时调用了三种方法，先后顺序分别为 onPause()、onStop()和 onDestroy()，如图 3-15 所示。

Level	Time	PID	TID	Application	Tag	Text
V	05-29 03:49:29.845	9008	9008	com.example.administrator.activitytest	Activity生命周期	onCreate()
V	05-29 03:49:29.846	9008	9008	com.example.administrator.activitytest	Activity生命周期	onStart()
V	05-29 03:49:29.846	9008	9008	com.example.administrator.activitytest	Activity生命周期	onResume()
V	05-29 03:49:31.594	9008	9008	com.example.administrator.activitytest	Activity生命周期	onPause()
V	05-29 03:49:32.255	9008	9008	com.example.administrator.activitytest	Activity生命周期	onStop()
V	05-29 03:49:33.638	9008	9008	com.example.administrator.activitytest	Activity生命周期	onRestart()
V	05-29 03:49:33.638	9008	9008	com.example.administrator.activitytest	Activity生命周期	onStart()
V	05-29 03:49:33.639	9008	9008	com.example.administrator.activitytest	Activity生命周期	onResume()
V	05-29 03:51:23.540	9008	9008	com.example.administrator.activitytest	Activity生命周期	onPause()
V	05-29 03:51:24.070	9008	9008	com.example.administrator.activitytest	Activity生命周期	onStop()
V	05-29 03:51:24.070	9008	9008	com.example.administrator.activitytest	Activity生命周期	onDestroy()

图 3-15　界面 1 结束运行时调用的三种方法

分析 Activity 的生命周期，可以得出以下结论。

- 不论 Activity 的是初始启动，还是暂停后重新开始，onResume()方法是 Activity 在显示前必经的阶段。因此，该阶段是屏幕数据刷新的最佳机会。正因如此，日记列表的数据刷新操作要在 onResume()方法中完成。
- 不论 Activity 是被隐藏，还是被杀死，onPause()方法是 Activity 离开显示后必经的阶段。因此，该阶段是保存状态的最佳机会。正因如此，保存日记的代码要在 onPause()方法中完成。
- onCreate()方法用于初始化 Activity，所有初始化代码都放入 onCreate()方法中。

此外，我们还可以发现，当系统内存不足时，系统可以随时杀死处于已停止或已暂停的 Activity，从而释放内存满足系统的需求。所以，在多数情况下，我们设计应用程序时不必考虑程序的自行退出情况，可以让操作系统结束程序的运行状态。但是，我们需要考虑在 Activity 进入暂停之前，要在 onPause()方法中保存当前的状态；在 Activity 对用户显示之前，要在 onResume()方法中恢复原先的状态。

Android 系统没有提供完全退出应用程序的方法，而 System.exit()方法也只是停止当前的 Activity。如果要完全退出应用程序，就必须想办法停止该应用程序的所有 Activity，这是一项比较复杂的工作，但通常情况下，不需要程序员来完成。

3.6　Intent

Intent 的单词原意为意图，即想要做什么，这个概念在 Android 系统开发过程中有十分重要的作用。

在 Android 系统中，Intent 用于协助应用程序之间的交互与通信，它负责对应用中某次操作的动作、动作涉及的数据及附加数据进行描述，Android 系统则根据 Intent 的描述，找到对应的调用组件并将 Intent 传递过去，进而完成组件的调用过程。Intent 不仅用于应用程序之间，也可用于应用程序内部的 Activity/Service 之间的交互。因此，Intent 在这里起中介的作用，专门提供组件互相调用的相关信息，实现调用者与被调用者之间的解耦。

Intent 的使用方式有以下三种。

- 直接通过 startActivity()方法或 Activity.startActivityForResult()方法启动一个 Activity。
- 通过 startService()方法启动一项服务，或者通过 bindService()方法和后台服务交互。
- 通过广播的方法（如 sendBroadcast()、sendOrderedBroadcast()或 sendStickyBroadcast()）发送给 Broadcast receivers。

例如，在本项目的 MainActivity.java 中，通过 startActivity()方法启动 DiaryListActivity：

```
Intent it = new Intent(MainActivity.this, DiaryListActivity.class);
startActivity(it);
```

Intent 分为两种：一种是显式 Intent，另一种是隐式 Intent。

3.6.1 显式 Intent

显式 Intent 在构造 Intent 对象时就指定接收者，因此具有确定的将要运行的对象。

例如，对于如下代码：

```
Intent it = new Intent(Activity1.this, Activity2.class);
```

其中，Activity1 是发送者，Activity2 是接收者，这里的 Intent 有明确的接收者，因此它是显式 Intent。

显式 Intent 的用途有以下三种，下面分别介绍。

1. 无数据传递 Activity 跳转

举例说明，从 Activity1 跳转到 Activity2，无数据传递。代码为：

```
Intent it = new Intent(Activity1.this, Activity2.class);
startActivity(it);
```

这种方式在本章的运动会报名项目的 MainActivity 中使用过。

2. 向下一个 Activity 传递数据

举例说明，从 Activity1 跳转到 Activity2 时，使用 Bundle 和 Intent.putExtras 将数据从 Activity1 传递到 Activity2，过程如下。

（1）发送方 Activity1 的代码如下：

```
Intent it = new Intent(Activity1.this, Activity2.class);
Bundle bundle=new Bundle();
bundle.putString("name", "This is from MainActivity!");
it.putExtras(bundle);
startActivity(it);
```

（2）接收方 Activity2 获取数据时可以采用如下代码：

```
Bundle bundle=getIntent().getExtras();
String name=bundle.getString("name");
```

注意：“name”是数据的 key，必须保证双方一致。

这种方式在本章的运动会报名项目的 DiaryListActivity 中使用过，即在跳转的同时向 DiaryEditActivity 传递数据。

3. 向上一个 Activity 返回结果

举例说明，从 Activity1 跳转到 Activity2 时，如果 Activity2 要向 Activity1 返回数据，这时需要使用 startActivityForResult(it,REQUEST_CODE)方法来启动 Activity2，过程如下。

（1）Activity 跳转，代码为：

```
Intent it = new Intent(Activity1.this, Activity2.class);
startActivityForResult (it, 1234);
```

（2）返回方（返回数据）Activity2 的代码为：

```
Intent intent=getIntent();
Bundle bundle2=new Bundle();
bundle2.putString("name", "This is from ShowMsg!");
intent.putExtras(bundle2);
setResult(RESULT_OK, intent);
```

（3）调用方（接收数据）Activity1 的代码为：

```
@Override
protected void onActivityResult(int requestCode, int resultCode, Intent data) {
    super.onActivityResult(requestCode, resultCode, data);
    if (requestCode==1234){
        if(resultCode==RESULT_CANCELED)
            setTitle("cancel");
```

```
        else if (resultCode==RESULT_OK) {
            String temp=null;
            Bundle bundle=data.getExtras();
            if(bundle!=null) temp=bundle.getString("name");
            setTitle(temp);
        }
    }
}
```

其中用到了回调上一个 Activity 的结果处理方法 onActivityResult()。

3.6.2 隐式 Intent

隐式 Intent 的发送者在构造 Intent 对象时，并不知道也不关心接收者是谁，有利于降低发送者和接收者之间的耦合。例如，对于如下代码：

```
Uri uri = Uri.parse("http://google.com");
Intent it = new Intent(Intent.ACTION_VIEW, uri);
```

其中，并没有指定接收 Intent 的 Activity，因此它是一个隐式 Intent。

隐式 Intent 的用途有以下几种，下面分别介绍。

1．显示网页

举例说明，用内置浏览器显示网页内容，代码如下：

```
Uri uri = Uri.parse("http://google.com");
Intent it = new Intent(Intent.ACTION_VIEW, uri);
startActivity(it);
```

本例中，接收Intent的Activity并不明确，系统会根据uri查找浏览器，启动浏览器的Activity来接收这个 Intent。如果设备上安装了多个浏览器应用程序，则系统会提示用户选择哪种浏览器来打开这个 uri。

2．显示地图

举例说明，用内置地图显示指定位置（如经纬度），代码如下：

```
Uri uri = Uri.parse("geo:31.498814,120.267279");
Intent it = new Intent(Intent.ACTION_VIEW, uri);
startActivity(it);
```

再比如，显示其他地理位置的 geo uri，代码如下：

```
geo:latitude,longitude
geo:latitude,longitude?z=zoom
geo:0,0?q=my_street_address
geo:0,0?q=business+near+city
google.streetview:cbll=lat,lng&cbp=1,yaw,,pitch,zoom&mz=mapZoom
```

3．行车导航规划

举例说明，显示两个指定位置之间的导航信息，代码如下：

```
Uri uri = Uri.parse("http://maps.google.com/maps?f=d&saddr=startLat%20startLng&daddr=
endLat%20endLng&hl=en");
Intent it = new Intent(Intent.ACTION_VIEW, uri);
startActivity(it);
```

4．拨打电话

拨打电话分以下两种情形，分别举例说明。

（1）调用拨号程序，代码如下：

```
Uri uri = Uri.parse("tel:0800000123");
Intent it = new Intent(Intent.ACTION_DIAL, uri);
startActivity(it);
```

（2）直接拨打电话，代码如下：

```
Uri uri = Uri.parse("tel:0800000123");
```

```
Intent it = new Intent(Intent.ACTION_CALL, uri);
startActivity(it);
```

请读者注意，这种方式需要在 AndroidManifest.xml 中添加权限设置：

```
<uses-permission id="android.permission.CALL_PHONE" />
```

5．传送手机短信/手机彩信

传送手机短信（SMS，Short Message Service）或手机彩信（MMS，Multimedia Messaging Service）分为以下两种情形。下面均以手机短信为例说明。

（1）调用手机短信程序，代码如下：

```
Intent it = new Intent(Intent.ACTION_VIEW, uri);
it.putExtra("sms_body", "The SMS text");
it.setType("vnd.android-dir/mms-sms");
startActivity(it);
```

（2）直接发送手机短信，代码如下：

```
Uri uri = Uri.parse("smsto://0800000123");
Intent it = new Intent(Intent.ACTION_SENDTO, uri);
it.putExtra("sms_body", "The SMS text");
startActivity(it);
```

6．传送电子邮件

传送电子邮件分以下两种情形，分别举例说明。

（1）调用电子邮件程序，代码如下：

```
Uri uri = Uri.parse("mailto:xxx@abc.com");
Intent it = new Intent(Intent.ACTION_SENDTO, uri);
startActivity(it);
```

（2）直接发送电子邮件，代码如下：

```
Intent it = new Intent(Intent.ACTION_SEND);
it.putExtra(Intent.EXTRA_EMAIL, "me@abc.com");
it.putExtra(Intent.EXTRA_TEXT, "The email body text");
it.setType("text/plain");
startActivity(Intent.createChooser(it, "Choose Email Client"));
```

7．播放多媒体

举例说明，播放多媒体的代码如下：

```
Uri uri = Uri.parse("file:///sdcard/song.mp3");
Intent it = new Intent(Intent.ACTION_VIEW, uri);
it.setType("audio/mp3");
startActivity(it);
Uri uri = Uri.withAppendedPath(MediaStore.Audio.Media.INTERNAL_CONTENT_URI, "1");
Intent it = new Intent(Intent.ACTION_VIEW, uri);
startActivity(it);
```

8．应用商店相关服务

若 Android 的应用程序在应用商店（Market）出售，可以通过 Intent 查询相关信息。

（1）在应用商店寻找某个应用程序，代码如下：

```
Uri uri = Uri.parse("market://search?q=pname:pkg_name");
Intent it = new Intent(Intent.ACTION_VIEW, uri);
startActivity(it);
```

注意：pkg_name 是应用程序的完整路径。

（2）显示某个应用程序的相关信息，代码如下：

```
Uri uri = Uri.parse("market://details?id=app_id");
Intent it = new Intent(Intent.ACTION_VIEW, uri);
startActivity(it);
```

注意：app_id 是应用程序的 ID，在应用商店中可以看到。

9．卸载应用程序

举例说明，卸载应用程序的代码如下：

```
Uri uri = Uri.fromParts("package", strPackageName, null);
Intent it = new Intent(Intent.ACTION_DELETE, uri);
startActivity(it);
```

3.6.3　Intent 的解析机制

Android 系统不需要对显式 Intent 进行解析，因为目标组件已经很明确；不过，Android 系统需要对隐式 Intent 进行解析。通过解析，系统将隐式 Intent 映射给可以处理该 Intent 的 Activity、IntentReceiver 或 Service。

Intent 的解析机制：通过查找已在 AndroidManifest.xml 中注册的所有 IntentFilter 及其中定义的 Intent，找到匹配的 Intent。在这个解析过程中，Android 系统通过 Intent 的 action、type 和 category 三个属性进行判断，判断方法如下。

- 如果 Intent 指定了 action，则目标组件里 IntentFilter 的 action 列表中就必须包含这个 action，否则不能匹配。
- 如果 Intent 没有提供 type，系统将从 data 中得到数据类型。和 action 一样，目标组件的数据类型列表中必须包含 Intent 的数据类型，否则不能匹配。
- 如果 Intent 中的数据不是 content: 类型的 uri，而且 Intent 也没有明确指定它的 type，系统将根据 Intent 中数据的 scheme （如 http: 或 mailto:） 进行匹配。同理，Intent 的 scheme 必须出现在目标组件的 scheme 列表中。
- 如果 Intent 指定了一个或多个 category，这些类别必须全部出现在组建的类别列表中。例如，Intent 中包含了两个类别：LAUNCHER_CATEGORY 和 ALTERNATIVE_CATEGORY，解析得到的目标组件必须至少包含这两个类别。

扫描旁边的二维码，了解 IntentFilter 的定义。

IntentFilter 的定义

3.7　Activity 之间传递数据

Activity 之间传递数据需要 Intent 的协助，传递的方法有多种，本节将举例说明。

3.7.1　直接传递

例如，从 Activity1 跳转到 Activity2，当传递少量简单的数据（基本数据类型及其数组）时，可以使用直接传递的方法。

（1）在 Activity1 中编写代码：

```
Intent intent = new Intent(Activity1.this,
            Activity2.class);
intent.putExtra("key", "a value");
startActivity(intent);
```

（2）在 Activity2 中编写代码：

```
Intent it = getIntent();
String str = it.getStringExtra("key");
```

3.7.2　使用 Bundle 类

例如，从 Activity1 跳转到 Activity2，当传递成组的数据或复杂的数据时，可以使用 Bundle 类。Bundle 的单词原意为束、捆、包等，因此 Bundle 类在 Android 系统中用于传递成组或复杂的数据。

（1）在 Activity1 中编写代码：

```
Intent intent = new Intent(Activity1.this,
            Activity2.class);
Bundle bundle = new Bundle();
bundle.putString("key", "a value");
intent.putExtras(bundle);
startActivity(intent);
```

（2）在 Activity2 中编写代码：

```
Intent it = getIntent();
Bundle bundle = it.getExtras();
String str = bundle.getString("key");
```

3.7.3 返回数据

例如，从 Activity1 跳转到 Activity2，当被调用的 Activity 需要向调用的 Activity 返回数据时，则应使用特别的技术来实现。

（1）在 Activity1 中编写代码：

```
Intent intent = new Intent(Activity1.this, Activity2.class);
startActivityForResult (intent,1234);
```

（2）在 Activity2 中编写代码：

```
Intent intent=getIntent();
Bundle bundle2=new Bundle();
bundle2.putString("name", "This is from ShowMsg!");
intent.putExtras(bundle2);
setResult(RESULT_OK, intent);
```

（3）在 Activity1 中编写代码：

```
@Override
protected void onActivityResult(int requestCode, int resultCode, Intent data) {
    super.onActivityResult(requestCode, resultCode, data);
    if (requestCode == 1234) {
        if (resultCode == RESULT_CANCELED){
            setTitle("cancel");
        }else if (resultCode == RESULT_OK) {
            String temp = null;
            Bundle bundle = data.getExtras();
            if (bundle != null){
                temp = bundle.getString("name");
            }
            setTitle(temp);
        }
    }
}
```

3.8 实训：完善运动会报名项目

围绕本章的运动会报名项目，请读者从以下几个方面考虑，进一步完善项目。

- 运动员报名的主要信息有四项（学号、姓名、性别、报名项目），这些信息比较分散，目前还没有进行统一管理。读者可以建立一个报名信息管理类，把上述四项信息作为报名信息管理类的属性，加入类中。
- 目前的报名信息为写入 SD 卡，读者可以利用串行化思想，将报名信息写入磁盘。此外，读者也可以在学习了第 4 章后，把报名信息写入数据库。

3.9　练习题

1．下列关于 Activity 的说法，不正确的是（　　）。
A．Activity 是为用户操作而展示的可视化用户界面
B．一个应用程序可以有若干个 Activity
C．Activity 可以通过一个别名进行访问
D．Activity 可以表现为一个悬浮的窗口
2．下列状态，不属于 Activity 的是（　　）。
A．暂停状态　　B．运行状态　　C．睡眠状态　　D．停止状态
3．下列生命周期方法，不属于 Activity 的是（　　）。
A．onCreate()　　B．onInit()　　C．onStart()　　D．onResume()
4．在 Activity 的生命周期中，当它从可见状态转向半透明状态时，它的（　　）方法必须被调用。
A．onStop()　　B．onPause()　　C．onRestart()　　D．onStart()
5．对一些资源及状态操作进行恢复，最好保存在 Activity 生命周期的（　　）方法中。
A．onStart()　　B．onPause()　　C．onCreate()　　D．onResume()
6．下列方法，不属于 Activity 启动方法的是（　　）。
A．startActivity()　　B．goToActivity()
C．startActivityForResult()　　D．startActivityFromChild()
7．关于 Intent 对象，说法错误的是（　　）。
A．在 Android 系统中，Intent 对象是用于传递信息的
B．Intent 对象可以把值传递给广播或 Activity
C．Intent 传递值时，可以传递一部分值类型
D．Intent 传递值时，它的 key 值可以是对象
8．在 Android 系统中，（　　）属于 Intent 的作用。
A．实现应用程序之间的数据共享
B．是一段长的生命周期，没有用户界面的程序，可以保持应用在后台运行，而不会因为切换页面而消失
C．可以实现界面之间的切换，可以包含动作和动作数据，连接四大组件的纽带
D．处理某个应用程序整体性的工作
9．Intent 传递数据时，（　　）类型的数据不可以被传递。
A．Serializable　　B．JSON 对象　　C．Bundle　　D．CharSequence
10．语句 Intent intent = new Intent(Intent. ACTION.VIEW,Uri.parse("http://mail.google.com")) 的作用是（　　）。
A．发送电子邮件　B．在浏览器里浏览该网址　C．发送手机短信　D．都不正确

3.10　作业

1．解释 Activity 生命周期的三种状态。
2．解释 Activity 生命周期的七种回调方法。
3．Android 系统是如何处理事件的？
4．Intent 有哪些作用？试举例说明。
5．在 Android 系统中，显式 Intent 和隐式 Intent 的区别是什么？
6．Activity 之间如何传递数据？如何返回数据？分别举例说明。
7．AndroidManifest.xml 有哪些作用？

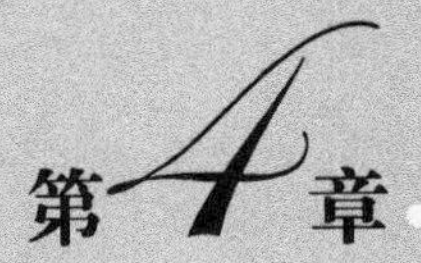

第4章 数据库与 ContentProvider——用户管理项目

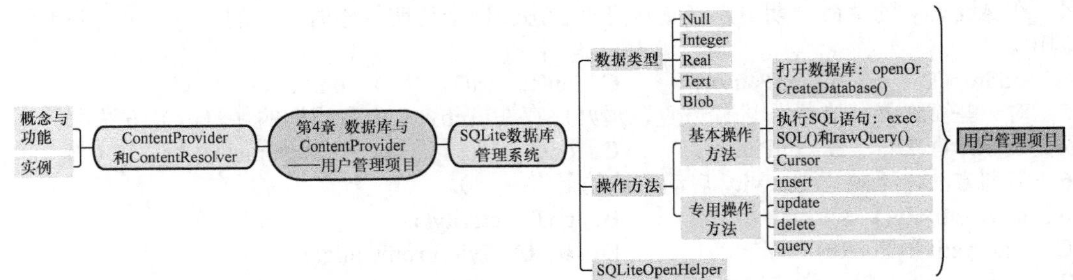

本章知识要点思维导图

在前面的章节中，我们对 Android 应用程序开发的基础知识进行了深入学习。读者应该掌握 Activity 生命周期和 Intent 的基本概念与使用方法，从而能开发简单的 Android 应用程序，并构造合理的图形界面。本章将主要介绍数据库技术在 Android 应用程序中的运用，以及通过 ContentProvider 实现应用程序间的数据共享。

4.1 需求分析

利用 ViewPager+Fragment 开发一个简单的用户管理项目，要求如下。

（1）本项目包含两个界面，界面 1 是用户登录界面，用户登录时，系统可以验证数据库中保存的用户名和用户名对应的密码是否匹配。

（2）界面 2 的下半部分，可以显示已注册过的用户信息。

（3）实现新用户注册功能。

（4）可以更新、删除选中的用户信息。

（5）可以根据用户名查询某条用户信息。

4.2 界面设计

用户管理项目的界面 1 为用户登录界面，如图 4-1（a）所示；界面 2 为用户信息管理与维护界面，如图 4-1（b）～图 4-1（d）所示。

- 用户登录界面：如果 usergl.db 数据库的 user 表里已保存现成的用户信息，则用户单击“登录”按钮后，系统直接进行验证流程，验证通过后，允许用户登录；用户单击“取消”按钮，则结束程序运行。此外，此处的“记住密码”和“忘记密码”两个复选框暂不编程，为第 5 章及后续练习做好准备。

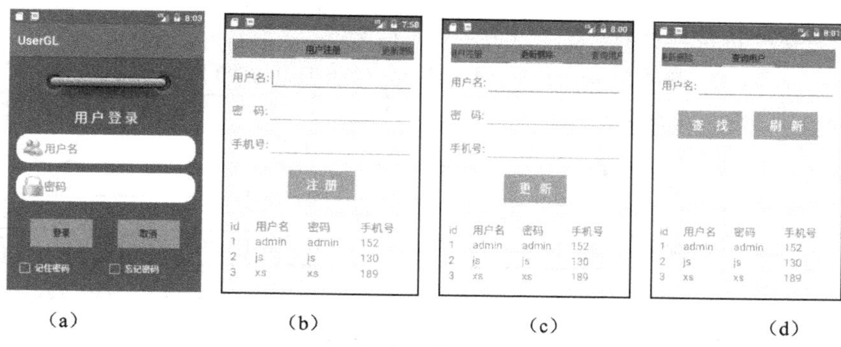

图 4-1　用户管理项目界面设计

- 用户信息管理与维护界面：该界面的上半部分采用 ViewPager 控件，在界面的顶部有三个 PagerTabStrip 控件，用户可以滑动切换。界面的下半部分采用 Fragment 控件，在这片区域中以 TextView+ListView 的方式显示数据表中已注册过的用户信息。

用户登录界面 activity_main.xml 中图形控件的 Text 及 ID 属性见表 4-1，用户信息管理与维护界面 activity_weihu.xml 中图形控件的 Text 及 ID 属性见表 4-2。

表 4-1　activity_main.xml 中图形控件的 Text 及 ID 属性

控　件	Text 属 性	ID 属 性	控　件	Text 属 性	ID 属 性
ImageView		imageView1			
TextView	用 户 登 录				
EditText		etPersonName			
EditText		etPassword			
Button	登录	btnLogin	Button	取消	btnCancel
CheckBox	记住密码	chkRemember	CheckBox	忘记密码	chkForget

表 4-2　activity_weihu.xml 中图形控件的 Text 及 ID 属性

控　件	Text 属 性	ID 属 性
ViewPager		viewpager
PagerTabStrip		pagertab
Fragment		fragment1

与 ViewPager 控件搭配的三个自定义布局文件分别为 layout1.xml、layout2.xml 和 layout3.xml。创建自定义布局文件的步骤如下：

在 res 目录中，右击 layout，在弹出的菜单中选择 New→Layout resource File，弹出如图 4-2 所示的 New Resource File 对话框，在该对话框中输入布局文件名称“layout1”，创建 layout1.xml 布局文件。

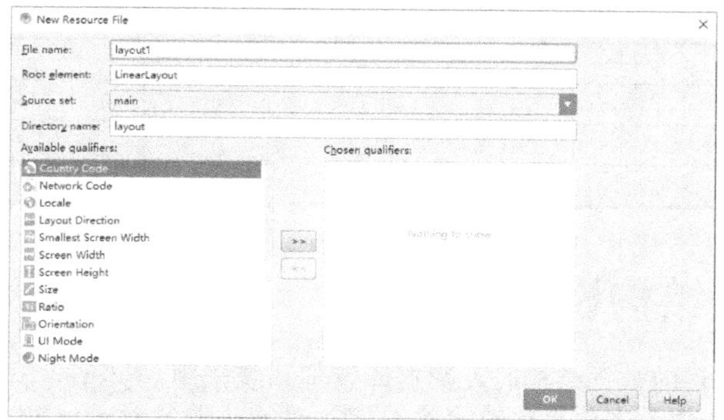

图 4-2　创建自定义布局文件

layout1.xml 中图形控件的 Text 及 ID 属性见表 4-3。

表 4-3 layout1.xml 中图形控件的 Text 及 ID 属性

控　件	Text 属 性	ID 属 性	控　件	Text 属 性	ID 属 性
TextView	用户名：	textView1	EditText		etpname1
TextView	密　码：	textView1	EditText		etpwd1
TextView	手机号：	textView1	EditText		etphone1
Button	注　册	btnRegister			

注意：此处以创建 layout1.xml 为例进行说明，其他布局文件的创建方法可参照 layout1.xml。此外，还要注意，布局文件的名称要用小写字母。

layout2.xml 和 layout3.xml 中图形控件的 Text 及 ID 属性分别见表 4-4 和 4-5。

表 4-4 layout2.xml 中图形控件的 Text 及 ID 属性

控　件	Text 属 性	ID 属 性	控　件	Text 属 性	ID 属 性
TextView	用户名：	textView1	EditText		etpname2
TextView	密　码：	textView1	EditText		etpwd2
TextView	手机号：	textView1	EditText		etphone2
Button	更　新	btnUpdate			

表 4-5 layout3.xml 中图形控件的 Text 及 ID 属性

控　件	Text 属 性	ID 属 性	控　件	Text 属 性	ID 属 性
TextView	用户名：	textView1	EditText		etpname3
Button	查　找	btnFind	Button	刷　新	btnRefresh

在界面 2 界面的下半部分，与 Fragment 控件搭配的自定义布局文件是 layout_data.xml，其图形控件的 Text 及 ID 属性见表 4-6。

表 4-6 layout_data.xml 中图形控件的 Text 及 ID 属性

控　件	Text 属 性	ID 属 性	控　件	Text 属 性	ID 属 性
TextView	id	TextView01	TextView	用户名	TextView02
TextView	密码	TextView03	TextView	手机号	TextView04
ListView		lstInfo			

同样，在界面 2 的下半部分，与 ListView（lstInfo）控件搭配的自定义布局文件是 lst_layout.xml，其图形控件的 Text 及 ID 属性见表 4-7。

表 4-7 lst_layout.xml 中图形控件的 Text 及 ID 属性

控　件	Text 属 性	ID 属 性
TextView	id	tvid
TextView	用户名	tvpname
TextView	密码	tvpwd
TextView	手机号	tvphone

4.3 数据结构设计

设计用户管理项目的数据结构时，数据库管理系统采用 SQLite，SQLite 是一个微型的关系型数据库管理系统。由于本项目比较简单，只需一张表，表名为 user，数据库名称为 usergl.db，用户信息见表 4-8。

表 4-8　用户信息表（user）

字 段 名	数 据 类 型	为 空 性	含　义
_id	integer	Not null	ID，主键，自动增加 autoincrement
pname	Text		用户名
pwd	Text		密码
phone	Text		手机号

4.4　实施

4.4.1　创建项目

创建一个名为 UserGL 的项目，默认的布局文件 activity_main.xml 和 MainActivity 类构成用户登录界面；再创建一个 WeihuActivity 类，构成用户信息管理与维护界面。

4.4.2　数据库相关代码

1. 封装数据库记录

创建一个类，用于封装数据库的记录，代码如下：

```
package com.example.administrator.usergl;

import java.io.Serializable;

public class User implements Serializable {
	/**
	 * 该类封装了一条记录的数据，用于传递数据库记录
	 * 该类实现了 Serializable 接口，这样才能在 Activity 之间传递
	 * 该类的三个有参构造方法是为了方便调用而添加的
	 */

	int _id;
	String pname;
	String pwd;
	String phone;

	public User() {
		super();
	}

	public User(int id) {
		super();
		this._id = _id;
	}

	public User(String pname, String pwd, String phone) {
		super();
		this.pname = pname;
		this.pwd = pwd;
		this.phone = phone;
	}

	public User(int _id, String pname, String pwd, String phone) {
		super();
```

```
            this._id = _id;
            this.pname = pname;
            this.pwd = pwd;
            this.phone = phone;
        }
}
```

2．数据库操作

SqlHelper 是数据库操作的核心类，该类封装了本项目需要的所有数据库操作，并继承了父类 SQLiteOpenHelper，SQLiteOpenHelper 是 Android 系统提供的 API 之一，详细内容将在本章的后续部分进行介绍。

提醒：建议读者先借助代码中的注释了解对应的方法，待学习完本章的后续内容，再进一步对代码深入分析，以便加深理解。

```
package com.example.administrator.usergl;

import android.content.ContentValues;
import android.content.Context;
import android.database.Cursor;
import android.database.sqlite.SQLiteDatabase;
import android.database.sqlite.SQLiteDatabase.CursorFactory;
import android.database.sqlite.SQLiteOpenHelper;

public class SqlHelper extends SQLiteOpenHelper{

    public SqlHelper(Context context) { //构造方法，根据提供的数据库名自动创建数据库
        super(context,"usergl.db", null, 1);
    }

    @Override
    public void onCreate(SQLiteDatabase sd) { //必须覆盖父类的此方法，作用是创建数据表
        String sql="create table user(_id integer primary key autoincrement,pname text,pwd
text,phone text)";
        sd.execSQL(sql);
        sd.execSQL("delete from user");
        sd.execSQL("insert into user(pname,pwd,phone) values('admin','admin','152')");
        sd.execSQL("insert into user(pname,pwd,phone) values('js','js','130')");
        sd.execSQL("insert into user(pname,pwd,phone) values('xs','xs','189')");
    }

    public Cursor select()     //查询所有记录
    {
        SQLiteDatabase sd=this.getReadableDatabase();
        Cursor cursor=sd.query("user", null, null, null, null, null, null);
        return cursor;
    }

    public long insert(User user){     //增加
        SQLiteDatabase sd=this.getWritableDatabase();
        ContentValues cv=new ContentValues();
        cv.put("pname",user.pname);
        cv.put("pwd",user.pwd);
        cv.put("phone",user.phone);
        long id=sd.insert("user", null, cv);
        return id;
    }

    public int update(User user){     //修改
```

```
            SQLiteDatabase sd=this.getWritableDatabase();
            ContentValues cv=new ContentValues();
            cv.put("pname",user.pname);
            cv.put("pwd",user.pwd);
            cv.put("phone",user.phone);
            int id=sd.update("user", cv,"_id='"+user._id+"'",null);
            return id;
      }

      public int delete(int _id){      //删除
            SQLiteDatabase sd=this.getWritableDatabase();
            int id=sd.delete("user","_id='"+_id+"'",null);
            return id;
      }

      public Cursor query(String pname){     //根据账号，查某条信息
            SQLiteDatabase sd=this.getReadableDatabase();
            Cursor cursor=sd.query("user",null, "pname='"+pname+"'",null,null,null,null);
            //也支持模糊查询，代码修改为"pname like '%"+pname+"%'"
            return cursor;
      }

      @Override
      public void onUpgrade(SQLiteDatabase arg0, int arg1, int arg2) {
            // 必须覆盖父类的此方法，作用是升级数据表

      }
}
```

项目运行后，打开 File Explorer 窗口，在 data→data→com.example.usergl→databases 目录下，可以看到生成的数据库文件。如图 4-3 所示。

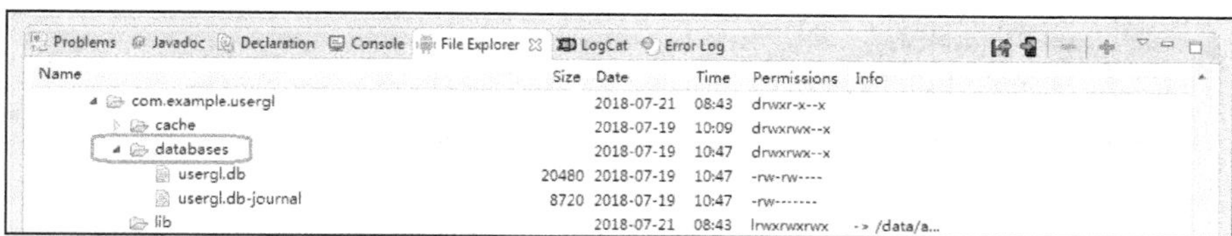

图 4-3　生成的数据库文件

4.4.3　界面实现

1．bg_et.xml

自定义布局文件 bg_et.xml 属于 Drawable resource file，该文件存放于 res→drawable 目录下。bg_et.xml 能够生成一个圆角矩形，作为用户登录界面中两个 EditText（etPersonName 和 etPassword）的文本框。

bg_et.xml 的代码如下：

```
<?xml version="1.0" encoding="utf-8"?>
<shape xmlns:android="http://schemas.android.com/apk/res/android"
      android:shape="rectangle" >

      <solid android:color="#FFFFFF" />

      <corners
            android:bottomLeftRadius="20dp"
            android:bottomRightRadius="20dp"
            android:topLeftRadius="20dp"
```

```
        android:topRightRadius="20dp" >
    </corners>

</shape>
```

2. activity_main.xml

用户登录界面的布局文件为 activity_main.xml，代码如下：

```
<LinearLayout xmlns:android="http://schemas.android.com/apk/res/android"
    xmlns:tools="http://schemas.android.com/tools"
    android:id="@+id/LinearLayout1"
    android:layout_width="match_parent"
    android:layout_height="match_parent"
    android:background="#394046"
    android:orientation="vertical"
    android:paddingBottom="@dimen/activity_vertical_margin"
    android:paddingLeft="@dimen/activity_horizontal_margin"
    android:paddingRight="@dimen/activity_horizontal_margin"
    android:paddingTop="@dimen/activity_vertical_margin"
    tools:context=".MainActivity" >

    <ImageView
        android:id="@+id/imageView1"
        android:layout_width="wrap_content"
        android:layout_height="wrap_content"
        android:layout_gravity="center"
        android:src="@drawable/top" />

    <TextView
        android:id="@+id/textView1"
        android:layout_width="wrap_content"
        android:layout_height="wrap_content"
        android:layout_gravity="center"
        android:layout_marginTop="15dp"
        android:text="用 户 登 录"
        android:textAppearance="?android:attr/textAppearanceLarge"
        android:textColor="#FFFFFF" />

    <EditText
        android:id="@+id/etPersonName"
        android:layout_width="match_parent"
        android:layout_height="45dp"
        android:layout_marginTop="15dp"
        android:background="@drawable/bg_et"
        android:drawableLeft="@drawable/username"
        android:ems="10"
        android:inputType="textPersonName"
        android:paddingLeft="10dp" >

        <requestFocus />
    </EditText>

    <EditText
        android:id="@+id/etPassword"
        android:layout_width="match_parent"
        android:layout_height="45dp"
        android:layout_marginTop="15dp"
        android:background="@drawable/bg_et"
        android:drawableLeft="@drawable/password"
```

```
            android:ems="10"
            android:inputType="textPassword"
            android:paddingLeft="10dp" >
        </EditText>

        <LinearLayout
            android:layout_width="wrap_content"
            android:layout_height="wrap_content"
            android:layout_gravity="center"
            android:layout_marginTop="40dp" >

            <Button
                android:id="@+id/btnLogin"
                android:layout_width="100dp"
                android:layout_height="wrap_content"
                android:background="#FF6100"
                android:text="登录" />

            <Button
                android:id="@+id/btnCancel"
                android:layout_width="100dp"
                android:layout_height="wrap_content"
                android:layout_marginLeft="40dp"
                android:background="#FF6100"
                android:text="取消" />
        </LinearLayout>

        <LinearLayout
            android:layout_width="match_parent"
            android:layout_height="wrap_content"
            android:layout_marginTop="35dp" >

            <CheckBox
                android:id="@+id/chkRemember"
                android:layout_width="wrap_content"
                android:layout_height="wrap_content"
                android:layout_weight="1"
                android:text="记住密码"
                android:textColor="#FFFFFF" />

            <CheckBox
                android:id="@+id/chkForget"
                android:layout_width="wrap_content"
                android:layout_height="wrap_content"
                android:layout_weight="1"
                android:text="忘记密码"
                android:textColor="#FFFFFF" />
        </LinearLayout>

    </LinearLayout>
```

3．activity_weihu.xml

用户信息管理与维护界面的布局文件为 activity_weihu.xml，代码如下：

```
<LinearLayout xmlns:android="http://schemas.android.com/apk/res/android"
    xmlns:tools="http://schemas.android.com/tools"
    android:id="@+id/LinearLayout1"
    android:layout_width="match_parent"
    android:layout_height="match_parent"
```

```
    android:orientation="vertical"
    android:paddingBottom="@dimen/activity_vertical_margin"
    android:paddingLeft="@dimen/activity_horizontal_margin"
    android:paddingRight="@dimen/activity_horizontal_margin"
    android:paddingTop="@dimen/activity_vertical_margin"
    tools:context=".WeihuActivity" >

    <android.support.v4.view.ViewPager
        android:id="@+id/viewpager"
        android:layout_width="match_parent"
        android:layout_height="0dp"
        android:layout_weight="2"
        android:layout_gravity="center" >

        <android.support.v4.view.PagerTabStrip
            android:id="@+id/pagertab"
            android:layout_width="wrap_content"
            android:layout_height="wrap_content"
            android:layout_gravity="top" >
        </android.support.v4.view.PagerTabStrip>
    </android.support.v4.view.ViewPager>

    <fragment
        android:id="@+id/fragment1"
        android:name="com.example.usergl.FragmentData"
        android:layout_weight="1"
        android:layout_width="match_parent"
        android:layout_height="0dp" />

</LinearLayout>
```

4．自定义布局文件

ViewPager 控件共分为三个 Pager，即需要三个 View（视图）。每个 View 都分别对应一个自定义布局文件，即 layout1.xml、layout2.xml 和 layout3.xml。

此外，与 Fragment 控件搭配的自定义布局文件是 layout_data.xml；与 ListView 控件搭配的自定义布局文件是 lst_layout.xml。

扫描下方的二维码，查看 layout1.xml、layout2.xml、layout3.xml、layout_data.xml 和 lst_layout.xml 的代码。

layout1.xml

layout2.xml

layout3.xml

layout_data.xml

lst_layout.xml

4.4.4 Java 代码

通过学习前面的章节，我们了解到 Activity 类主要负责图形界面的交互操作。而本节将会介绍 Activity 类的另一项重要功能，即与数据库之间通信。通常，Activity 类与数据库通信时会调用 SqlHelper 类的相关方法，SqlHelper 类的出现极大地方便了在 Activity 类中对数据库进行操作。

在本章的用户管理项目中，数据库操作的代码被封装在用户自定义类 SqlHelper.java 文件中。该自定义类封装了八种方法，作用分别如下。

- SqlHelper (Context context)：用于创建数据库。
- onCreate(SQLiteDatabase sd)：用于创建数据表，并输入初始数据。
- select()：查询 user 表中的所有记录。

- insert(User user)：增加信息。
- update(User user)：修改记录。
- delete(int _id)：根据 id 删除记录。
- query(String pname)：根据用户名返回对应的一条记录。
- onUpgrade(SQLiteDatabase arg0, int arg1, int arg2)：用于升级数据库。

数据库的基本操作主要有四种：增、删、改、查。

select()、insert(User user)、update(User user)、delete(int_id)和 query(String pname)这五个用户自定义方法，调用了能够创建与管理数据库的 SQLiteOpenHelper()方法、以读写方式打开数据库的 getReadableDatabase()方法和 getWritableDatabase()方法。此外，五种用户自定义方法又调用了 SQLiteDatabase()的 insert、delete、update、query 方法，实现了增、删、改、查功能。

1. MainActivity.java

MainActivity 的主要功能为验证用户在界面 1 中输入的用户名、密码是否与数据库中的信息一致。如果用户名、密码与数据库中的信息一致，则提示“登录成功”，同时跳转到界面 2。如果信息不一致，则提示“用户名或密码错误”。

用户登录界面的 MainActivity.java 代码如下：

```
package com.example.administrator.usergl;

import android.os.Bundle;
import android.support.v7.app.AppCompatActivity;
import android.content.Intent;
import android.database.Cursor;
import android.util.Log;
import android.view.Menu;
import android.view.View;
import android.view.View.OnClickListener;
import android.widget.Button;
import android.widget.EditText;
import android.widget.Toast;

public class MainActivity extends AppCompatActivity implements OnClickListener {
    EditText etPersonName,etPassword;
    Button btnLogin,btnCancel;
    String pname,pwd;
    SqlHelper helper=new SqlHelper(this);

    @Override
    protected void onCreate(Bundle savedInstanceState) {
        super.onCreate(savedInstanceState);
        setContentView(R.layout.activity_main);

        etPersonName=(EditText) this.findViewById(R.id.etPersonName);
        etPassword=(EditText) this.findViewById(R.id.etPassword);

        btnLogin=(Button) this.findViewById(R.id.btnLogin);
        btnLogin.setOnClickListener(this);
        btnCancel=(Button) this.findViewById(R.id.btnCancel);
        btnCancel.setOnClickListener(this);
    }


    @Override
    public void onClick(View   v) {
        Cursor cursor=helper.query(etPersonName.getText().toString());
        while (cursor.moveToNext()) {
```

```
            pname = cursor.getString(1);// 获取第二列的值
            pwd = cursor.getString(2);// 获取第三列的值

        System.out.println(pname+","+pwd);
            Log.v("user", pname+","+pwd);
        }

        switch(v.getId()){
        case R.id.btnLogin:
            if(pname.equals(etPersonName.getText().toString()) &&
                    pwd.equals(etPassword.getText().toString()))
            {
                //提示“登录成功”，然后跳转到界面 2
                Toast.makeText(MainActivity.this, "登录成功",1).show();
                Intent i=new Intent(MainActivity.this,WeihuActivity.class);
                startActivity(i);
            }
            else
            {
                //账号或密码错误，并清除 EditText
                Toast.makeText(MainActivity.this, "用户名或密码错误",1).show();
                etPersonName.setText("");
                etPassword.setText("");
            }
            break;
        case R.id.btnCancel:
            System.exit(0);
            break;
        }
    }
}
```

2．FragmentData.java

FragmentData.java 继承 Fragment，可以通过代码动态加载自定义布局文件 layout_data.xml。FragmentData.java 的代码如下：

```
package com.example.administrator.usergl;

import android.os.Bundle;
import android.support.v4.app.Fragment;
import android.view.LayoutInflater;
import android.view.View;
import android.view.ViewGroup;

public class FragmentData extends Fragment{

    @Override
    public View onCreateView(LayoutInflater inflater, ViewGroup container,
            Bundle savedInstanceState) {
        return inflater.inflate(R.layout.layout_data, null);
    }
}
```

3．WeihuActivity.java

WeihuActivity 主要使用 ViewPager 和 Fragment 控件，把界面 2 分割为上、下两部分。上半部分为 ViewPager 控件，可实现数据的增、删、改、查功能。下半部分为 Fragment 控件，借助 ListView 显示 user 中的所有记录。

请注意以下两点内容。

- 为 ListView 增加长按监听器（setOnItemLongClick Listener），通过获取当前记录的对应 ID，在对话框中删除记录，如图 4-4 所示。
- 通过自定义方法 isValidPhone(String phone)，检测手机号码的格式是否正确。例如，手机号码的第 1 位必须为 1，第 2 位是 3 或 5 或 8，第 3 位是 0~9 之间的任意数字，代码如下：

图 4-4　删除记录

```
public boolean isValidPhone(String phone) {
        if (phone.length() == 3 && phone.matches("1[358]\\d{1}"))
            return true;
        else
            return false;
    }
```

用户信息管理与维护界面的 WeihuActivity.java 代码如下：

```
package com.example.administrator.usergl;

import java.util.ArrayList;
import java.util.List;
import android.app.Activity;
import android.app.AlertDialog;
import android.content.DialogInterface;
import android.database.Cursor;
import android.os.Bundle;
import android.support.v4.app.FragmentActivity;
import android.support.v4.view.PagerAdapter;
import android.support.v4.view.PagerTabStrip;
import android.support.v4.view.ViewPager;
import android.support.v4.widget.SimpleCursorAdapter;
import android.view.LayoutInflater;
import android.view.View;
import android.view.View.OnClickListener;
import android.view.ViewGroup;
import android.widget.AdapterView;
import android.widget.AdapterView.OnItemClickListener;
import android.widget.AdapterView.OnItemLongClickListener;
import android.widget.Button;
import android.widget.EditText;
import android.widget.ListView;
import android.widget.Toast;

public class WeihuActivity extends FragmentActivity {
    ViewPager viewPager;
    View view1, view2, view3;
    List<View> viewList;
    PagerTabStrip pagerTabStrip;
    List<String> titleList;
    ListView lstInfo;
    Cursor cursor;
    SqlHelper helper;
    EditText etpname1, etpwd1, etphone1;
    Button btnRegister;
    EditText etpname2, etpwd2, etphone2;
    Button btnUpdate,btnFind,btnRefresh;
    int _id;
    EditText etpname3;

    @Override
    protected void onCreate(Bundle savedInstanceState) {
```

```
        super.onCreate(savedInstanceState);
        setContentView(R.layout.activity_weihu);

        helper = new SqlHelper(this);
        initView();
        lstInfo = (ListView) findViewById(R.id.lstInfo);
        lstInfo.setOnItemClickListener(new OnItemClickListener(){

            @Override
            public void onItemClick(AdapterView<?> av, View arg1, int position,
                    long arg3) {
                cursor.moveToPosition(position);
                _id=cursor.getInt(0);
                etpname2.setText(cursor.getString(1));
                etpwd2.setText(cursor.getString(2));
                etphone2.setText(cursor.getString(3));
            }});

        lstInfo.setOnItemLongClickListener(new OnItemLongClickListener(){
            @Override
            public boolean onItemLongClick(AdapterView<?> av, View arg1,
                    int position, long arg3) {
                cursor.moveToPosition(position);
                _id=cursor.getInt(0);
                AlertDialog.Builder builder=new AlertDialog.Builder(WeihuActivity.this);
                builder.setTitle("删除").setMessage("确定要删除吗？ ");
                builder.setPositiveButton("确定", new DialogInterface.OnClickListener() {
                    @Override
                    public void onClick(DialogInterface arg0, int arg1) {
                            helper.delete(_id);
                            DataFresh();
                            etpname2.setText("");
                            etpwd2.setText("");
                            etphone2.setText("");
                    }
                });

                builder.setNegativeButton("取消", new DialogInterface.OnClickListener(){
                    @Override
                    public void onClick(DialogInterface arg0, int arg1) {
                        // TODO Auto-generated method stub

                    }});

                AlertDialog dialog=builder.create();
                dialog.show();
                return false;
            }});


        etpname1 = (EditText) view1.findViewById(R.id.etpname1);
        etpwd1 = (EditText) view1.findViewById(R.id.etpwd1);
        etphone1 = (EditText) view1.findViewById(R.id.etphone1);
        btnRegister = (Button) view1.findViewById(R.id.btnRegister);
        btnRegister.setOnClickListener(new OnClickListener() {

            @Override
            public void onClick(View arg0) {
```

```
                if(isValidPhone(etphone1.getText().toString()))
                {
                helper.insert(new User(2, etpname1.getText().toString(), etpwd1
                        .getText().toString(), etphone1.getText().toString()));
                Toast.makeText(WeihuActivity.this, "添加成功", 1).show();
                etpname1.setText("");
                etpwd1.setText("");
                etphone1.setText("");
                DataFresh();
                }else
                    Toast.makeText(WeihuActivity.this, "手机号码不正确", 1).show();
        }
    });

    etpname2 = (EditText) view2.findViewById(R.id.etpname2);
    etpwd2 = (EditText) view2.findViewById(R.id.etpwd2);
    etphone2 = (EditText) view2.findViewById(R.id.etphone2);
    btnUpdate = (Button) view2.findViewById(R.id.btnUpdate);
    btnUpdate.setOnClickListener(new OnClickListener() {

        @Override
        public void onClick(View arg0) {
            if(isValidPhone(etphone2.getText().toString()))
            {
            helper.update(new User(_id, etpname2.getText().toString(), etpwd2
                    .getText().toString(), etphone2.getText().toString()));
            Toast.makeText(WeihuActivity.this, "已更新", 1).show();
            etpname2.setText("");
            etpwd2.setText("");
            etphone2.setText("");
            DataFresh();
            }else
                Toast.makeText(WeihuActivity.this, "手机号码不正确", 1).show();
        }
    });

    etpname3 = (EditText) view3.findViewById(R.id.etpname3);
    btnFind = (Button) view3.findViewById(R.id.btnFind);
    btnFind.setOnClickListener(new OnClickListener(){

        @Override
        public void onClick(View arg0) {
            Cursor cursor3=helper.query(etpname3.getText().toString());
            SimpleCursorAdapter adapter = new SimpleCursorAdapter(
                    view3.getContext(), R.layout.lst_layout, cursor3, new String[] {
                            "_id", "pname", "pwd", "phone" }, new int[] {
                            R.id.tvid, R.id.tvpname, R.id.tvpwd, R.id.tvphone });
            lstInfo.setAdapter(adapter);
        }});

    btnRefresh = (Button) view3.findViewById(R.id.btnRefresh);
    btnRefresh.setOnClickListener(new OnClickListener(){

        @Override
        public void onClick(View arg0) {
            DataFresh();

        }});
```

```
	}

	public boolean isValidPhone(String phone) {
		if (phone.length() == 3 && phone.matches("1[358]\\d{1}"))
			return true;
		else
			return false;
	}

	@Override
	protected void onResume() {
		super.onResume();
		DataFresh();
	}

	private void DataFresh() {
		cursor = helper.select();
		SimpleCursorAdapter adapter = new SimpleCursorAdapter(
				view1.getContext(), R.layout.lst_layout, cursor, new String[] {
						"_id", "pname", "pwd", "phone" }, new int[] {
						R.id.tvid, R.id.tvpname, R.id.tvpwd, R.id.tvphone });
		lstInfo.setAdapter(adapter);
	}

	private void initView() {
		viewPager = (ViewPager) this.findViewById(R.id.viewpager);
		pagerTabStrip = (PagerTabStrip) this.findViewById(R.id.pagertab);

		pagerTabStrip
				.setBackgroundColor(getResources().getColor(R.color.azure));
		pagerTabStrip.setTabIndicatorColor(getResources()
				.getColor(R.color.gold));

		view1 = this.findViewById(R.layout.layout1);
		view2 = this.findViewById(R.layout.layout2);
		view3 = this.findViewById(R.layout.layout3);

		LayoutInflater lf = getLayoutInflater().from(this);
		view1 = lf.inflate(R.layout.layout1, null);
		view2 = lf.inflate(R.layout.layout2, null);
		view3 = lf.inflate(R.layout.layout3, null);

		viewList = new ArrayList<View>();
		viewList.add(view1);
		viewList.add(view2);
		viewList.add(view3);

		titleList = new ArrayList<String>();
		titleList.add("用户注册");
		titleList.add("更新删除");
		titleList.add("查询用户");

		PagerAdapter adapter = new PagerAdapter() {

			@Override
			public int getCount() {
				return viewList.size();
			}
```

```
                @Override
                public boolean isViewFromObject(View arg0, Object arg1) {
                    return arg0 == arg1;
                }

                @Override
                public Object instantiateItem(ViewGroup container, int position) {
                    container.addView(viewList.get(position));
                    return viewList.get(position);
                }

                @Override
                public CharSequence getPageTitle(int position) {
                    return titleList.get(position);
                }

                @Override
                public void destroyItem(ViewGroup container, int position,
                        Object object) {
                    container.removeView(viewList.get(position));

                }

                @Override
                public int getItemPosition(Object object) {

                    return super.getItemPosition(object);
                }
            };

            viewPager.setAdapter(adapter);
        }
    }
```

4.5　SQLite 数据库管理系统

本章的用户管理项目使用了数据库管理系统，下面详细介绍。

4.5.1　SQLite 概述

SQLite 是一款非常流行的嵌入式数据库，它占用内存少且性能较好，符合 SQL-92 标准的基本要求。SQLite 是一个相对完整的 SQL 系统，拥有完整的触发器和事务处理功能等。但 SQLite 不支持一些标准的 SQL 功能，特别是外键约束（Foreign Key Constrain）、事务的嵌套、外连接，也不支持 Alter Table 功能。

Android 系统运行环境包含了完整的 SQLite 数据库管理系统，因此不需要另外安装 SQLite，可以直接使用。

4.5.2　数据类型

SQLite 的数据类型共有五种，分别如下。

- Null：空值。
- Integer：带符号的整型数值。
- Real：浮点型数值，一般存储为 8-byte IEEE 浮点数。

- Text：字符串型文本。
- Blob：二进制对象。

SQLite 和其他数据库最大的区别就是对数据类型的支持有所相同。比如创建一张表时，可以在 Create Table 语句中指定某列的数据类型，但实际上可以把任意数据类型放入任意列中。当把某个值插入数据库时，SQLite 将检查该值的数据类型。如果该数据类型与对应列的数据类型不匹配，则 SQLite 将尝试把该值转换为该列要求的数据类型。如果该值不能转换，则将该值按自身数据类型存储。例如，可以把一个字符串（String）放入 Integer 列。

创建数据库和表时，应注意以下几点。

- 所有表都应该有主键，使用关键字 Primary Key 定义，主键通常为自增的整数。
- 图像和音频文件等不应存储在数据库中。

Android 系统的数据库都存储在设备的“/data/data/<package_name>/databases”文件夹中。默认情况下，所有数据库都是私有的，只能由创建它们的应用程序访问。

4.5.3 基本操作方法

由于 JDBC[①]会消耗太多的系统资源，所以它并不适合手机这类内存有限的设备。因此，Android 系统提供了一些新的 API 来使用 SQLite 数据库管理系统，这些 API 可以完成所有的 SQL 操作。

1. 打开数据库

打开数据库的方式有两种：第一种，打开数据库，如果数据库不存在则会抛出异常；第二种，先判断数据库是否存在，如果不存在，则创建新数据库，然后再打开数据库。通常情况下，我们使用第二种方式，比如：

```
SQLiteDatabase sqlDB = openOrCreateDatabase(DATABASE_NAME,
SQLiteDatabase.CREATE_IF_NECESSARY, null);
```

2. 执行 SQL 语句

SQLiteDatabase 提供了两种执行 SQL 语句的方法：execSQL()和 rawQuery()，前者用于执行一条非查询 SQL 语句，后者执行一条查询 SQL 语句并返回查询的结果。

execSQL()方法的举例，代码如下：

```
sqlDB.execSQL("insert into person(name, age) values('王五', 23)");
sqlDB.execSQL("update person set name= '李丽' where id=2");
```

rawQuery()方法的举例，代码如下：

```
Cursor cursor = sqlDB.rawQuery("select * from person", null);
```

利用这两种方法就能完成数据表的创建，数据的增、删、改、查等功能。

3. Cursor

在前面的章节，Cursor（光标）已经出现过，它用于接受 select 查询的返回值，包含 0~*n* 条记录的数据。此外，rawQuery()方法会把查询的结果包装在一个 Cursor 的对象中返回，从而对结果集进行向前、向后或随机访问。Cursor 的常用方法见表 4-9。

表 4-9 Cursor 的常用方法

常用方法	说明
int getCount()	返回 Cursor 中的行数
int getColumnCount()	返回 Cursor 中的列数
boolean moveToFirst()	移动光标到第一行
boolean moveToLast()	移动光标到最后一行
boolean moveToNext()	移动光标到下一行
boolean moveToPrevious()	移动光标到上一行

① JDBC（Java DataBase Connectivity）即 Java 数据库连接，它是一种用于执行 SQL 语句的 Java API。Java 程序开发人员一般使用 JDBC 访问后台数据库。

续表

常 用 方 法	说　明
boolean moveToPosition(int position)	移动光标到一个绝对的位置
boolean isAfterLast()	在最后一行之后
boolean isBeforeFirst()	在第一行之前
boolean isFirst()	是否为第一行
boolean isLast()	是否为最后一行
String getColumnName(int columnIndex)	从给定的索引返回列名
getString(int columnIndex)	根据列索引值，返回该列的值（字符串型）
getInt(int columnIndex)	根据列索引值，返回该列的值（整型）
getXxx(int columnIndex)	根据列索引值，返回该列的值（×××类型，如长整型、浮点型等）
boolean isNull(int columnIndex)	该列的值是否为 Null
void close()	关闭光标，释放资源
boolean isClosed()	是否已关闭

对 Cursor 操作时，需要注意以下几点。

- Cursor 是数据行的集合，使用循环从第一行遍历到最后一行。
- 使用 moveToFirst()方法可以将Cursor 定位到第一行，使用moveToNext方法()将 Cursor 移到下一行，如果 moveToNext()方法返回值为 False，则表示没有下一行；或者使用 isAfterLast()方法进行判断，是否已到最后一行之后。
- 通过列下标获得列的值，如果不知道列下标，则通过 getColumnIndex()方法查询列名对应的列下标。

【例 4.1】Cursor 应用实例。

本例综合了数据库的创建，数据表的创建，数据的增、删、改、查等功能，代码如下：

```
private static final String DATABASE_NAME = "MyContacts.db";
private static final String TABLE_NAME = "person";
// Declare an object of SQLiteDatabase class
private SQLiteDatabase sqlDB;

public void createDB(View v) {
        sqlDB = openOrCreateDatabase(DATABASE_NAME,
                    SQLiteDatabase.CREATE_IF_NECESSARY, null);
        Log.v(TAG,"数据库已创建");

        sqlDB.execSQL("CREATE TABLE " + TABLE_NAME
                    + "(id integer primary key autoincrement, name text, age integer)");
        Log.v(TAG,"表已创建");

        sqlDB.execSQL("insert into person(name, age) values('张三', 21)");
        sqlDB.execSQL("insert into person(name, age) values('李四', 22)");
        sqlDB.execSQL("insert into person(name, age) values('王五', 23)");
        Log.v(TAG,"插入三行");

        sqlDB.execSQL("update person set name= '李丽' where id=2");
        Log.v(TAG,"更新一行");

        sqlDB.execSQL("delete from person where id=3");
        Log.v(TAG,"删除一行");

        Cursor cursor = sqlDB.rawQuery("select * from person", null);
        while (cursor.moveToNext()) {
            int id = cursor.getInt(0); // 获取第一列的值，第一列的索引值从 0 开始
            String name = cursor.getString(1);// 获取第二列的值
```

```
            int age = cursor.getInt(2);// 获取第三列的值

            Log.v(TAG, "id=" + id + ", name=" + name + ", age=" + age);
        }
        cursor.close();
        sqlDB.close();
    }
```

4.5.4 专用操作方法

对数据库进行操作时，直接使用 SQL 语句有许多不便。为此，SQLiteDatabase 专门定义了一些与 insert、update、delete 和 query 有关的专用操作方法，使编写程序更方便，这些专用操作方法见表 4-10。

表 4-10 SQLiteDatabase 的专用操作方法

专用操作方法	说 明
public long insert (String table, String nullColumnHack, ContentValues initialValues);	• table：需要插入数据的表名。 • nullColumnHack：该参数需要传入一个列名。SQL 标准并不允许插入一行所有列均为空的数据，所以当传入的 initialValues 值为空或者为 0 时，用 nullColumnHack 参数指定的列会被插入值为 null 的数据，然后再将此行插入表中。 • initalValues：用来描述要插入行数据的 ContentValues 对象，即列名和列值的映射。 • 返回值：新插入行的行 ID。如果有错误发生返回-1
public int update (String table, ContentValues values, String whereClause,String[] whereArgs);	• table：需要更新数据的表名。 • values：用来描述更新后的行数据的 ContentValues 对象，即列名和列值的映射。whereClause，可选的 where 语句（不包括 WHERE 关键字），用来指定需要更新的行。若传入 null 则表中所有的行均会被更新。 • whereArgs：where 语句中表达式的占位参数列表，参数只能为 String 类型。返回值：被更新的行的数量
public int delete (String table, String whereClause, String[] whereArgs);	• table：需要删除行的表名。 • whereClause：可选的 where 语句（不包括 WHERE 关键字），用来指定需要删除的行。若传入 null 则会删除所有的行。 • whereArgs：where 语句中表达式的占位参数列表，参数只能为 String 类型。 • 返回值：若传入了正确的 where 语句则被删除的行数会被返回。若传入 null，则会返回 0。若要删除所有行并且返回删除的行数，则需要在 where 语句的位置传入字符串 1
public Cursor query (String table, String[] columns, String selection, String[] selectionArgs, String groupBy, String having, String orderBy, String limit);	• table：需要检索的表名。 • columns：由需要返回列的列名所组成的字符串数组，传入 null 会返回所有的列。selection，指定需要返回的行的 where 语句（不包括 WHERE 关键字）。传入 null 则返回所有行。 • selectionArgs：where 语句中表达式的占位参数列表，参数只能为 String 类型。 • groupBy：对结果集进行分组的 group by 语句（不包括 GROUP BY 关键字）。传入 null 将不对结果集进行分组。 • having：对分组结果集设置条件的 having 语句（不包括 HAVING 关键字）。必须配合 groupBy 参数使用，传入 null 将不对分组结果集设置条件。 • orderBy：对结果集进行排序的 order by 语句（不包括 ORDER BY 关键字）。传入 null 将对结果集使用默认的排序。 • limit：对返回的行数进行限制的 limit 语句（不包括 LIMIT 关键字）。传入 null 将不限制返回的行数。 • distinct：如果希望结果集没有重复的行传入 true，否则传入 false。 • CursorFactory：使用 CursorFactory 构造返回的 Cursor 子类对象，传入 null 将使用默认的 CursorFactory。 • 返回值：指向第一行数据之前的 Cursor 子类对象

在用户管理项目的 SqlHelper 类中，主要采用上述部分方法实现增、删、改功能。

【例 4.2】专用操作方法实例。

以实现 4.5.3 节中例 4.1 的数据的增、删、改功能为目标，本例将介绍另一种方法，代码如下：

```
//sqlDB.execSQL("insert into person(name, age) values('张三', 21)");
ContentValues values = new ContentValues();
values.put("name","张三");
values.put("age",21);
sqlDB.insert("person", "id", values); // ID 列为空，由 SQLite 自动赋值

//sqlDB.execSQL("insert into person(name, age) values('李四', 22)");
long newId = sqlDB.insert("person", "id", null); // 插入空列，然后再对它进行修改
if(newId!=-1){
      values.clear();
      values.put("name","李四");
      values.put("age",22);
      sqlDB.update("person", values, "id = "+newId, null); // newId 是新插入行的行 ID
}

//sqlDB.execSQL("insert into person(name, age) values('王五', 23)");
values.clear();
values.put("name","王五");
values.put("age",23);
newId = sqlDB.insert("person", "id", values); // ID 列为空，由 SQLite 自动赋值

Log.v(TAG,"插入三行");

if(newId!=-1){
      sqlDB.delete("person","id=?",new String[] {newId+""});
      Log.v(TAG,"删除一行");
}

//Cursor cursor = sqlDB.rawQuery("select * from person", null);
// 改为查询某条记录
String[] columns = { "id", "name", "age"};
String[] parms = { "张三" };
Cursor cursor = sqlDB.query("person", columns, "name=?", parms, null,
            null, null);
while (cursor.moveToNext()) {
      int id = cursor.getInt(0); // 获取第一列的值，第一列的索引值从 0 开始
      String name = cursor.getString(1);// 获取第二列的值
      int age = cursor.getInt(2);// 获取第三列的值

      Log.v(TAG, "id=" + id + ", name=" + name + ", age=" + age);
}
```

4.5.5 SQLiteOpenHelper

为了进一步提高编程效率，Android 提供了 SQLiteOpenHelper 类。SQLiteOpenHelper 类是一个抽象类，定义了创建、打开和升级数据库的方法。程序员在开发程序时，需要创建一个 SQLiteOpenHelper 类的子类，实现 SQLiteOpenHelper 类的构造方法、两个抽象方法，以及一些管理数据的方法。详细说明见以下代码的注释：

```
import android.content.Context;
import android.database.sqlite.SQLiteDatabase;
import android.database.sqlite.SQLiteOpenHelper;
```

```
public class SqlDemo extends SQLiteOpenHelper {
  /**
   *构造方法，根据提供的数据库名自动创建数据库。并根据数据库的版本编号维护不同的数据库
   *
   * @param context
   */
    public SqlDemo(Context context) {
        // 数据库名是 mydatabase.db，数据库的版本编号是 1
        super(context, "mydatabase.db", null, 1);
    }

   /**
    * 必须实现的方法，创建数据表
    */
    @Override
    public void onCreate(SQLiteDatabase db) {
        // 根据数据结构的设计，创建数据表的语句
    }

    /**
     * 必须实现的方法，升级数据表
     */
    @Override
    public void onUpgrade(SQLiteDatabase db, int oldVersion, int newVersion) {
    // 一种方法是删除旧数据表，直接创建新版本的数据表
    // 另一种方法是创建新数据表后将数据从旧数据表中导入新数据表中，再删除旧数据表
    }

    /**
     * 以下代码是用于自定义添加的方法，必须包括对本数据库进行增、删、改、查操作的方法
     */
    // public Cursor select() {
    // public void delete(int id) {
}
```

提示：完整的例子请参考本章的用户管理项目中的代码（SqlHelper.java）。

4.5.6 SQLite 数据库的管理

对 SQLite 数据库进行管理，需要用到相应的管理类工具软件。手机和 AVD 中缺少这类软件，但是，可以在计算机中运行的此类软件较多，比如 SQLite Expert Personal，它是一款开源软件，其界面如图 4-5 所示。

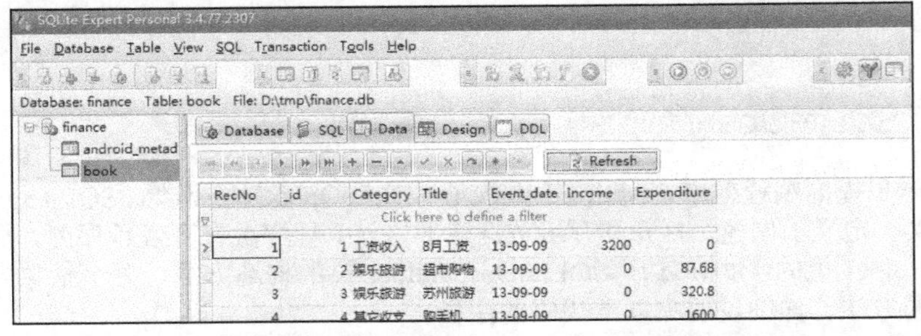

图 4-5 SQLite Expert Personal 的界面

SQLite Expert Personal 的使用方法如下。

（1）打开 Eclipse 的 File Explorer 文件夹，将手机或 AVD 中的 SQLite 数据库文件（如 finance.db）下载到计算机中。

（2）使用 SQLite Expert Personal 打开刚下载的数据库文件，可对其进行管理操作（增、删、改、查）；还可以将修改后的数据库上传到手机或 AVD 中。

此外，编写 SQL 语句时，也可以在 SQLite Expert Personal 中进行测试。

4.6　ContentProvider 和 ContentResolver

4.6.1　概念与功能

为了提高 Android 系统的安全性，Android 系统对数据的共享进行了严格限制：应用程序之间不能自由交换数据，而只能使用统一的工具——ContentProvider（内容提供者）和 ContentResolver（内容接收者）实现数据的交换。这种做法的优点是统一了数据的访问方式，提高了 Android 应用程序的安全性。

当应用程序 A 需要将数据共享给应用程序 B 时，此时的应用程序 A 是 ContentProvider，应用程序 B 是 ContentResolver，则应用程序 A 需要完成以下任务。

- 创建 ContentProvider 的子类，并实现其中的抽象方法（增、删、改、查）以及提供数据的功能。
- 将 ContentProvider 的子类在配置文件 AndroidManifest.xml 中进行注册，以便 Content Resolver 能够找到它。

与此同时，应用程序 B 需要完成以下任务。

- 使用 ContentResolver 类的增、删、改、查方法实现对应用程序 A 的数据进行访问。其中，ContentResolver 类的增、删、改、查方法的签名与 ContentProvider 类的对应方法的签名是完全相同的。

ContentProvider 共享的数据不仅包括数据库中的数据，也包括后续章节要讨论的其他类型的数据。

4.6.2　实例代码

例如，访问本章的用户管理项目中的数据，原来的 UserGL 项目为内容提供者，即应用程序 A，新建一个名为 UserGLShare 的项目作为内容接收者，即应用程序 B，实现步骤如下。

1．创建 ContentProvider 的子类

在内容提供者 UserGL 项目（应用程序 A）中创建一个 ContentProvider 的子类 ShareData，使其继承 ContentProvider，代码如下：

```
package com.example.administrator.usergl;

import android.content.ContentProvider;
import android.content.ContentValues;
import android.database.Cursor;
import android.net.Uri;

public class ShareData extends ContentProvider{
    /**
     * 作为一个简单的 ContentProvider 的演示程序，仅支持对数据库进行查询操作
     * 不支持对数据库进行增、删、改操作，故相关方法的返回值为空或 0
     */
```

```
    @Override
    public int delete(Uri arg0, String arg1, String[] arg2) {
        // TODO Auto-generated method stub
        return 0;
    }

    @Override
    public String getType(Uri arg0) {
        // TODO Auto-generated method stub
        return null;
    }

    @Override
    public Uri insert(Uri arg0, ContentValues arg1) {
        // TODO Auto-generated method stub
        return null;
    }

    @Override
    public boolean onCreate() {
        // TODO Auto-generated method stub
        return false;
    }

    /*
     * 演示数据库的查询和共享过程
     */
    @Override
    public Cursor query(Uri uri, String[] arg1, String arg2, String[] arg3,
            String arg4) {
        // 为简单起见，忽略传入的 uri 等参数
        SqlHelper helper = new SqlHelper(this.getContext());
        return helper.select();
    }

    @Override
    public int update(Uri arg0, ContentValues arg1, String arg2, String[] arg3) {
        // TODO Auto-generated method stub
        return 0;
    }
}
```

2. 在 AndroidManifest.xml 中注册 ContentProvider 的子类

在内容提供者 UserGL 项目（应用程序 A）的配置文件中注册第一步创建的子类，代码如下：

```
<application ...>
<!-- 原有的配置... -->
    <provider
        android:name=".ShareData"
        android:authorities="com.example.administrator.usergl"
        android:exported="true" >
    </provider>
</application>
```

注意： Android 4.2 及以后版本的代码中要增加 android:exported 属性。

3．使用 ContentResolver 类访问共享的数据

在内容接收者 UserGLShare 项目（应用程序 B）中，访问 UserGL 项目的 user 表内所有行的数据，结果显示在 TableLayout 布局中，如图 4-6 所示。

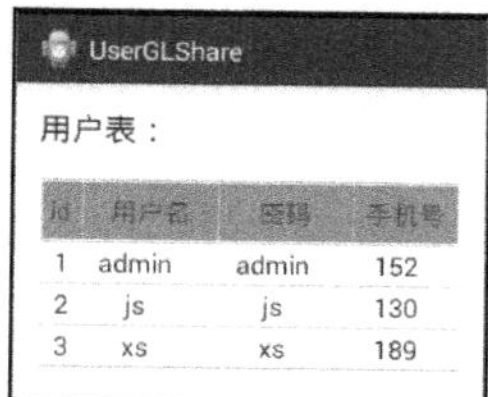

图 4-6　访问共享的数据

UserGLShare 项目的界面布局文件为 activity_main.xml，代码如下：

```
<LinearLayout xmlns:android="http://schemas.android.com/apk/res/android"
    xmlns:tools="http://schemas.android.com/tools"
    android:id="@+id/LinearLayout1"
    android:layout_width="match_parent"
    android:layout_height="match_parent"
    android:orientation="vertical"
    android:paddingBottom="@dimen/activity_vertical_margin"
    android:paddingLeft="@dimen/activity_horizontal_margin"
    android:paddingRight="@dimen/activity_horizontal_margin"
    android:paddingTop="@dimen/activity_vertical_margin"
    tools:context=".MainActivity" >

    <LinearLayout
        android:layout_width="match_parent"
        android:layout_height="wrap_content" >

        <TextView
            android:id="@+id/tvid"
            android:layout_width="100dp"
            android:layout_height="wrap_content"
            android:text="用户表："
            android:textAppearance="?android:attr/textAppearanceLarge" />
    </LinearLayout>

    <ScrollView
        android:id="@+id/scroll"
        android:layout_width="match_parent"
        android:layout_height="match_parent"
        android:layout_marginTop="20dp" >

        <TableLayout
            android:id="@+id/tableLayout1"
            android:layout_width="match_parent"
            android:layout_height="wrap_content"
             android:divider="@drawable/table_v_divider"
             android:orientation="vertical"
        android:showDividers="middle|beginning|end"
            android:layout_margin="1dp" >

            <TableRow
                android:layout_width="match_parent"
                android:layout_margin="1dp"
                android:background="#2691f2"
```

```
            android:divider="@drawable/table_h_divider"
            android:orientation="horizontal"
            android:showDividers="middle|beginning|end" >

            <TextView
                android:layout_height="40dp"
                android:layout_weight="1"
                android:gravity="center"
                android:paddingBottom="2dp"
                android:paddingLeft="5dp"
                android:paddingRight="8dp"
                android:paddingTop="2dp"
                android:text="id"
                android:textSize="18sp" />

            <TextView
                android:layout_height="40dp"
                android:layout_weight="1"
                android:gravity="center"
                android:paddingBottom="2dp"
                android:paddingLeft="8dp"
                android:paddingRight="8dp"
                android:paddingTop="2dp"
                android:text="用户名"
                android:textSize="18sp" />

            <TextView
                android:layout_height="40dp"
                android:layout_weight="1"
                android:gravity="center"
                android:paddingBottom="2dp"
                android:paddingLeft="8dp"
                android:paddingRight="8dp"
                android:paddingTop="2dp"
                android:text="密码"
                android:textSize="18sp" />

            <TextView
                android:layout_height="40dp"
                android:layout_weight="1"
                android:gravity="center"
                android:paddingBottom="2dp"
                android:paddingLeft="8dp"
                android:paddingRight="8dp"
                android:paddingTop="2dp"
                android:text="手机号"
                android:textSize="18sp" />
        </TableRow>
    </TableLayout>
</ScrollView>

</LinearLayout>
```

MainActivity.java 的代码如下：

```
package com.example.administrator.userglshare;

import android.annotation.SuppressLint;
import android.annotation.TargetApi;
import android.support.v7.app.AppCompatActivity;
```

```
import android.content.ContentResolver;
import android.database.Cursor;
import android.graphics.Color;
import android.graphics.drawable.ColorDrawable;
import android.net.Uri;
import android.os.Build;
import android.os.Bundle;
import android.view.Gravity;
import android.widget.TableLayout;
import android.widget.TableRow;
import android.widget.TextView;

public class MainActivity extends AppCompatActivity {
    private Cursor cursor;
    public static final Uri CONTENT_URI = Uri
            .parse("content://com.example.usergl/usergl");

    @TargetApi(Build.VERSION_CODES.HONEYCOMB)
    @SuppressLint("NewApi")
    @Override
    protected void onCreate(Bundle savedInstanceState) {
        super.onCreate(savedInstanceState);
        setContentView(R.layout.activity_main);

        ContentResolver cr = getContentResolver();
        cursor = cr.query(CONTENT_URI, null, null, null, null);
        TableLayout tableLayout1 = (TableLayout) this
                .findViewById(R.id.tableLayout1);

        while (cursor.moveToNext()) {
            TableRow row = new TableRow(this);
            android.widget.TableRow.LayoutParams lp=new android.widget.
TableRow.LayoutParams();
          lp.setMargins(2, 2, 2, 2);   //left, top, right, bottom

            for (int j = 0; j < cursor.getColumnCount(); j++) {
                // 创建显示的内容，这里创建的是一列
                TextView text = new TextView(this);
                // 设置显示内容
                if(j==0)
                    text.setText(String.valueOf(cursor.getInt(j)));
                else
                    text.setText(cursor.getString(j)+"            ");
                text.setPadding(10, 2, 10, 2);
                text.setTextColor(Color.BLACK); // 字体颜色
                text.setTextSize(18); // 字体大小
                text.setGravity(Gravity.CENTER); // 居中显示
                // 添加到 row
                row.addView(text);
            }
            // 将一行数据添加到表格中
            tableLayout1.addView(row,lp);
        }
    }
}
```

提醒：本例仅展示了如何对数据库进行查询，在查询过程中也没有通过 query()方法传递有关的查询参数。而实际应用中的情况要比本例复杂得多，建议读者参考相关 API 文档。

4.7 使用内置的 ContentProvider

Android 系统自带了许多内置的应用程序，如电话号码簿、短信、通话记录、日历等。在开发其他应用程序时，有时需要访问上述程序中的数据，唯一的解决办法就是通过 ContentProvider 进行访问。Android 系统自带的应用程序提供了内置的 ContentProvider，极大地方便了开发人员，有关 ContentProvider 的介绍在 Android API 中有详细说明。如图 4-7 所示为内置的 ContentProvider 之电话号码簿。

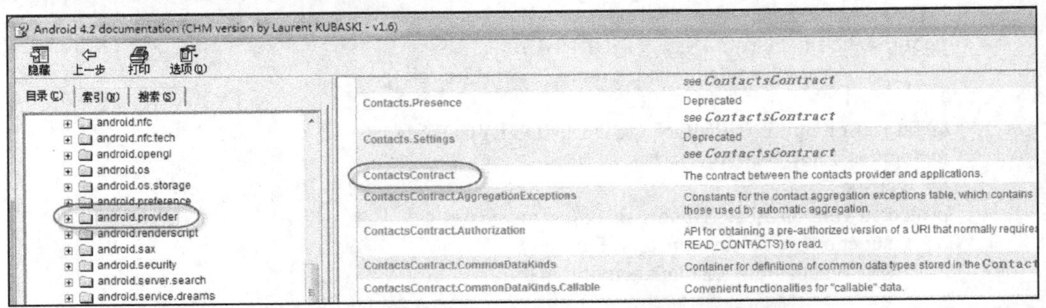

图 4-7 内置的 ContentProvider 之电话号码簿

例如，通过电话号码簿的 ContentProvider，即 ContactsContract.Contacts.CONTENT_URI，可以查询手机上所有联系人的电话号码，代码如下：

```
public class MainActivity extends Activity {

    @Override
    protected void onCreate(Bundle savedInstanceState) {
        super.onCreate(savedInstanceState);
        // setContentView(R.layout.main); //不使用 xml 界面文件

        // 取得内置的 ContentProvider
        ContactsContract.Contacts.CONTENT_URI（联系人）
        ContentResolver cr = getContentResolver();
        Cursor cursor = cr.query(ContactsContract.Contacts.CONTENT_URI, null,
                null, null, null);

        // 从返回的光标中取得联系人的信息
        String allContact = "";
        while (cursor.moveToNext()) {
            // 取得联系人的名字索引值
            int nameIndex = cursor.getColumnIndex(PhoneLookup.DISPLAY_NAME);
            String contact = cursor.getString(nameIndex);
            allContact += (contact + ": ");

            // 取得联系人的 ID 索引值
            String contactId = cursor.getString(cursor
                    .getColumnIndex(ContactsContract.Contacts._ID));
            // 查询该联系人的电话号码，类似地，可以查询 email、photo
            // 每位联系人的电话号码也是一个光标，因为一个人可能有多个电话号码
```

```
            Cursor phone = cr.query(
                    ContactsContract.CommonDataKinds.Phone.CONTENT_URI, null,
                    ContactsContract.CommonDataKinds.Phone.CONTACT_ID + " = "
                            + contactId, null, null);
            // 一位联系人的多个电话号码
            while (phone.moveToNext()) {
                String strPhoneNumber = phone
                        .getString(phone
                                .getColumnIndex(ContactsContract.CommonDataKinds.
Phone.NUMBER));
                allContact += (strPhoneNumber + "; ");
            }
            allContact += "\r\n";
            phone.close();
        }
        cursor.close();

        // 直接显示内容
        TextView tv = new TextView(this);
        tv.setText(allContact);
        setContentView(tv); // 不通过布局文件，直接显示到屏幕上

        // 或者显示到 LogCat 上，然后复制到计算机中进行保存
        Log.v("电话号码簿", allContact);

    }
}
```

另外，还需要在配置文件 AndroidManifest.xml 中设置访问权限（在 application 元素之前），代码如下：

```
<uses-permission android:name="android.permission.READ_CONTACTS" />
<application ....
```

如果对联系人的信息进行增、删、改操作，则还需要设置写的权限，代码如下：

```
<uses-permission android:name="android.permission.WRITE_CONTACTS" />
```

如果在真实的手机上运行程序，就能读取手机上的所有电话号码；如果在虚拟仿真器上运行程序，则需要事先在虚拟仿真器上创建联系人及相应的电话号码，否则，无法找到联系人的信息。

4.8　实训：完善用户管理项目

在本章的用户管理项目的基础上，增加以下需求。

- 当前的用户管理项目，用户的手机号码为 3 位，在实际生活中，用户的手机号码是 11 位，如何将用户的手机号码由 3 位改为 11 位？
- 当前的用户管理项目，用户的基本信息有三项：用户名、密码、手机号，请在此基础上，增加昵称、邮箱、头像等基本信息。

4.9　实训：商品选购界面

请读者使用 TabHost 选项卡+FrameLayout 帧布局，设计一个简单的商品选购界面，运行效果如图 4-8 所示，项目要求如下。

(a)

(b)

(c)

图 4-8　商品选购界面运行效果

- 本项目的名称为 Shopping。
- 本项目只有一个界面（activity_main.xml），且该界面分为两部分，上半部分为 ImageView 控件，下半部分为 TabHost 选项卡。
- TabHost 有三个选项卡，每个选项卡都是垂直线性布局，其中仅包含一个 ListView 控件。此外，利用 FrameLayout 帧布局控制三个选项卡的切换效果。
- ListView 控件采用自定义布局（moban.xml），实现思路如下：外框为水平线性布局，里面嵌套了两个垂直线性布局，布局内的控件见表 4-11。

表 4-11　垂直线性布局内的控件

左　侧	右　侧
商品图片 ImageView（imgPhoto1）	商品图片 ImageView（imgPhoto2）
商品名称 TextView（tvTitle1）	商品名称 TextView（tvTitle2）
商品价格 TextView（tvPrice1）	商品价格 TextView（tvPrice2）

- 本项目中的商品基本信息主要有三项：商品图片、商品名称和商品价格。所以，相应的三个属性被封装在用户自定义 Goods 类中，参考代码如下：

```
package com.example.administrator.shopping;

public class Goods {
    int photo;
    String title;
    String price;

    public Goods() {
        super();
    }

    public Goods(int photo, String title, String price) {
        super();
        this.photo = photo;
        this.title = title;
```

```
            this.price = price;
        }
    }
```

- 本项目中的商品分为三类：牛奶、饮料和咖啡。所以，用三个 LinkedList<Goods>容器存放这三类商品信息。第一个容器中有 5 种商品，第二个和第三个容器中均有 6 种商品。
- 设置一个用户自定义适配器 MyAdapter.java 类，可以实现显示方式为两列的 ListView 填充效果。扫描旁边的二维码，查看相关代码。

MyAdapter.java

如图 4-9 所示为商品数量为奇数和偶数时的显示效果。

（a）　　（b）

图 4-9　商品数量为奇数和偶数时的显示效果

- 在 MainActivity.java 中编写代码。当切换选项卡时，当前选项卡上的标题文字的显示效果为红色，字体大小为 16 号；非当前选项卡上的标题文字的显示效果为黑色，字体大小为 12 号。参考代码如下：

```
/**更新文字大小、颜色
* @param tabHost
*/
    private void updateTab(TabHost tabHost) {
            for (int i = 0; i < tabHost.getTabWidget().getChildCount(); i++) {
                TextView tv = (TextView) tabHost.getTabWidget().
getChildAt(i).findViewById(android.R.id.title);
                if (tabHost.getCurrentTab() == i) {//选中
                  tv.setTextSize(16);
                      tv.setTextColor(Color.RED);
                } else {//不选中
                  tv.setTextSize(12);
                      tv.setTextColor(Color.BLACK);
                }
            }
        }
    });
```

读者可以在本项目的基础上，再增加以下需求。

- 本项目的商品信息存储在类和List中，请读者改为使用SQLite数据库或MySQL数据库。
- 为本项目增加购物车环节，即用户选购商品后，所选商品会被自动添加到购物车中。用户在购物车中可以看到所选的商品，删除购物车中的商品，修改购物车中商品的数量，查看购物车中所有商品的总金额等。

4.10 练习题

1．开发手机应用程序时，常用的数据库是（　）。

A．SQLite　B．Oracle　C．SQL Server　D．MySQL

2．使用SQLiteOpenHelper类时，（　）方法可以实现版本升级。

A．onCreate()　B．onCreateView()　C．onUpdate()　D．onUpgrade()

3．下列关于SQLite数据库的说法，不正确的是（　）。

A．SqliteOpenHelper类主要用于创建数0据库和更新数据库

B．SqliteDatabase 类用于操作数据库

C.在每次调用SqliteDatabase的getWritableDatabase()方法时，系统会执行SqliteOpenHelper的onCreate()方法

D．当数据库版本发生变化时，可以自动更新数据库结构

4．如果数据源为SQLite数据库，那么查询信息时最合适的适配器是（　）。

A．SimpleAdapter　B．SimpleCursorAdapter　C．ArrayAdapter　D．ListAdapter

5．程序员Jane创建包含四个活动的应用程序时，需要使用SQLite数据库，并且应用程序中的所有活动都需要在Details表中插入记录，则她应使用（　）类创建Insert语句。

A．SQLiteOpenHelper　B．SQLiteStatement

C．SQLiteDatabase　D．SQLiteQueryBuilder

6．SQLiteOpenHelper的子类不需要实现（　）方法。

A．构造函数　B．onCreate()　C．onUpgrade()　D．openDatabase()

7．继承ContentProvider不需要实现（　）方法。

A．add　B．delete　C．update　D．query

8．在多个应用中读取共享存储数据时，需要用到（　）对象的query()方法。

A．ContentResolver　B．ContentProvider　C．Cursor　D．SQLiteHelper

9．在Android系统中使用SQLiteOpenHelper辅助类，读取数据应采用（　）方法。

A．getDatabase()　B．openDatabase()

C．getReadableDatabase()　D．getWritableDatabase()

10．下列关于ContentProvider的说法，错误的是（　）。

A．ContentProvider的作用是实现数据共享和交换

B．要访问ContentProvider，只需调用ContentProvider的增、删、改、查相关方法

C．ContentProvider提供的URI必须以“content://”开头

D．对于系统中的音频、视频、图像、通讯录等，Android提供了内置的ContentProvider

4.11 作业

1．比较SQLite与SQL Server的不同之处。

2．SQLite有哪几种数据类型？在SQLite中，日期型（年-月-日或时-分-秒）数据应该用哪种类型？需要进行怎样的处理？

3．SQLiteDatabase 类的作用是什么？试举例说明。
4．SQLiteOpenHelper 类的作用是什么？试举例说明。
5．ContentProvider 的作用是什么？它向谁提供内容？
6．编写一个 ContentProvider 需要哪些方法？请举例说明。
7．在 Android 系统中，如何访问自定义 ContentProvider?
8．请简述通过 ContentResolver 获取 ContentProvider 内容的基本步骤。
9．试写出一个内置的 ContentProvider 的示例（只需写出关键代码）。

第5章

数据存储——简易相册项目

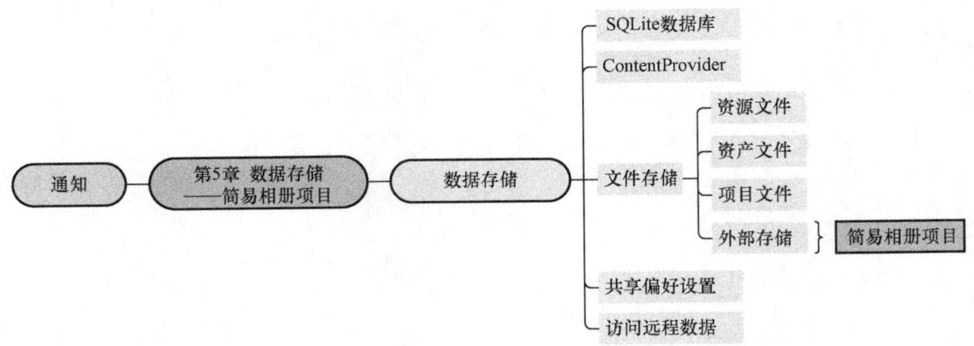

本章知识要点思维导图

本章主要介绍数据存储的相关内容，包括文件存储（访问 SD 卡）、共享偏好设置（SharedPreferences）、通知（Notification）、远程数据访问等。

5.1 需求分析

开发一个简易相册项目，可以实现预览相册功能，设计要求如下。

- 本项目包含两个界面。
- 以网格方式（GridView）读取 SD 卡中所有的图片文件，文件格式包括 jpg、bmp、gif、png 等，并在界面 1 中显示。
- 在 GridView 中设置监听器（setOnItemClickListener()），当用户单击监听器后，系统会携带两个参数——所有图片文件的路径（ArrayList<String> imgPath）和单击位置（position），跳转至界面 2。
- 界面 2 只有一个 ImageView 控件，用于显示 position 对应的图片。
- 在 ImageView 控件中设置触摸监听器（setOnTouchListener()），用于在屏幕上左右滑动，预览所有图片，当预览第一张图片时，给出提示“已经是第一张了！”当预览最后一张图片时，给出提示“已经是最后一张了！”。

5.2 界面设计

简易相册项目有两个界面，界面 1 使用网格方式，需要注意，为了突出填充效果，建议读者选择的图片数量避免是 3 的倍数；界面 2 用于显示用户所选的图片，用户既可以放大图片进行观看，也可以左右滑动屏幕切换图片。本项目的界面如图 5-1 所示。

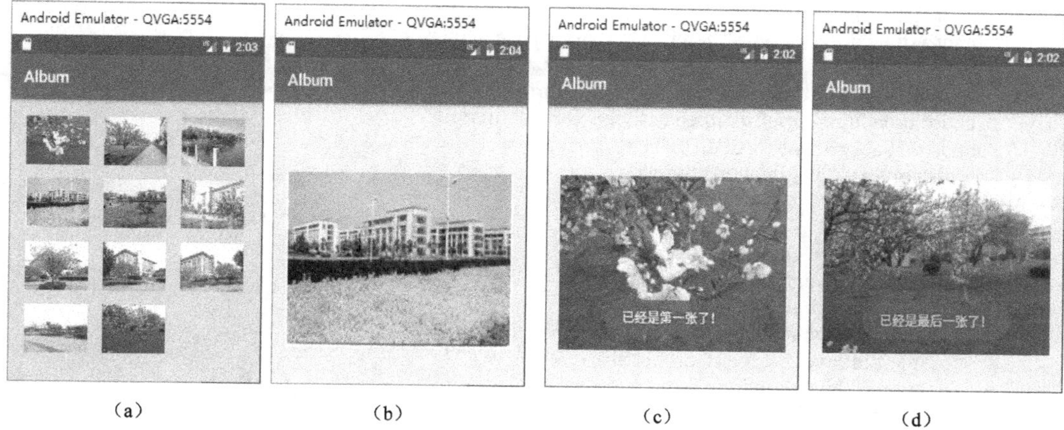

（a）　（b）　（c）　（d）

图 5-1　简易相册项目的界面

5.3 实施

5.3.1 创建项目

创建一个名为 Album 的项目，包含 MainActivity 和 BigImageActivity。

5.3.2 界面实现

简易相册项目有两个界面，界面 1 是主界面，界面 2 是大图显示界面。此外，由于简易相册项目还涉及 SD 卡的读取权限问题，所以要单独设置。

（1）界面 1 的布局文件为 activity_main.xml，代码如下：

```
<RelativeLayout xmlns:android="http://schemas.android.com/apk/res/android"
    xmlns:tools="http://schemas.android.com/tools"
    android:layout_width="match_parent"
    android:layout_height="match_parent"
    android:paddingBottom="@dimen/activity_vertical_margin"
    android:paddingLeft="@dimen/activity_horizontal_margin"
    android:paddingRight="@dimen/activity_horizontal_margin"
    android:paddingTop="@dimen/activity_vertical_margin"
    android:background="#B0E2FF"
    tools:context=".MainActivity" >

   <GridView
        android:id="@+id/gridView1"
        android:layout_width="match_parent"
        android:layout_height="match_parent"
        android:numColumns="3"
         android:horizontalSpacing="10dip"
        android:verticalSpacing="10dip" />

</RelativeLayout>
```

（2）界面 2 的布局文件为 activity_big_image.xml，代码如下：

```
<RelativeLayout xmlns:android="http://schemas.android.com/apk/res/android"
    xmlns:tools="http://schemas.android.com/tools"
    android:layout_width="match_parent"
```

```
    android:layout_height="match_parent"
    android:paddingBottom="@dimen/activity_vertical_margin"
    android:paddingLeft="@dimen/activity_horizontal_margin"
    android:paddingRight="@dimen/activity_horizontal_margin"
    android:paddingTop="@dimen/activity_vertical_margin"
    android:background="#EBEBEB"
    tools:context=".BigImageActivity" >

    <ImageView
        android:id="@+id/bigimage"
        android:layout_width="match_parent"
        android:layout_height="300dp"
        android:layout_alignParentTop="true"
        android:layout_centerHorizontal="true"
        android:layout_marginTop="40dp"
        android:src="@drawable/ic_launcher" />

</RelativeLayout>
```

（3）测试项目时，SD 卡中的图片可以是任意格式的图片文件（如 jpg、bmp、gif、png 等）。读者可以将目标图片放入 File Explorer→storage→sdcard--Pictures 目录下，如图 5-2 所示。

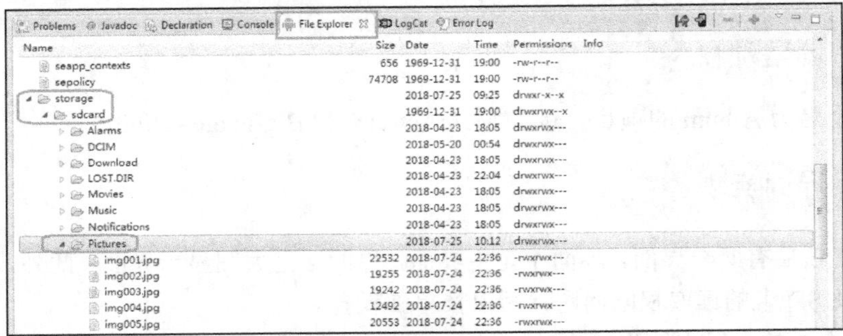

图 5-2　放入 SD 卡中的图片

（4）因为本项目要访问 SD 卡，所以请打开 SD 卡的读取权限，即在 AndroidManifest.xml 文件中输入如下代码：

```
<uses-permission android:name="android.permission.MOUNT_UNMOUNT_FILESYSTEMS" />
<uses-permission android:name="android.permission.READ_EXTERNAL_STORAGE"/>
<uses-permission android:name="android.permission.WRITE_EXTERNAL_STORAGE"/>
```

5.3.3　Java 代码

1．MainActivity.java

界面 1 是主界面，设计界面 1 时，有两大难点：第一，访问 SD 卡，并把 SD 卡中所有图片的完整路径写入 ArrayList<String>。第二，自定义适配器 BaseAdapter 用于填充前台控件 GridView。此外，还需要在 GridView 控件上添加监听功能，当用户单击某张图片时，系统携带参数，跳转至界面 2。

界面 1 的 MainActivity.java 完整代码如下：

```
package com.example.administrator.album;

import java.io.File;
import java.util.ArrayList;
import java.util.HashMap;
import java.util.List;
import java.util.Map;
```

```
import android.os.AsyncTask;
import android.os.Bundle;
import android.os.Environment;
import android.support.v7.app.AppCompatActivity;
import android.content.Intent;
import android.graphics.Bitmap;
import android.graphics.BitmapFactory;
import android.view.Menu;
import android.view.View;
import android.view.ViewGroup;
import android.widget.AdapterView;
import android.widget.AdapterView.OnItemClickListener;
import android.widget.BaseAdapter;
import android.widget.GridView;
import android.widget.ImageView;
import android.widget.SimpleAdapter;

public class MainActivity extends AppCompatActivity{
    ArrayList<String> imgPath = new ArrayList<String>(); // 定义一个数组用于保存文件路径
    String[] imageFormat = new String[] { "jpg", "bmp", "gif" ,"png"}; // 定义图片格式
    String filePath = Environment.getExternalStorageDirectory().getPath()
            + "/Pictures/";
    GridView gridView1;

    @Override
    protected void onCreate(Bundle savedInstanceState) {
        super.onCreate(savedInstanceState);
        setContentView(R.layout.activity_main);

        gridView1 = (GridView) findViewById(R.id.gridView1);
        getSdCardImgFile(filePath);

        if (imgPath.size() < 1) {// 如果不存在图片文件
            return;
        }

        BaseAdapter adapter = new BaseAdapter() {
            // 获取当前选项的 ID
            @Override
            public long getItemId(int position) {
                return position;
            }

            // 获取当前选项
            @Override
            public Object getItem(int position) {
                return position;
            }

            // 获得数量
            @Override
            public int getCount() {
                return imgPath.size();
            }

            @Override
            public View getView(int position, View convertView, ViewGroup parent) {
                ImageView imageView;// 声明 ImageView 的对象
```

```
                if (convertView == null) {
                    imageView = new ImageView(MainActivity.this);// 实例化 ImageView 的对象
                    // 设置对象的宽度和高度
                    imageView.setAdjustViewBounds(true);
                    imageView.setMaxWidth(150);
                    imageView.setMaxHeight(95);
                    imageView.setPadding(3, 3, 3, 3);// 设置 ImageView 的内边距
                } else {
                    imageView = (ImageView) convertView;
                }
                // 为 ImageView 设置要显示的图片
                Bitmap bm = BitmapFactory.decodeFile(imgPath.get(position));
                imageView.setImageBitmap(bm);
                return imageView;
            }
        };

        gridView1.setAdapter(adapter);// 将适配器与 GridView 关联
        gridView1.setOnItemClickListener(new OnItemClickListener(){

            @Override
            public void onItemClick(AdapterView<?> av, View arg1, int position,
                    long arg3) {
                Intent i = new Intent(MainActivity.this,BigImageActivity.class);
                Bundle bundle = new Bundle();
                bundle.putStringArrayList("imgPath", imgPath);
                bundle.putInt("position", position);
                i.putExtras(bundle);
                startActivity(i);
            }});
    }

    private boolean isImageFile(String path) { // 判断是否为图片文件的方法
        for (String format : imageFormat) {
            if (path.contains(format)) { // 如果文件名称包含定义的格式后缀，则返回 true
                return true;
            }
        }
        return false;
    }

    private void getSdCardImgFile(String url) { //获取指定路径下的图片文件，传入路径参数
        File files = new File(url);          //新定义一个文件，路径是传入的 url
        File[] file = files.listFiles();     //遍历该文件所有的子文件夹，生成文件夹数组
        for (File f : file) {                //for 循环遍历文件数组
            if (f.isDirectory()) {           //如果是文件夹，则递归调用此方法遍历子文件夹
                getSdCardImgFile(f.getAbsolutePath());     //递归调用
            } else {
                if (isImageFile(f.getPath())) {       //如果文件是图片文件

                    imgPath.add(f.getAbsolutePath());//获取绝对路径，返回定义好的数组
                }
            }
        }
    }
}
```

2．BigImageActivity.java

界面 2 是大图显示界面，主要功能有两项：第一，对界面 1 携带的两个参数进行解析，并以大图的方式显示被用户单击的图片。第二，在 ImageView 控件上添加触摸监听，以便用户在屏幕上可以左右滑动。

界面 2 的 BigImageActivity.java 完整代码如下：

```
package com.example.administrator. album;

import java.util.ArrayList;

import android.os.Bundle;
import android.annotation.SuppressLint;
import android.support.v7.app.AppCompatActivity;
import android.content.Intent;
import android.graphics.Bitmap;
import android.graphics.BitmapFactory;
import android.view.Menu;
import android.view.MotionEvent;
import android.view.View;
import android.view.View.OnTouchListener;
import android.widget.ImageView;
import android.widget.Toast;

public class BigImageActivity extends AppCompatActivity implements OnTouchListener {
    ArrayList<String> imgPath;
    int position;
    ImageView bigimage;

    @Override
    protected void onCreate(Bundle savedInstanceState) {
        super.onCreate(savedInstanceState);
        setContentView(R.layout.activity_big_image);

        bigimage = (ImageView) this.findViewById(R.id.bigimage);
        bigimage.setClickable(true);  //这行代码很重要，请务必牢记并正确输入

        Intent it = getIntent();
        Bundle bundle = it.getExtras();
        imgPath = bundle.getStringArrayList("imgPath");
        position = bundle.getInt("position");
        Bitmap bm = BitmapFactory.decodeFile(imgPath.get(position));
        bigimage.setImageBitmap(bm);

        bigimage.setOnTouchListener(this);
    }

    float x0 = 0, x1;

    @Override
    public boolean onTouch(View arg0, MotionEvent arg1) {
        switch (arg1.getAction()) {
        case MotionEvent.ACTION_DOWN:
            x0 = arg1.getX();
            break;
        case MotionEvent.ACTION_UP:
            x1 = arg1.getX();
            float w;
            w = x1 - x0;
```

```
                if (w > 80)
                    viewPrePhoto();
                if (-w > 80)
                    viewNextPhoto();
                break;
            }
            return false;
        }

        private void viewPrePhoto() {
                if (position== 0)
                    Toast.makeText(this, "已经是第一张了！ ", 0).show();
                else {
                    Bitmap bm = BitmapFactory.decodeFile(imgPath.get(position- 1));
                    bigimage.setImageBitmap(bm);
                    position=position-1;
                }
        }

        private void viewNextPhoto() {
                if (position== imgPath.size()-1)
                    Toast.makeText(this, "已经是最后一张了！ ", 0).show();
                else {
                    Bitmap bm = BitmapFactory.decodeFile(imgPath.get(position+1));
                    bigimage.setImageBitmap(bm);
                    position=position+1;
                }
        }
}
```

5.3.4 运行测试

通过模拟器或真实手机运行项目，测试简易相册的功能。

5.4 数据存储概述

一个完整的应用程序，通常会涉及数据存储操作。因此，Android 系统提供了五种数据存储（访问）方式，分别如下。

- 使用 SQLite 数据库存储数据。
- 使用 ContentProvider 存储数据。
- 使用文件存储数据。
- 使用共享偏好设置存储数据。
- 访问远程数据。

在第 4 章中，我们已经详细介绍过 SQLite 数据库和 ContentProvider，本章主要讨论文件存储、共享偏好设置及访问远程数据。

5.5 文件存储

文件是一种存储数据的重要方式，出于安全考虑，Android 系统将可访问的文件分为四种类型并给予不同的访问控制方式，这四种文件类型分别是资源文件（Resource File）、资产文件（Asset File）、项目文件（Data File）和外部存储（Storage File）。

5.5.1 资源文件

资源文件包括 Android 应用程序使用的各种资源，如布局、菜单、图片、字符串等，详细内容见表 5-1。在项目的“/res”目录下的文件都是资源文件，这些文件都会被赋予 ID 值，从而在程序中被引用。编译程序时，资源文件被打包到 apk 文件中；安装程序时，资源文件随之被安装到目标设备中。因此，资源文件为只读状态，程序运行过程中不能修改它们。

表 5-1 资源文件

资 源 类 型	目录/文件名	说 明
布局	/res/layout	图形界面的布局文件（xml 格式）
菜单	/res/menu	定义菜单项的菜单文件（xml 格式）
图片	/res/drawable	直接复制到该目录，不同分辨率的图片归于不同的目录
颜色值	/res/values/color.xml	以 resources 为 Root 的 xml 文件
字符串	/res/values/string.xml	以 resources 为 Root 的 xml 文件
单位资源	/res/values/demens.xml	以 resources 为 Root 的 xml 文件
样式和主题	/res/values/style.xml	以 resources 为 Root 的 xml 文件
原始文件	/res/raw	直接复制到这个文件夹

表 5-1 中的所有资源在编译的时都会被适当处理：系统先为每个资源赋予一个 ID 值，并生成 R.java 文件，使应用程序能够高效地引用这些资源；然后系统将 xml 文件编译为二进制形式。但是对于“/res/raw”目录中的内容，系统只赋予其 ID 值并在 R.java 中引用，不会对文件进行任何处理。

使用资源文件时，需要注意以下事项。

- 同一文件夹下的资源文件，其名称不能重复，即不能用文件扩展名进行区别，因为 R.java 中只保留资源的文件名而不保留扩展名。例如，有两个图片文件，一个是 icon.png，另一个是 icon.jpg，那么在 R.java 中只有一个 R.drawable.icon，便会引发问题。
- 资源文件的名字必须符合 Java 变量的命名规则，并且不能有“大写字母”，只能是“小写字母”或“小写字母+数字”，其中“数字”的范围指 0~9 的整数，否则会有编译错误，因为 R.java 中的 ID 变量名要与资源中的文件一一对应，即用资源文件名作为 ID 变量名，命名时要遵守上述规则。
- 除 Android SDK 规定的目录名称外，不能有其他的目录名或者子目录。虽然不会出现编译错误，但是，其他的目录名或子目录会被完全忽略。例如，在“/res/layout”目录下建一个子目录“/activity”，那么在生成的 R.java 中是看不到“/activity”和其中的内容的。
- Android SDK 说明文档对资源文件的大小有限制，建议单个资源文件的小于 1MB。

5.5.2 资产文件

资产文件是一些与项目共存但又不希望在编译时被处理的文件。在项目的“/assets”目录下的文件都是资产文件，这些文件没有 ID 值，与普通文件相似，但在编译程序时，这些文件会被打包到 apk 文件中，安装程序时，资产文件随之被安装到目标设备中。因此，资产文件也是只读状态，程序运行过程中不能修改它们。

资产文件和资源文件一样，在编译过程中不会被处理，但它们有所区别。

- 资源 ID 映射：虽然“/res/raw”目录中的文件不会被编译为二进制形式，但是这些文件会被映射到 R.java 中，便于通过资源 ID 的形式进行访问。
- 子目录结构：虽然“/res/raw”目录中允许有子目录，但是子目录是无效的，也不会被打包到 apk 中；而“/assets”中允许有子目录，并且可以访问其中的文件，此外，这些子目录会被打包到 apk 中，安装程序后，子目录中的文件仍然存在，且和源码包中的一样。

在本章的简易相册项目中，封面图片就是一个资产文件，保存在“/assets”目录中，它在编译后与apk文件一同安装到AVD或手机上。注意，引用该文件的代码如下：

```
InputStream jpgFile = getAssets().open("photoViewer.jpg");
```

5.5.3 项目文件

在安装应用程序时，Android系统会为每个应用程序分配一个私有的存储空间——私有文件夹，用来保存与项目有关的文件，如数据库文件、共享偏好设置文件等。私有文件夹的位置为“/data/data/包名”，其中，包名指创建项目时输入的包名，可作为应用程序的唯一标识。

应用程序对其私有文件夹拥有独立的操作权限，如创建子目录、读取和写入等。但是，私有文件夹不能被其他应用程序访问，如果要将私有文件夹共享给其他应用程序，则只能通过内容提供者机制实现。

对于Android系统而言，项目文件是普通的文件，可以通过标准的Java I/O流读取或写入数据。例如，在第3章的运动会报名项目中，将所有已报名的学生其学号及详细信息，分别写入SD卡的两个文件（Sno.txt和Submit.txt）中，路径是“File Explorer→storage→sdcard”。

5.5.4 外部存储

外部存储指SD卡的存储空间，这里的SD卡是广义概念，包括内置存储设备，也包括可拔插型SD卡。SD卡用于保存文件，其存储空间是公开的，所有应用程序都可以自由地访问，常用的访问操作包括：读取、写入、创建、修改和删除。

注意：在AVD上使用SD卡时，需要在AVD的配置选项中指定SD卡的容量，然后再启动AVD。

【例5.1】在SD卡中读写文本文件，代码如下：

```
package com.example.administrator.testsdcard;

import java.io.BufferedReader;
import java.io.File;
import java.io.FileNotFoundException;
import java.io.FileReader;
import java.io.FileWriter;
import java.io.IOException;
import android.support.v7.app.AppCompatActivity;
import android.os.Bundle;
import android.os.Environment;
import android.util.Log;
import android.widget.Toast;

public class MainActivity extends AppCompatActivity{
	String sdPath = Environment.getExternalStorageDirectory().getPath();

	@Override
	protected void onCreate(Bundle savedInstanceState) {
		super.onCreate(savedInstanceState);
		setContentView(R.layout.activity_main);

		//①写文件
		String state = Environment.getExternalStorageState();
		if (Environment.MEDIA_MOUNTED.equals(state)) {
			try {
				FileWriter fw = new FileWriter(sdPath + "/Test.txt");
				fw.write("this is a test text file.");
				fw.flush();
```

```
                        fw.close();
System.out.println("The file written.");
                        } catch (IOException e) {
                        e.printStackTrace();
                    }
            } else if (Environment.MEDIA_MOUNTED_READ_ONLY.equals(state)) {
                Log.v("SD 卡状态", "已经插入了 SD 卡，但是只能读取");
            } else {
                Log.v("SD 卡状态", "不能使用（不存在，或其他问题）" + state);
            }

            //②读文件
            try {
                BufferedReader br = new BufferedReader(new FileReader(sdPath+ "/Test.txt"));
                String result = "";
                while ((result = br.readLine()) != null)
                    System.out.println(result);
            } catch (FileNotFoundException e) {
                e.printStackTrace();
            } catch (IOException e) {
                e.printStackTrace();
            }
        }
}
```

注意，本例需要访问 SD 卡，请设置 SD 卡的访问权限。本例的运行结果如图 5-3 所示。

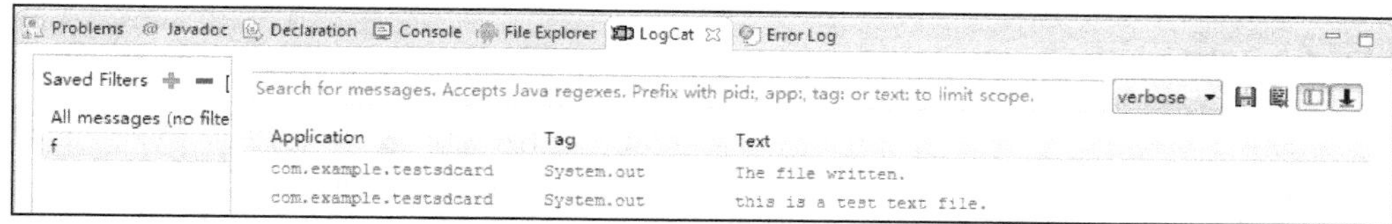

图 5-3　【例 5.1】运行结果

5.6　共享偏好设置

共享偏好设置（SharedPreferences）是一种轻量级机制，用于存储基本数据类型的键-值对，如布尔型、字符串型、浮点型、长整型、整型等。共享偏好设置是一种快速存储默认值、类实例变量、当前用户界面状态及用户偏好设置的理想方法。

偏好配置文件用于保存用户使用程序时的一些偏好配置参数，如喜欢的字体、字号、颜色等，再次运行程序时，系统自动恢复用户所选的配置参数。

保存到共享偏好设置的代码，一般写入 onPause()方法。即当用户离开程序界面前，系统把重要的配置参数信息保存到“File Explorer→data→data→项目包名→shared_prefs”下的某个 xml 文件中。

从共享偏好设置中读取的代码，一般写入 onCreate()方法。即当用户再次返回程序界面时，从已有的 info.xml 文件中把配置参数信息读取出来。

例如，在第 4 章的用户管理项目中，用户登录界面里的“记住密码”复选框就用到了共享偏好设置。具体而言，如果用户不选中“记住密码”复选框，则离开登录界面时，系统会忽略已输入的用户名和密码；反之，如果用户选中“记住密码”复选框，系统则会把用户名和密码写入“项目包名→shared_prefs”下的 info.xml 文件，如图 5-4 所示，当用户再次回到用户登录界面时，系统直接读取 info.xml 文件，并自动填写信息，用户就不必重新输入用户名和密码了。

Name	Size	Date	Time	Permissions	Info
com.example.training8		2018-07-26	03:30	drwxr-x--x	
cache		2018-07-26	03:28	drwxrwx--x	
databases		2018-07-26	03:28	drwxrwx--x	
usergl.db	20480	2018-07-26	03:28	-rw-rw----	
usergl.db-journal	8720	2018-07-26	03:28	-rw-------	
lib		2018-07-26	03:27	lrwxrwxrwx	-> /data/a...
shared_prefs		2018-07-26	03:30	drwxrwx--x	
info.xml	148	2018-07-26	03:30	-rw-rw----	

图 5-4　将用户名和密码写入 info.xml 文件

其中，onPause()方法用于保存配置参数信息，关键代码如下：

```
if (chkRemember.isChecked()) {
            SharedPreferences share = getSharedPreferences("info",Activity.MODE_PRIVATE);
            SharedPreferences.Editor editor = share.edit();
            pname = etPersonName.getText().toString();
            pwd = etPassword.getText().toString();
            editor.putString("pname", pname);
            editor.putString("pwd", pwd);
            editor.commit();
    }
```

onCreate()方法用于读取配置参数信息，关键代码如下：

```
SharedPreferences share = getSharedPreferences("info",Activity.MODE_PRIVATE);
etPersonName.setText(share.getString("pname", pname));
etPassword.setText(share.getString("pwd", pwd));
```

5.7　通知

在 Android 系统中，一般使用通知（Notification）提示用户，通知具有全局效果，它以图标的形式在屏幕顶端显示，当用户自屏幕顶部向下滑动时，通知会显示具体内容。

通常情况下，应用程序通过 NotificationManager 服务发送通知，即通过 NotificationManager 服务发送一个 Notification 对象。NotificationManager 是一个重要的系统级服务，它位于应用程序的框架层中，应用程序可以通过它向系统发送全局的通知。理论上，系统需要创建一个 Notification 对象，用于承载通知的内容。但是，在实际的使用过程中，系统一般不会直接构建 Notification 对象，而是使用它的一个内部类 NotificationCompat.Builder（Android3.0 以下版本使用 Notification.Builder）来实例化一个对象，并设置通知的各种属性，最后，通过 NotificationCompat.Builder.build()方法得到一个 Notification 对象。当获得 Notification 对象后，可以使用 NotificationManager.notify()方法发送通知。

例如，在第 4 章的用户管理项目中，用户登录界面里的“忘记密码”复选框就用到了通知。具体而言，如果用户选中“忘记密码”复选框，则界面会弹出“找回密码”对话框，接下来，用户输入注册时的手机号，单击“发送”按钮，则新密码会以通知的形式发送给用户，如图 5-5 所示。

图 5-5　“找回密码”通知

5.8　访问远程数据

在日常生活中，使用 Android 平台接入互联网访问远程数据的案例比比皆是。一方面，用户可以通过互联网读取远程主机上的数据，比如获取每天的外汇汇率、了解当天的天气预报等；另一方面，用户也可以通过互联网将数据存储在远程主机上。

【例 5.2】读取谷歌网站首页的内容，代码如下：

```
@Override
protected void onCreate(Bundle savedInstanceState) {
    super.onCreate(savedInstanceState);
    setContentView(R.layout.activity_main);

    new Thread(new Runnable() {
        @Override
        public void run() {
            BufferedReader in = null;
            try {
                URL url = new URL("http://www.google.com.hk/"); // 创建 URL
                in = new BufferedReader(new InputStreamReader(
                        url.openStream())); // 获得输入流，并加入缓冲功能
                String str;
                while ((str = in.readLine()) != null) { // 从输入流中逐行读取
                    System.out.println(str); // 输出到控制台
                }
            } catch (MalformedURLException e) {
                e.printStackTrace();
            } catch (IOException e) {
                e.printStackTrace();
            } finally { // 关闭语句，写在 finally 块中
                try {
                    if (in != null) {
                        in.close(); // 结束时关闭
                    }
                } catch (IOException e) {
                    e.printStackTrace();
                }
            }

        }
    }).start();
}
<uses-permission android:name="android.permission.INTERNET" />
```

5.9　实训：完善简易相册项目

观察真实手机的相册功能，完善本章的简易相册项目，增加以下需求。

- 增加手势识别操作，实现照片的缩小、放大功能。提醒读者，可以使用手势监听器 OnGestureListener 类，以便识别双触点操作。
- 对相册中的照片进行分类。

5.10　实训：进一步完善用户管理项目

在第 4 章的用户管理项目的基础上，增加共享偏好设置和通知功能。项目的登录信息来自 SQLite 数据库，用户登录界面和主界面如图 5-6 所示。

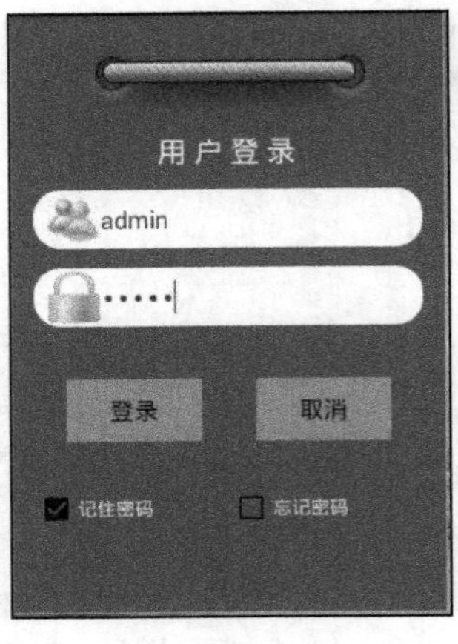

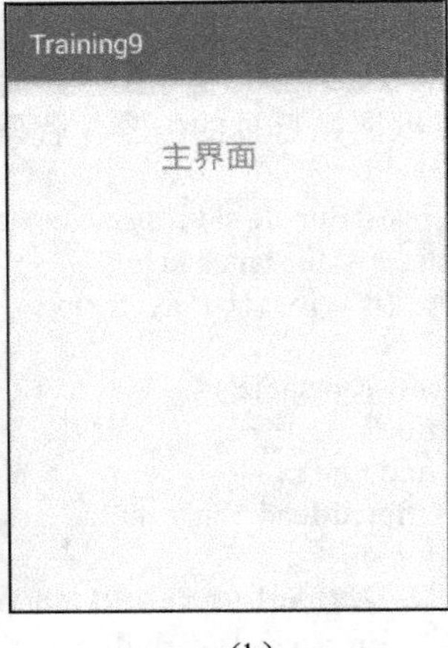

（a）　（b）

图 5-6　用户登录界面和主界面

（1）完成界面设计的布局文件。

（2）通知功能的自定义布局文件为 dialog_find.xml，代码如下：

```
<?xml version="1.0" encoding="utf-8"?>
<LinearLayout xmlns:android="http://schemas.android.com/apk/res/android"
    android:layout_width="match_parent"
    android:layout_height="match_parent"
    android:orientation="vertical" >

    <LinearLayout
        android:layout_width="match_parent"
        android:layout_height="wrap_content" >

        <TextView
            android:id="@+id/textView1"
            android:layout_width="wrap_content"
            android:layout_height="wrap_content"
            android:text="手机号："
            android:textAppearance="?android:attr/textAppearanceMedium" />

        <EditText
            android:id="@+id/etPhoneFind"
            android:layout_width="wrap_content"
            android:layout_height="wrap_content"
            android:layout_weight="1"
            android:ems="10"
            android:inputType="phone" >

            <requestFocus />
        </EditText>
    </LinearLayout>
</LinearLayout>
```

（3）完成登录界面的 MainActivity.java，主要代码如下：

```
package com.example.administrator.raining8;

import android.os.Bundle;
import android.annotation.SuppressLint;
import android.support.v7.app.AppCompatActivity;
import android.app.AlertDialog;
import android.app.Notification;
```

```
import android.app.NotificationManager;
import android.content.DialogInterface;
import android.content.Intent;
import android.content.SharedPreferences;
import android.database.Cursor;
import android.util.Log;
import android.view.LayoutInflater;
import android.view.View;
import android.view.View.OnClickListener;
import android.view.Window;
import android.view.WindowManager;
import android.widget.Button;
import android.widget.CheckBox;
import android.widget.EditText;
import android.widget.Toast;

public class LoginActivity extends AppCompatActivity implements OnClickListener {
        EditText etPersonName, etPassword;
        Button btnLogin, btnCancel;
        CheckBox chkRemember, chkForget;
        String pname, pwd;
        SqlHelper helper = new SqlHelper(this);
        String phoneFind,pwdFind;

        @Override
        protected void onCreate(Bundle savedInstanceState) {
                super.onCreate(savedInstanceState);
                setContentView(R.layout.activity_login);

                etPersonName = (EditText) this.findViewById(R.id.etPersonName);
                etPassword = (EditText) this.findViewById(R.id.etPassword);

                btnLogin = (Button) this.findViewById(R.id.btnLogin);
                btnLogin.setOnClickListener(this);
                btnCancel = (Button) this.findViewById(R.id.btnCancel);
                btnCancel.setOnClickListener(this);

                chkRemember = (CheckBox) this.findViewById(R.id.chkRemember);
                chkRemember.setOnClickListener(this);

                chkForget = (CheckBox) this.findViewById(R.id.chkForget);
                chkForget.setOnClickListener(new OnClickListener() {

                        @Override
                        public void onClick(View arg0) {
                                AlertDialog.Builder builder_Find = new AlertDialog.Builder(
                                                LoginActivity.this);
                                LayoutInflater inflater_Find = getLayoutInflater();
                                final View layout_Find = inflater_Find.inflate(
                                                R.layout.dialog_find, null);// 获取自定义布局
                                builder_Find.setView(layout_Find);
                                builder_Find.setIcon(R.drawable.key);// 设置标题图标
                                builder_Find.setTitle("找回密码");// 设置标题内容
                                final EditText etPhoneFind = (EditText) layout_Find
                                                .findViewById(R.id.etPhoneFind);

                                // 发送（确认）按钮
                                builder_Find.setPositiveButton("发送",
```

```
                                        new DialogInterface.OnClickListener() {

                                            @SuppressLint("NewApi")
                                            @Override
                                            public void onClick(DialogInterface arg0, int arg1) {
                                                Toast.makeText(layout_Find.getContext(),
                                                        "密码已发送", 1).show();
                                                Notification.Builder builder_Noti = new
Notification.Builder(
                                                        LoginActivity.this);
                                                builder_Noti.setSmallIcon(R.drawable.sms); // 设置图标
                                                builder_Noti.setContentTitle("找回密码"); // 设置标题

                                                phoneFind = etPhoneFind.getText().toString();
                                                Cursor cursor = helper.queryByphone(phoneFind);
                                                while (cursor.moveToNext()) {
                                                    pwdFind = cursor.getString(2);// 获取第三列的值
                                                }

                                                builder_Noti.setContentText("您好！密码是："+ pwdFind);
                                                //设置内容
                                                builder_Noti.setWhen(System.currentTimeMillis());
                                                //设置启动时间

                                                Notification msg = builder_Noti.build(); // 生成通知
                                                msg.defaults = Notification.DEFAULT_SOUND;
                                                NotificationManager manager = (NotificationManager)
getSystemService(NOTIFICATION_SERVICE); // 获取系统的 Notification 服务
                                                manager.notify(1, msg); // 发出通知
                                            }
                                        });

                    AlertDialog dlg_Find = builder_Find.create();
                    Window window_Find = dlg_Find.getWindow();
                    WindowManager.LayoutParams lp_Find = window_Find
                            .getAttributes();
                    lp_Find.alpha = 0.8f;
                    window_Find.setAttributes(lp_Find);
                    dlg_Find.show();
                }
            });
        }

        @Override
        public void onClick(View v) {
            Cursor cursor = helper.queryBypname(etPersonName.getText().toString());
            while (cursor.moveToNext()) {
                pname = cursor.getString(1);// 获取第二列的值
                pwd = cursor.getString(2);// 获取第三列的值

                System.out.println(pname + "," + pwd);
                Log.v("user", pname + "," + pwd);
            }

            switch (v.getId()) {
            case R.id.btnLogin:
                if (pname.equals(etPersonName.getText().toString())
```

```
                    && pwd.equals(etPassword.getText().toString())) {
                // 提示"登录成功"，然后再跳转到界面 2
                Toast.makeText(LoginActivity.this, "登录成功", 1).show();
                Intent i = new Intent(LoginActivity.this, MainActivity.class);
                startActivity(i);
            } else {
                // 用户名或密码错误，并清除 EditText
                Toast.makeText(LoginActivity.this, "用户名或密码错误", 1).show();
                etPersonName.setText("");
                etPassword.setText("");
            }
            break;
        case R.id.btnCancel:
            System.exit(0);
            break;
        }
    }

    @Override
    protected void onResume() {
        super.onResume();
        SharedPreferences share = getSharedPreferences("info",
                Activity.MODE_PRIVATE);
        etPersonName.setText(share.getString("pname", pname));
        etPassword.setText(share.getString("pwd", pwd));
        }

    @Override
    protected void onPause() {
        super.onPause();
        if (chkRemember.isChecked()) {
            SharedPreferences share = getSharedPreferences("info",
                    Activity.MODE_PRIVATE);
            SharedPreferences.Editor editor = share.edit();
            pname = etPersonName.getText().toString();
            pwd = etPassword.getText().toString();
            editor.putString("pname", pname);
            editor.putString("pwd", pwd);
            editor.commit();
        }
    }
}
```

5.11　练习题

1．下列关于 AlertDialog 的描述，不正确的是（　　）。
A．使用 new 关键字创建 AlertDialog 的实例
B．显示对话框需要调用 show()方法
C．setPositiveButton()方法用于添加“确定”按钮
D．setNegativeButton()方法用于添加“取消”按钮
2．（　　）方法在触摸按钮时会自动调用。
A．onClick()　　B．OnTouch()　　C．onActionDown()　　D．onTouchEvent()
3．Android 项目工程中的“/assets”目录的作用是（　　）。
A．放置一些文件资源，这些文件会被原封不动地打包到 apk 中

B．放置用到的图片资源
C．放置字符串、颜色、数组等常量数据
D．放置一些与 UI 相应的布局文件

4．（　）不是 Android 的数据存储方式。
A．SharedPreferences　B．SQLite　C．ContentProvider　D．ListView

5．存储在 SD 卡中（　）文件夹内的文件对 Android 系统是可见的。
A．res　B．system　C．data　D．sdcard

6．SharedPreferences 存放的数据类型不支持（　）。
A．boolean　B．int　C．String　D．double

7．下列关于 SharedPreferences 的描述，正确的是（　）。
A．SharedPreferences pref = new SharedPreferences()
B．Editor editor = new Editor()
C．SharedPreferences 对象用于读取和存储常用数据类型
D．Editor 对象存储数据时，最后都要调用 commit()方法

8．下列关于 SharedPreferences 保存文件的路径和扩展名的说法，正确的是（　）。
A．路径为“/data/data/shared_prefs”，扩展名为*.txt
B．路径为“/data/data/package name/shared_prefs”，扩展名为*.xml
C．路径为“/mnt/sdcard/指定文件夹”，扩展名为指定扩展名
D．路径为任意路径，扩展名为指定扩展名

9．下列关于 Notification 的描述，不正确的是（　）。
A．Notification 需要 NotificatinManager 进行管理
B．使用 NotificationManager 的 notify()方法显示 Notification 消息
C．显示 Notification 时，可以设置默认发声、震动等
D．Notification 中存在可以清除消息的方法

10．下列某个（些）类中，用于实现状态栏通知的是（　）。
A．Notification　B．NotificationManager
C．Dialog　D．Notification 和 NotificationManager

5.12 作业

1．在 Android 系统中，文件处理有哪几类？它们的文件各自保存在什么文件夹中？
2．什么是资源文件和资产文件？它们各有什么特点？
3．如何访问“/res/raw”和“/assets”目录下的文件？有什么区别？
4．Android 系统中的五种数据存储方式分别是什么？它们分别有什么特点？
5．简述 SharedPreferences 存储方式与 SQLite 数据库的区别？
6．在 Android 系统中，常见的 Notification 表现形式有哪些？

第6章

Service——MP3音乐播放器项目

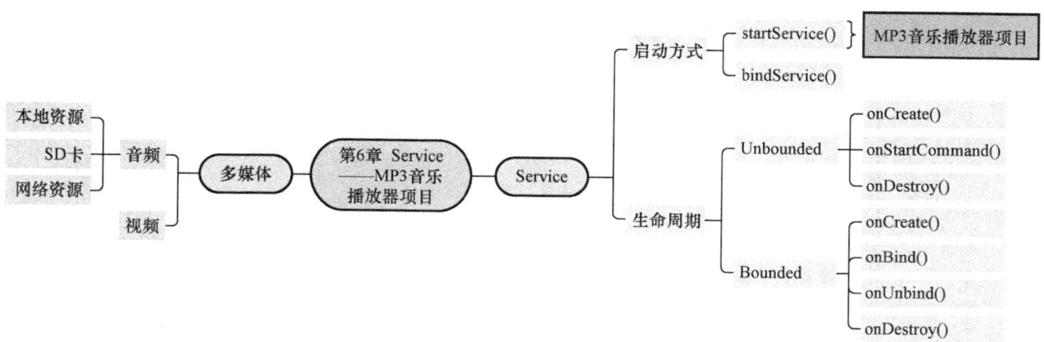

本章知识要点思维导图

本章将通过MP3音乐播放器项目介绍Android系统中的重要组件——Service（服务），它用于控制在后台运行的程序，如音乐播放程序等，此外，还将讨论多媒体（音频、视频）的相关内容。

6.1 需求分析

开发一个MP3音乐播放器项目，要求MediaPlayer的音乐播放流程放在后台的Service中进行控制，设计要求如下。

- 将每首歌曲的基本信息——singer（歌手）、song（歌曲名）、path（路径）、duration（时长）和size（大小）封装在歌曲类Song中。
- 自定义音乐工具类MusicUtils：通过ContentResolver读取设备（模拟器或真实手机）中的音频文件，并添加到容器类ArrayList<Song>中。
- 自定义适配器MyAdapter (extends BaseAdapter)：用于填充前台界面（界面布局文件为activity_main.xml）中的ListView控件。
- 自定义类CircleImageView (extends ImageView)：制作一个外观为圆角矩形的播放控制台，包括“上一首”“暂停和播放”“下一首”“停止”功能按钮。
- 单击ListView控件，实现点拨歌曲功能。
- 程序中的各种控制命令，由前台界面的MainActivity通过Intent将值传给后台的Service，由Service中的MediaPlayer来实现。
- 在播放控制台的左侧加入图片，使其能够360°旋转，形成动画。
- 在播放控制台的上方，以跑马灯的形式呈现歌曲名。

6.2 界面设计

MP3 音乐播放器项目只有一个界面，其布局文件为 activity_main.xml，后台 Service 不涉及屏幕操作，本项目的界面如图 6-1 所示。

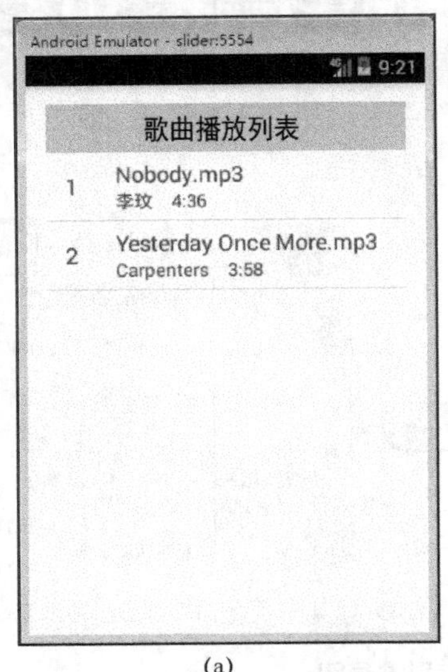

(a)

(b)

图 6-1 MP3 音乐播放器界面

6.3 实施

6.3.1 创建项目

创建一个名为 MusicPlayer 的项目，把准备好的歌曲放入 storage→sdcard→Music 目录下。

6.3.2 界面实现

在界面实现过程中，需要设计以下布局文件：ListView 控件对应的自定义布局文件（item_music_listview.xml）、主界面布局文件（activity_main.xml）、动画控制布局文件（animation.xml 和 img_animation.xml）及菜单布局文件（main.xml）。

1．item_music_listview.xml

ListView 控件对应的自定义布局文件为 item_music_listview.xml，显示效果如图 6-2 所示。

图 6-2 ListView 控件的显示效果

item_music_listview.xml 的代码如下：

```
<?xml version="1.0" encoding="utf-8"?>
<LinearLayout xmlns:android="http://schemas.android.com/apk/res/android"
    android:layout_width="match_parent"
    android:layout_height="110dp"
```

```
    android:gravity="center_vertical"
    android:orientation="horizontal"
    android:padding="5dp" >

    <TextView
        android:id="@+id/item_mymusic_position"
        android:layout_width="wrap_content"
        android:layout_height="wrap_content"
        android:layout_margin="10dp"
        android:gravity="center"
        android:text="1"
        android:textColor="#0d0c0c"
        android:textSize="18sp" />

    <RelativeLayout
        android:layout_width="wrap_content"
        android:layout_height="wrap_content"
        android:layout_marginLeft="20dp" >

        <TextView
            android:id="@+id/item_mymusic_song"
            android:layout_width="wrap_content"
            android:layout_height="wrap_content"
            android:text="歌曲名"
            android:textColor="#0d0c0c"
            android:textSize="18dp" />

        <TextView
            android:id="@+id/item_mymusic_singer"
            android:layout_width="wrap_content"
            android:layout_height="wrap_content"
            android:layout_below="@+id/item_mymusic_song"
            android:text="歌手"
            android:textColor="#0d0c0c"
            android:textSize="15sp" />

        <TextView
            android:id="@+id/item_mymusic_duration"
            android:layout_width="wrap_content"
            android:layout_height="wrap_content"
            android:layout_below="@+id/item_mymusic_song"
            android:layout_marginLeft="15dp"
            android:layout_toRightOf="@+id/item_mymusic_singer"
            android:text="时间"
            android:textColor="#0d0c0c"
            android:textSize="15sp" />
    </RelativeLayout>

</LinearLayout>
```

2．activity_main.xml

主界面的布局文件为 activity_main.xml，代码如下：

```
<RelativeLayout xmlns:android="http://schemas.android.com/apk/res/android"
    xmlns:tools="http://schemas.android.com/tools"
    android:layout_width="match_parent"
    android:layout_height="match_parent"
    android:paddingBottom="@dimen/activity_vertical_margin"
    android:paddingLeft="@dimen/activity_horizontal_margin"
```

```
android:paddingRight="@dimen/activity_horizontal_margin"
android:paddingTop="@dimen/activity_vertical_margin"
tools:context=".MainActivity" >

<TextView
    android:id="@+id/main_textview"
    android:layout_width="match_parent"
    android:layout_height="40dp"
    android:gravity="center"
    android:background="#87CEFF"
    android:text="歌曲播放列表"
    android:textSize="22sp"
    android:textColor="#FFFFFF" />
<ListView
    android:id="@+id/main_listview"
    android:layout_below="@+id/main_textview"
    android:layout_width="match_parent"
    android:layout_height="match_parent"
     android:divider="#40FF0000"
      android:dividerHeight="1dp"/>

 <RelativeLayout
    android:id="@+id/main_control_rl"
    android:layout_width="match_parent"
    android:layout_height="90dp"
    android:layout_alignParentBottom="true"
     android:background="@drawable/circle"
    android:visibility="invisible">

    <com.example.musicplayer.CircleImageView
        android:id="@+id/control_imageview"
        android:layout_width="60dp"
        android:layout_height="60dp"
        android:layout_centerVertical="true"
        android:layout_marginLeft="5dp"
        android:src="@drawable/music" />

    <TextView
        android:id="@+id/control_singer"
        android:layout_width="wrap_content"
        android:layout_height="wrap_content"
        android:layout_marginLeft="30dp"
        android:layout_marginTop="5dp"
        android:layout_toRightOf="@+id/control_imageview"
        android:text="歌手"
        android:textSize="15sp" />

    <TextView
        android:id="@+id/control_song"
        android:layout_width="150dp"
        android:layout_height="wrap_content"
        android:layout_marginLeft="15dp"
        android:layout_marginTop="5dp"
        android:singleLine="true"
        android:ellipsize="marquee"
        android:marqueeRepeatLimit="marquee_forever"
        android:layout_toRightOf="@+id/control_singer"
        android:text="歌曲名"
```

```
                android:textSize="16sp" />

        <RelativeLayout
            android:layout_width="match_parent"
            android:layout_height="50dp"
            android:layout_alignParentBottom="true"
            android:layout_marginBottom="10dp"
            android:layout_marginLeft="5dp"
            android:layout_toRightOf="@+id/control_imageview">

            <Button
                android:id="@+id/playing_btn_previous"
                android:layout_width="40dp"
                android:layout_height="40dp"
                android:background="@drawable/previous"
                android:layout_alignParentLeft="true"
                android:onClick="control" />
            <Button
                android:id="@+id/playing_btn_pause"
                android:layout_width="40dp"
                android:layout_height="40dp"
                android:background="@drawable/pause"
                android:layout_marginLeft="50dp"
                android:onClick="control" />

            <Button
                android:id="@+id/playing_btn_next"
                android:layout_width="40dp"
                android:layout_height="40dp"
                android:background="@drawable/next"
                android:layout_marginLeft="100dp"
                android:onClick="control" />

            <Button
                android:id="@+id/playing_btn_stop"
                android:layout_width="40dp"
                android:layout_height="40dp"
                android:background="@drawable/stop"
                android:layout_marginLeft="150dp"
                android:onClick="control" />

        </RelativeLayout>
    </RelativeLayout>

</RelativeLayout>
```

3．animation.xml 和 img_animation.xml

动画控制布局文件包括 animation.xml 和 img_animation.xml。

其中，animation.xml 的代码如下：

```
<?xml version="1.0" encoding="utf-8"?>
<set xmlns:android="http://schemas.android.com/apk/res/android" >

    <rotate
        android:duration="1000"
        android:fromDegrees="0"
        android:pivotX="50%"
        android:pivotY="50%"
        android:repeatMode="reverse"
```

```
            android:toDegrees="360" />

    <scale
        android:duration="1000"
        android:fromXScale="0"
        android:fromYScale="0"
        android:pivotX="50%"
        android:pivotY="50%"
        android:repeatMode="reverse"
        android:toXScale="1"
        android:toYScale="1" />

    <translate
        android:duration="1000"
        android:fromXDelta="-300"
        android:repeatMode="reverse"
        android:toXDelta="0" />

</set>
```

img_animation.xml 的代码如下：

```
<?xml version="1.0" encoding="utf-8"?>
<set xmlns:android="http://schemas.android.com/apk/res/android" >

    <rotate
        android:duration="10000"
        android:fromDegrees="0"
        android:interpolator="@android:anim/linear_interpolator"
        android:pivotX="50%"
        android:pivotY="50%"
        android:repeatMode="restart"
        android:repeatCount="-1"
        android:toDegrees="360" />

</set>
```

4. main.xml

菜单布局文件指 menu 目录下的 main.xml，代码如下：

```
<menu xmlns:android="http://schemas.android.com/apk/res/android" >

    <item
        android:id="@+id/item_stop"
        android:orderInCategory="100"
        android:showAsAction="never"
        android:icon="@drawable/stop"
        android:title="退出系统"/>

</menu>
```

6.3.3 Java 代码

1. Song.java

把每首歌曲的基本信息封装到歌曲类 Song 中，Song.java 的代码如下：

```
package com.example.administrator.musicplayer;

public class Song {
    public String singer;   //歌手
    public String song;   //歌曲名
```

```
    public String path;   //歌曲的路径
    public int duration;   //歌曲的时长
    public long size;   //歌曲的大小
}
```

2．MusicUtils.java

自定义音乐工具类 MusicUtils，通过 ContentResolver 读取设备（模拟器或真实手机）中的音频文件，并添加到容器类 ArrayList<Song>中，MusicUtils.java 的代码如下：

```
package com.example.administrator.musicplayer;

import android.content.Context;
import android.database.Cursor;
import android.provider.MediaStore;
import java.util.ArrayList;
import java.util.List;

public class MusicUtils {     //音乐工具类

    public static List<Song> getMusicData(Context context) {
        List<Song> list = new ArrayList<Song>();
        // 媒体库查询语句（自定义音乐工具类 MusicUtils）
        Cursor cursor = context.getContentResolver().query(
                MediaStore.Audio.Media.EXTERNAL_CONTENT_URI, null, null,
                null, MediaStore.Audio.AudioColumns.IS_MUSIC);
        if (cursor != null) {
            while (cursor.moveToNext()) {
                Song song = new Song();
                song.song = cursor.getString(cursor.getColumnIndexOrThrow
(MediaStore.Audio.Media.DISPLAY_NAME));
                song.singer = cursor.getString(cursor.getColumnIndexOrThrow
(MediaStore.Audio.Media.ARTIST));
                song.path = cursor.getString(cursor.getColumnIndexOrThrow
(MediaStore.Audio.Media.DATA));
                song.duration = cursor.getInt(cursor.getColumnIndexOrThrow
(MediaStore.Audio.Media.DURATION));
                song.size = cursor.getLong(cursor.getColumnIndexOrThrow
(MediaStore.Audio.Media.SIZE));
                if (song.size > 1000 * 800) {
// 下面的 if 语句用于切割标题，分离歌曲名和歌手（本地媒体库读取的歌曲信息不规范）
                    if (song.song.contains("-")) {
                        String[] str = song.song.split("-");
                        song.singer = str[0];
                        song.song = str[1];
                    }
                    list.add(song);
                }
            }
            // 释放资源
            cursor.close();
        }

        return list;
    }

    public static String formatTime(int time) {   //定义一个方法用于格式化获取的时间信息
        if (time / 1000 % 60 < 10) {
            return time / 1000 / 60 + ":0" + time / 1000 % 60;
```

```
            } else {
                return time / 1000 / 60 + ":" + time / 1000 % 60;
            }
        }
}
```

3．MyAdapter.java

自定义适配器 MyAdapter(extends BaseAdapter)，其功能是填充前台界面（界面布局文件为 activity_main.xml）中的 ListView 控件。MyAdapter.java 的代码如下：

```
package com.example.administrator.musicplayer;

import android.content.Context;
import android.graphics.Color;
import android.view.View;
import android.view.ViewGroup;
import android.widget.BaseAdapter;
import android.widget.TextView;

import java.util.List;

public class MyAdapter extends BaseAdapter {
        private Context context;
        private List<Song> list;

        public MyAdapter(MainActivity mainActivity, List<Song> list) {
                this.context = mainActivity;
                this.list = list;
        }

        @Override
        public int getCount() {
                return list.size();
        }

        @Override
        public Object getItem(int i) {
                return list.get(i);
        }

        @Override
        public long getItemId(int i) {
                return i;
        }

        @Override
        public View getView(int i, View view, ViewGroup viewGroup) {
                ViewHolder holder = null;
                if (view == null) {
                        holder = new ViewHolder();
                        // 引入布局
                        view = View.inflate(context, R.layout.item_music_listview, null);
                        // 实例化对象
                        holder.song = (TextView) view.findViewById(R.id.item_mymusic_song);
                        holder.singer = (TextView) view
                                        .findViewById(R.id.item_mymusic_singer);
                        holder.duration = (TextView) view
                                        .findViewById(R.id.item_mymusic_duration);
                        holder.position = (TextView) view
```

```
                .findViewById(R.id.item_mymusic_position);
            view.setTag(holder);
        } else {
            holder = (ViewHolder) view.getTag();
        }
        // 给控件赋值
        holder.song.setText(list.get(i).song.toString());
        holder.singer.setText(list.get(i).singer.toString());
        // 需要转换时间
        int duration = list.get(i).duration;
        String time = MusicUtils.formatTime(duration);
        holder.duration.setText(time);
        holder.position.setText(i + 1 + "");

        return view;
    }

    class ViewHolder {
        TextView song;// 歌曲名
        TextView singer;// 歌手
        TextView duration;// 时长
        TextView position;// 序号
    }
}
```

4．CircleImageView.java

自定义类 CircleImageView(extends ImageView)，用于制作一个外观为圆角矩形的播放控制台，包括“上一首”“暂停和播放”“下一首”“停止”功能按钮。CircleImageView.java 的代码如下：

```
package com.example.administrator.musicplayer;

import android.content.Context;
import android.content.res.TypedArray;
import android.graphics.Bitmap;
import android.graphics.BitmapShader;
import android.graphics.Canvas;
import android.graphics.Color;
import android.graphics.Matrix;
import android.graphics.Paint;
import android.graphics.RectF;
import android.graphics.Shader;
import android.graphics.drawable.BitmapDrawable;
import android.graphics.drawable.ColorDrawable;
import android.graphics.drawable.Drawable;
import android.util.AttributeSet;
import android.widget.ImageView;

public class CircleImageView extends ImageView {

    private static final ScaleType SCALE_TYPE = ScaleType.CENTER_CROP;
    private static final Bitmap.Config BITMAP_CONFIG = Bitmap.Config.ARGB_8888;
    private static final int COLORDRAWABLE_DIMENSION = 1;
    private static final int DEFAULT_BORDER_WIDTH = 0;
    private static final int DEFAULT_BORDER_COLOR = 0x00000000;
    private final RectF mDrawableRect = new RectF();
    private final RectF mBorderRect = new RectF();
    private final Matrix mShaderMatrix = new Matrix();
```

```
private final Paint mBitmapPaint = new Paint();
private final Paint mBorderPaint = new Paint();
private int mBorderColor = DEFAULT_BORDER_COLOR;
private int mBorderWidth = DEFAULT_BORDER_WIDTH;
private Bitmap mBitmap;
private BitmapShader mBitmapShader;
private int mBitmapWidth;
private int mBitmapHeight;
private float mDrawableRadius;
private float mBorderRadius;
private boolean mReady;
private boolean mSetupPending;

public CircleImageView(Context context) {
      super(context);
}

public CircleImageView(Context context, AttributeSet attrs) {
      this(context, attrs, 0);
}

public CircleImageView(Context context, AttributeSet attrs, int defStyle) {
      super(context, attrs, defStyle);
      super.setScaleType(SCALE_TYPE);

      TypedArray a = context.obtainStyledAttributes(attrs,
                  R.styleable.CircleImageView, defStyle, 0);

      mBorderWidth = a.getDimensionPixelSize(
                  R.styleable.CircleImageView_border_width, DEFAULT_BORDER_WIDTH);
      mBorderColor = a.getColor(R.styleable.CircleImageView_border_color,
                  DEFAULT_BORDER_COLOR);

      a.recycle();
      mReady = true;

      if (mSetupPending) {
            setup();
            mSetupPending = false;
      }
}

@Override
public ScaleType getScaleType() {
      return SCALE_TYPE;
}

@Override
public void setScaleType(ScaleType scaleType) {
      if (scaleType != SCALE_TYPE) {
            throw new IllegalArgumentException(String.format(
                        "ScaleType %s not supported.", scaleType));
      }
}

@Override
protected void onDraw(Canvas canvas) {
      if (getDrawable() == null) {
            return;
```

```
        }

        canvas.drawCircle(getWidth() / 2, getHeight() / 2, mDrawableRadius,
                mBitmapPaint);
        canvas.drawCircle(getWidth() / 2, getHeight() / 2, mBorderRadius,
                mBorderPaint);
    }

    @Override
    protected void onSizeChanged(int w, int h, int oldw, int oldh) {
        super.onSizeChanged(w, h, oldw, oldh);
        setup();
    }

    public int getBorderColor() {
        return mBorderColor;
    }

    public void setBorderColor(int borderColor) {
        if (borderColor == mBorderColor) {
            return;
        }

        mBorderColor = borderColor;
        mBorderPaint.setColor(mBorderColor);
        invalidate();
    }

    public int getBorderWidth() {
        return mBorderWidth;
    }

    public void setBorderWidth(int borderWidth) {
        if (borderWidth == mBorderWidth) {
            return;
        }

        mBorderWidth = borderWidth;
        setup();
    }

    @Override
    public void setImageBitmap(Bitmap bm) {
        super.setImageBitmap(bm);
        mBitmap = bm;
        setup();
    }

    @Override
    public void setImageDrawable(Drawable drawable) {
        super.setImageDrawable(drawable);
        mBitmap = getBitmapFromDrawable(drawable);
        setup();
    }

    @Override
    public void setImageResource(int resId) {
        super.setImageResource(resId);
        mBitmap = getBitmapFromDrawable(getDrawable());
```

```
        setup();
    }

    private Bitmap getBitmapFromDrawable(Drawable drawable) {
        if (drawable == null) {
            return null;
        }

        if (drawable instanceof BitmapDrawable) {
            return ((BitmapDrawable) drawable).getBitmap();
        }

        try {
            Bitmap bitmap;

            if (drawable instanceof ColorDrawable) {
                bitmap = Bitmap.createBitmap(COLORDRAWABLE_DIMENSION,
                        COLORDRAWABLE_DIMENSION, BITMAP_CONFIG);
            } else {
                bitmap = Bitmap.createBitmap(drawable.getIntrinsicWidth(),
                        drawable.getIntrinsicHeight(), BITMAP_CONFIG);
            }

            Canvas canvas = new Canvas(bitmap);
            drawable.setBounds(0, 0, canvas.getWidth(), canvas.getHeight());
            drawable.draw(canvas);
            return bitmap;
        } catch (OutOfMemoryError e) {
            return null;
        }
    }

    private void setup() {
        if (!mReady) {
            mSetupPending = true;
            return;
        }

        if (mBitmap == null) {
            return;
        }

        mBitmapShader = new BitmapShader(mBitmap, Shader.TileMode.CLAMP,
                Shader.TileMode.CLAMP);

        mBitmapPaint.setAntiAlias(true);
        mBitmapPaint.setShader(mBitmapShader);

        mBorderPaint.setStyle(Paint.Style.STROKE);
        mBorderPaint.setAntiAlias(true);
        mBorderPaint.setColor(mBorderColor);
        mBorderPaint.setStrokeWidth(mBorderWidth);

        mBitmapHeight = mBitmap.getHeight();
        mBitmapWidth = mBitmap.getWidth();

        mBorderRect.set(0, 0, getWidth(), getHeight());
        mBorderRadius = Math.min((mBorderRect.height() - mBorderWidth) / 2,
                (mBorderRect.width() - mBorderWidth) / 2);
```

```
        mDrawableRect.set(mBorderWidth, mBorderWidth, mBorderRect.width()
                - mBorderWidth, mBorderRect.height() - mBorderWidth);
        mDrawableRadius = Math.min(mDrawableRect.height() / 2,
                mDrawableRect.width() / 2);

        updateShaderMatrix();
        invalidate();
    }

    private void updateShaderMatrix() {
        float scale;
        float dx = 0;
        float dy = 0;

        mShaderMatrix.set(null);

        if (mBitmapWidth * mDrawableRect.height() > mDrawableRect.width()
                * mBitmapHeight) {
            scale = mDrawableRect.height() / (float) mBitmapHeight;
            dx = (mDrawableRect.width() - mBitmapWidth * scale) * 0.5f;
        } else {
            scale = mDrawableRect.width() / (float) mBitmapWidth;
            dy = (mDrawableRect.height() - mBitmapHeight * scale) * 0.5f;
        }

        mShaderMatrix.setScale(scale, scale);
        mShaderMatrix.postTranslate((int) (dx + 0.5f) + mBorderWidth,
                (int) (dy + 0.5f) + mBorderWidth);

        mBitmapShader.setLocalMatrix(mShaderMatrix);
    }
}
```

5．MainActivity.java

主界面的 MainActivity.java 代码如下：

```
package com.example.administrator.musicplayer;

import java.util.ArrayList;
import java.util.List;
import android.support.v7.app.AppCompatActivity;
import android.content.Intent;
import android.graphics.Color;
import android.os.Bundle;
import android.os.Parcelable;
import android.view.Menu;
import android.view.MenuItem;
import android.view.View;
import android.view.animation.Animation;
import android.view.animation.AnimationUtils;
import android.view.animation.LinearInterpolator;
import android.widget.AdapterView;
import android.widget.Button;
import android.widget.ImageView;
import android.widget.ListView;
import android.widget.TextView;
import android.widget.Toast;
```

```
public class MainActivity extends AppCompatActivity{

    private ListView mListView;
    private List<Song> list;
    private MyAdapter adapter;
    private int playPosition; // 当前播放歌曲的序号
    private boolean IsPlay = false; // 是否有歌曲在播放
    private Button playPause; // “暂停和播放”按钮
    private TextView song; // 歌曲名
    private TextView singer; // 歌手
    private ImageView imageView; // 播放控制台的图片
    private Animation animation; // 动画
    private String filePath;
    private String fileName;

    @Override
    protected void onCreate(Bundle savedInstanceState) {
        super.onCreate(savedInstanceState);
        setContentView(R.layout.activity_main);

        // 初始化
        list = new ArrayList<Song>();
        initView();
        setListener();
    }

    /**
     * ListView 控件的监听事件
     */
    private void setListener() {
        mListView.setOnItemClickListener(new AdapterView.OnItemClickListener() {

                @Override
                public void onItemClick(AdapterView<?> av, View view, int position,
                        long arg1) {
                    playPosition = position;
                    String posPath = list.get(position).path; // 长路径（绝对路径）
                    filePath = posPath.substring(0, posPath.lastIndexOf("/") + 1); // 路径的前半部分
                    fileName = list.get(position).song;
                    IsPlay = true;

                    Intent it = new Intent(MainActivity.this, MyService.class);
                    Bundle b = new Bundle();
                    // b.putStringArray("filesList", filesList);
                    b.putString("filePath", filePath);
                    b.putString("fileName", fileName);
                    b.putInt("playPosition", playPosition);
                    it.putExtras(b);
                    it.putExtra("key", 0);
                    startService(it);

                    // 设置当前项的背景色
                    if (((ListView) av).getTag() != null) {
                        ((View) ((ListView) av).getTag())
                                .setBackgroundDrawable(null);
                    }
                    ((ListView) av).setTag(view);
```

```
                view.setBackgroundColor(Color.rgb(230, 230, 250));

                // 单击 item，让播放控制台显示出来
                findViewById(R.id.main_control_rl).setVisibility(View.VISIBLE);
                //“暂停和播放”按钮图片变成播放状态
                playPause.setBackgroundResource(R.drawable.pause);

                setText(); // 设置歌曲名和歌手
                imageView.startAnimation(animation); // 图片开始旋转，形成动画
            }
        });
    }

    /**
     * 初始化 view
     */
    private void initView() {
        mListView = (ListView) findViewById(R.id.main_listview);
        // 把扫描获取的音乐赋值给 list
        list = MusicUtils.getMusicData(this);
        adapter = new MyAdapter(this, list);
        mListView.setAdapter(adapter);

        playPause = (Button) findViewById(R.id.playing_btn_pause);
        song = (TextView) findViewById(R.id.control_song);
        singer = (TextView) findViewById(R.id.control_singer);
        imageView = (ImageView) findViewById(R.id.control_imageview);
        // 动画
        animation = AnimationUtils.loadAnimation(this, R.anim.img_animation);
        LinearInterpolator lin = new LinearInterpolator(); // 设置动画匀速播放
        animation.setInterpolator(lin);
    }

    /**
     * 底部控制栏的单击事件
     *
     * @param view
     */
    public void control(View view) {
        Intent it = new Intent(MainActivity.this, MyService.class);
        switch (view.getId()) {
        case R.id.playing_btn_previous: // 上一首歌曲
        // 若当前歌曲的序号≤0 时，用户单击“上一首”按钮，提示当前是第一首歌曲
            if (playPosition <= 0) {
                Toast.makeText(MainActivity.this, "已经是第一首歌了",
                        Toast.LENGTH_SHORT).show();
                playPosition = 0;
            } else {
                // 让歌曲的序号减 1
                playPosition--;
            }
            // 播放
            playPause.setBackgroundResource(R.drawable.pause);
            String posPath_previous = list.get(playPosition).path; // 长路径（绝对路径）
            filePath = posPath_previous.substring(0,
                    posPath_previous.lastIndexOf("/") + 1); // 路径的前半部分
```

```
            fileName = list.get(playPosition).song;
            Bundle b_previous = new Bundle();
            b_previous.putString("filePath", filePath);
            b_previous.putString("fileName", fileName);
            b_previous.putInt("playPosition", playPosition);
            it.putExtras(b_previous);
            it.putExtra("key", 0);
            break;
        case R.id.playing_btn_pause: // 暂停和播放
            if (IsPlay == false) {
                //"暂停和播放"按钮图片变成播放状态
                playPause.setBackgroundResource(R.drawable.pause);
                it.putExtra("key", 1);
                imageView.startAnimation(animation);
                IsPlay = true; // 判断是否在播放，并为其赋值 true
                animation.start();
                Toast.makeText(MainActivity.this,
                        "播放" + list.get(playPosition).song, Toast.LENGTH_SHORT)
                        .show();
            } else {
                // "暂停和播放"按钮图片变成暂停状态
                playPause.setBackgroundResource(R.drawable.play);
                // 暂停歌曲
                it.putExtra("key", 2);
                imageView.clearAnimation(); // 停止动画
                IsPlay = false; // 判断是否在播放，并为其赋值 false
                Toast.makeText(MainActivity.this,
                        "暂停" + list.get(playPosition).song, Toast.LENGTH_SHORT)
                        .show();
            }
            break;
        case R.id.playing_btn_next: // 下一首歌曲
            // 若当前歌曲的序号≥歌曲播放列表的数量-1 时，提示当前是最后一首歌曲
            if (playPosition >= list.size() - 1) {
                Toast.makeText(MainActivity.this, "已经是最后一首歌了",
                        Toast.LENGTH_SHORT).show();
                playPosition = list.size() - 1;
            } else {
                // 单击"下一首"按钮，歌曲的序号加 1
                playPosition++;
            }
            //"暂停和播放"按钮图片变成播放状态
            playPause.setBackgroundResource(R.drawable.pause);
            String posPath_next = list.get(playPosition).path; // 长路径（绝对路径）
            filePath = posPath_next.substring(0,
                    posPath_next.lastIndexOf("/") + 1); // 路径的前半部分
            fileName = list.get(playPosition).song;
            Bundle b_next = new Bundle();
            b_next.putString("filePath", filePath);
            b_next.putString("fileName", fileName);
            b_next.putInt("playPosition", playPosition);
            it.putExtras(b_next);
            it.putExtra("key", 0);
            break;
        case R.id.playing_btn_stop: // 停止服务
            it.putExtra("key", 3);
```

```
int playPosition;

@Override
public IBinder onBind(Intent arg0) {
    return null;
}

@TargetApi(Build.VERSION_CODES.HONEYCOMB_MR1)
@SuppressLint("NewApi")
@Override
public int onStartCommand(Intent intent, int flags, int startId) {

    Bundle b = intent.getExtras();
    filePath=b.getString("filePath");
    fileName=b.getString("fileName");
    playPosition=b.getInt("playPosition");
    String path="file://"+filePath+fileName;

    int m = intent.getIntExtra("key", 0);
   switch (m) {
   case 0:        //点播
        player.reset();
    playing(path);
    player.start();
        break;
        case 1:        //播放
          player.start();
            break;
        case 2:      // 暂停
            if (player != null && player.isPlaying())
                player.pause();
            break;
      case 3:     // 停止
       if(player!=null){
                player.stop();      //停止播放歌曲
                try{
                    player.prepare();//停止播放后，下次播放歌曲先要进入 prepared 状态
                    player.seekTo(0); //须将播放时间设置为 0；下次播放歌曲时才能
                                  重新开始，否则会继续播放上次退出时的歌曲
                    }catch(IOException e){
                    e.printStackTrace();
                    }
        }
            break;
   }
    return super.onStartCommand(intent, flags, startId);
}

@Override
public void onDestroy() {
    super.onDestroy();
player.stop();        // 服务销毁时，停止播放
player.release();    //释放资源
}


void playing(String path) {
```

```
        player.reset();
            try {
                player.setDataSource(path);
                player.prepare();
            } catch (Exception e) {
                e.printStackTrace();
            }
            player.start();
        }
    }
```

6.3.4 注册

Service 不涉及屏幕操作，和 Activity 有所不同。因此，需要在 AndroidManifest.xml 中手动注册 Service，代码如下：

```
<application …… >
    ……
    <service android:name=".MyService" android:process="system" />
</application>
```

6.3.5 SD 卡的访问权限

MP3 音乐播放器项目要读取 SD 卡中的信息，需开启 SD 卡的访问权限。

6.3.6 运行测试

按照需求分析中提出的要求，进行测试。

首先，播放歌曲；其次，单击 ListView 控件，查看歌曲能否切换，背景色是否修改；然后，分别测试“上一首”“暂停和播放”“下一首”“停止”功能按钮；最后，单击“菜单”按钮，停止服务并退出程序。

6.4 Service

6.4.1 Service 概述

有些应用程序很少涉及与用户的交互操作，这些应用程序通常在后台运行，而且它们运行时一般不影响其他应用程序正常运行。例如，媒体播放程序可以转入后台运行，并继续播放歌曲；文件下载程序可以在后台执行文件下载任务。

为了处理这种后台进程，Android 系统引入了 Service 的概念。Service 是一种长生命周期的组件，它不能用于实现任何用户界面。

Service 是一个在后台运行的应用组件，它被启动后就在后台持续运行，直到外界停止它或自己停止。Service 的启动方式和停止方式如下。

- 启动方式：外界调用 Context.startService()方法和 Context.bindService()方法。
- 停止方式：外界调用 Context.stopService()方法，或者 Service 自己调用 Service.stopSelf()方法/ Service.stopSelfResult()方法。

注意：不论调用了多少次 startService()方法，只需调用一次 stopService()方法，即可停止 Service。

Service 主要用于完成一些耗时任务，如查询升级信息，播放音乐等。

6.4.2 Service 的启动方式

1．通过 startService()方法启动

通过 startService()方法启动 Service，则 Service 独立于开启它的 Activity。Activity 退出后，Service 仍然运行。具体的启动、运行和停止过程如下。

（1）启动过程。Activity 通过 startService()方法启动 Service，Service 会经历 onCreate()方法和 onStart()方法。

（2）Service 持续运行。

（3）停止过程。Activity 通过 stopService()方法停止 Service，这时 Service 会经历 onDestroy()方法，Service 通过 stopSelf()方法停止运行。此外，Activity 再次启动后也能通过 stopService()方法停止 Service。

2．通过 bindService()方法启动

通过 bindService()方法启动 Service，则 Service 与开启它的 Activity 绑定在一起。Activity 退出时，Service 会同时退出。具体的启动、运行和停止过程如下。

（1）启动过程。Activity 通过 bindService()方法启动 Service，Service 会经历 onCreate()方法和 onBind()方法。

（2）Service 在 Activity 运行期间持续运行。

（3）停止过程。Activity 退出时，Service 会经历 onUnbind()方法和 onDestroy()方法。

读者可以这样理解：两者绑定意味着它们共存亡，正因如此，两者间可以传输信息。

6.4.3 生命周期

在 6.4.2 节中，我们介绍了 Service 的两种启动方式，这两种启动方式下的 Service，其生命周期有所不同，如图 6-3 所示。

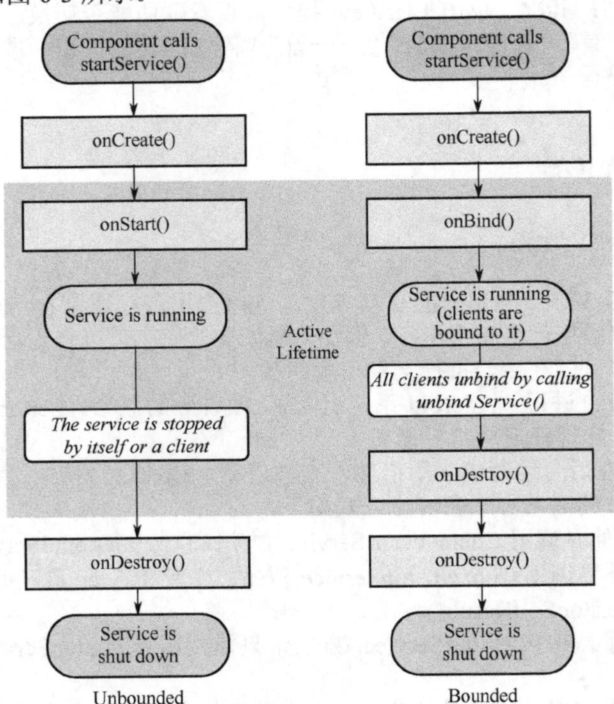

图 6-3 两种启动方式下的 Service 生命周期

根据如图 6-3 所示的流程，我们可以合理地利用相关方法实现 Service 的功能。

6.5　多媒体

Android 系统支持图片（包括 png、gif、jpg、bmp 等文件格式）、音频（包括 mp3、m4a、mid 等文件格式）、视频（mp4、3gp、mxf 等文件格式）等多媒体文件。在前面的章节中，我们已经讨论过使用 ImageView 控件显示图片的方法，本节将介绍使用 MediaPlayer 类播放音频，以及使用 VideoView 类播放视频。

MediaPlayer 类和 VideoView 类可以抓取、解码并播放音/视频资源，这些资源主要包括：

- 本地资源，如项目中的资源；
- 内部 URI，如 SD 卡中的资源；
- 外部 URI，如网络资源（从互联网的流服务器获得的文件）。

6.5.1　音频

【例 6.1】播放本地资源中的音频文件，代码如下：

```
MediaPlayer player = MediaPlayer.create(this, R.raw.start);
player.start();
```

【例 6.2】播放 SD 卡中的音频文件，代码如下：

```
String url = "file://" + Environment.getExternalStorageDirectory().toString() + "/Music/001.mp3";
MediaPlayer player = new MediaPlayer();
player.setAudioStreamType(AudioManager.STREAM_MUSIC);
player.setDataSource(url);
player.prepare();
player.start();
```

【例 6.3】播放网络资源中的音频文件，代码如下：

```
String url = "http://www.music.net/download/001.mp3"; // 与播放 SD 卡中的音频有区别，其 URL 不同
MediaPlayer player = new MediaPlayer();
player.setAudioStreamType(AudioManager.STREAM_MUSIC);
player.setDataSource(url);
player.prepare();
player.start();
```

注意：使用网络资源时要在配置文件中加入访问互联网的权限，代码如下：

```
<uses-permission android:name="android.permission.INTERNET"></uses-permission>
```

6.5.2　视频

视频播放窗口要在屏幕中布局，因此需要在布局文件内添加一个 VideoView 控件，VideoView 控件的位置和大小就是视频播放窗口的位置和大小。所以，应该给 VideoView 控件预留尽可能大的空间，在有条件的情况下，可以让视频全屏显示。

【例 6.4】新建一个项目，测试视频播放功能。

（1）设计布局文件，代码如下：

```
<LinearLayout xmlns:android="http://schemas.android.com/apk/res/android"
    xmlns:tools="http://schemas.android.com/tools"
    android:id="@+id/LinearLayout1"
    android:layout_width="match_parent"
    android:layout_height="match_parent"
    android:orientation="vertical"
    android:paddingBottom="0dp"
    android:paddingLeft="0dp"
    android:paddingRight="0dp"
    android:paddingTop="0dp"
```

```
    tools:context=".MainActivity" >

    <VideoView
        android:id="@+id/videoView1"
        android:layout_width="match_parent"
        android:layout_height="wrap_content" />

</LinearLayout>
```

（2）在配置文件中进行设置，代码如下：

```
<application
    android:allowBackup="true"
    android:icon="@drawable/ic_launcher"
    android:label="@string/app_name"
    android:theme="@android:style/Theme.NoTitleBar.Fullscreen" >
```

这样就能实现全屏显示了，需要注意，当前的界面没有标题。

（3）Activity 类的代码如下：

```
@Override
protected void onCreate(Bundle savedInstanceState) {
    super.onCreate(savedInstanceState);
    setContentView(R.layout.activity_main);

    String url = "file://"
            + Environment.getExternalStorageDirectory().toString()
            + "/Movies/big_buck_bunny.mp4";
    VideoView videoView = (VideoView) findViewById(R.id.videoView1);
    MediaController controller = new MediaController(this);
    videoView.setVideoPath(url);
    videoView.setMediaController(controller);
    videoView.start();
}
```

6.6 实训：完善 MP3 音乐播放器项目

本节的实训任务为针对 MP3 音乐播放器项目，发现问题并提出完善意见。

- 如果歌曲名包含汉字，则播放列表中的歌曲名就会显示为乱码，请尝试修改。
- 增加歌曲的循环播放功能，即播放完歌曲播放列表中的所有歌曲后，播放器自动从第一首歌曲开始播放。
- 单击 ListView 控件时，当前项的背景色会改变；再单击“上一首”“下一首”按钮时，播放功能是正常的，但 ListView 控件对应项的背景却没有改变，请尝试修改。
- 在模拟器中，歌曲播放完毕，用户单击“菜单”按钮，结束服务并退出程序。当用户再次进入音乐播放器程序并播放歌曲时，便会报错，请尝试修改。
- 增加播放选定歌曲的功能，即只播放歌曲播放列表中选定的部分歌曲，而不是播放全部歌曲。
- 增加歌曲播放的进度条控件（SeekBar），播放歌曲时，用户可以在 SeekBar 上看到歌曲播放进度，此外，用户还可以通过拖动 SeekBar 控件，使音乐快进或后退。

6.7 实训：制作音乐盒项目

本节的实训任务为制作音乐盒项目，与 MP3 音乐播放器项目相比，主要有三大区别：第一，掌握第三方 jar 包的使用方法；第二，使用两种不同的启动方式（startService()方法与

bindService()方法）启动 Service，并比较两种启动方式的效果；第三，学会使用 SharedPreferences。

提醒：本节的音乐盒项目虽然开发规模较小，但可以实现音乐点播、暂停与播放、停止播放、循环播放、退出等功能，非常灵活。

音乐盒的界面如图 6-4 所示。

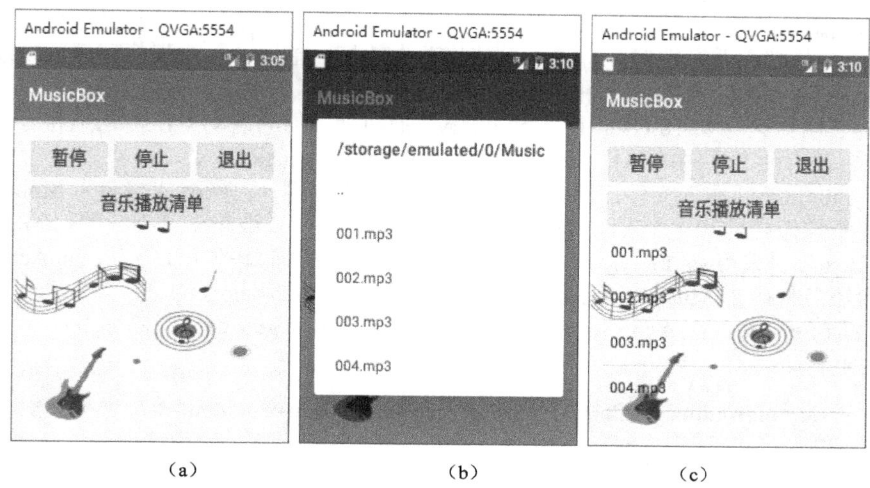

图 6-4　音乐盒界面

（1）本项目的制作要求如下。

① “音乐播放清单”按钮：单击该按钮，弹出文件选择列表，加载用户要播放的歌曲清单。

②文件选择列表：用于实现功能的第三方 jar 包（FileDialog-0.1.jar）由素材提供，将该 jar 包放在 libs 目录下，然后右击 FileDialog-0.1.jar，在弹出的菜单中选择“Add As Library…”选项，弹出如图 6-5 所示的 Create Library 对话框。单击“OK”按钮，即可将此 jar 包加入当前项目类库中。该 jar 包中封装了 FileDialog 的 FileSelectedListener 监听器接口。任意选择一个 mp3 格式的音频文件，则该目录下所有的 mp3 文件都会被加入 String 数组的 filesArr（文件数组）中，并用 filesArr 填充前台的 ListView 控件。

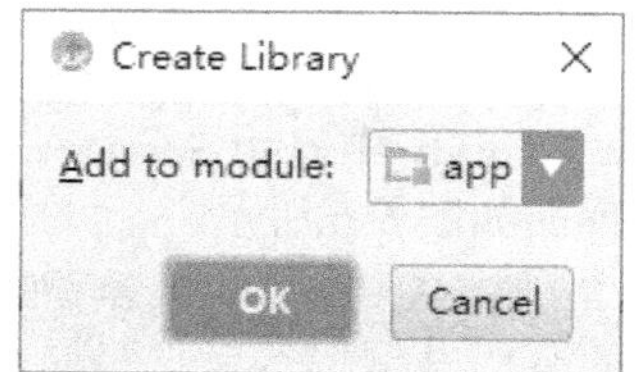

图 6-5　Create Library 对话框

③SharedPreferences：用于记录 path 和 filesArr。在离开程序之前，MainActivity 类的 onPause()方法能够将当前的 path 和 filesArr 保存到 SharedPreferences 中；当下次进入程序或返回 MainActivity 时，从 onResume()方法的 SharedPreferences 中读取记录的 path 和 filesArr。

④PermissionUtils 类：如果模拟器对应的 API 太高（如 API24），在访问 SD 卡时，就需要添加 PermissionUtils 类，该类具有 verifyStoragePermissions 功能，用于对 SD 卡的访问进行 Allow 授权。使用 PermissionUtils 类需两个步骤，第一，将素材中的相关文件复制到项目包下；第二，在 MainActivity 类的 onCreate()方法中添加以下代码：

```
PermissionUtils.verifyStoragePermissions(this);
```

⑤Service：自定义类 MusicService(extends Service)，用于控制 MediaPlayer 的播放和停止功能。

⑥音乐点播、暂停与播放、停止播放、循环播放等功能。

- 音乐点播功能：用户单击如图 6-4（c）下方的 ListView 控件时，可点播歌曲。
- 暂停与播放功能：当一首歌曲正在播放时，用户单击“暂停”按钮，歌曲暂停播放，同时“暂停”按钮变为“播放”按钮。当用户再单击“播放”按钮时，歌曲继续播放。
- 停止播放功能：用户单击“停止”按钮时，会强制中止正在播放的歌曲。
- 循环播放功能：当播放完一首歌曲时，会自动从数组中读取下一首歌曲进行播放。当播放完最后一首歌曲时，会播放歌曲清单中的第一首歌曲，实现循环播放功能。

⑦结束项目：单击“退出”按钮，停止服务，结束项目。

（2）为了明显区分 startService()与 bindService()两种不同的启动方法，这里将音乐盒项目（MusicBox）分为两个不同的项目分别实现（MusicBox 与 MusicBox2）。

activity_main.xml 布局文件：音乐盒项目仅有一个主界面，主界面的类名为 MainActivity，界面中有 4 个按钮控件和 1 个 ListView 控件。请扫描下方的二维码浏览相应的代码。

MusicBox 项目采用 startService()方法启动 Service，对应的前端控制代码为 MainActivity.java，对应的后台服务代码为 MusicService.java，请扫描下方的二维码进行浏览。

（3）注册 Service。在 AndroidManifest.xml 中注册 MusicService，代码如下：

```
<application …… >
    ……
    <service android:name=".MusicService" android:process="system" />
</application>
```

提醒： MusicBox2 项目的布局文件、注册服务与 MusicBox 项目是相同的，此处不再赘述。

MusicBox2 项目采用 bindService()方法启动 Service，对应的前端控制代码为 MainActivity2.java，对应的后台服务代码为 MusicService2.java，请扫描下方的二维码进行浏览。

activity_main.xml

MainActivity.java

MusicService.java

PermissionUtils.java

MainActivity2.java

MusicService2.java

6.8 实训：Service 练习

本节的实训任务为编写一个 Service，其内容是每秒向 LogCat 输出信息“Service is running”，共输出 120 次，即运行 2 分，然后自动停止。要求用 startService()方法启动，并且可以在 Service 启动后的 2 分内，从 Activity 中停止 Service。

项目的名称为 ServiceDemo，界面如图 6-6 所示。

图 6-6 ServiceDemo 界面

（1）MainActivity.java 的参考代码如下：

```
package com.example.administrator.servicedemo;

import android.os.Bundle;
import android.support.v7.app.AppCompatActivity;
import android.content.Intent;
import android.view.Menu;
import android.view.View;
import android.view.View.OnClickListener;
import android.widget.Toast;
```

```
public class MainActivity extends AppCompatActivity implements OnClickListener{

    @Override
    protected void onCreate(Bundle savedInstanceState) {
        super.onCreate(savedInstanceState);
        setContentView(R.layout.activity_main);

        this.findViewById(R.id.btn_Start).setOnClickListener(this);
        this.findViewById(R.id.btn_Stop).setOnClickListener(this);
    }

    @Override
    public void onClick(View v) {
        Intent intent=new Intent(this,ServiceDemo.class);
        switch(v.getId()){
        case R.id.btn_Start:
            Toast.makeText(this, "startService", Toast.LENGTH_SHORT).show();
            startService(intent);
            break;
        case R.id.btn_Stop:
            Toast.makeText(this, "stopService", Toast.LENGTH_SHORT).show();
            stopService(intent);
            break;
        }

    }
}
```

（2）ServiceDemo.java 的参考代码如下：

```
package com.example.administrator.servicedemo;

import android.app.Service;
import android.content.Intent;
import android.os.IBinder;
import android.util.Log;

public class ServiceDemo extends Service{
    boolean isRunning=false;

    @Override
    public IBinder onBind(Intent arg0) {
        // TODO Auto-generated method stub
        return null;
    }

    @Override
    public int onStartCommand(Intent intent, int flags, int startId) {
        new Thread(new Runnable(){
            @Override
            public void run() {
                isRunning=true;
                for(int i=0;i<120;i++){
                    Log.v("Demo","Service is Running"+i);
                    try {
                        Thread.sleep(1000);
                    } catch (InterruptedException e) {
                        // TODO Auto-generated catch block
                        e.printStackTrace();
```

```
                        }
                        if(!isRunning){
                                break;
                        }
                    }
                }}).start();
            return 0;
        }

        @Override
        public void onDestroy() {
            super.onDestroy();
            isRunning=false;
        }
}
```

（3）在 AndroidManifest.xml 中注册 Service，代码如下：

```
<service android:name=".ServiceDemo"></service>
```

6.9 练习题

1．下面选项中，不是 Android 系统四大组件的是（　　）。

A．Activity　　B．Intent　　C．Service　　D．ContentProvider

2．在 Android 系统的四大组件中，（　）可以在后台长时间运行，且不需要提供用户界面。

A．Service　　B．Activity　　C．BroadcastReceiver　　D．ContentProvider

3．MediaPlayer 播放资源前，需要调用（　　）方法完成准备工作。

A．setDataSource()　　B．prepare()　　C．reset()　　D．release()

4．在 Android 平台中，下列有关视频播放的说法不正确的是（　　）。

A．可以使用 SurfaceView 控件播放视频

B．可以使用 VideoView 控件播放视频

C．VideoView 控件可以控制视频播放窗口的位置和大小

D．使用 VideoView 控件播放视频，视频可以是 3gp 格式

5．使用 MediaPlayer 播放 SD 卡中的 mp3 文件时，下列说法中正确的是（　　）。

A．需要使用 MediaPlayer.create() 方法创建 MediaPlayer

B．直接调用 new MediaPlayer() 方法即可

C．需要调用 setDataSource() 方法设置文件源

D．直接调用 start() 方法，无须设置文件源

6．下列关于 Android 系统中 MediaPlayer 的说法错误的是（　　）。

A．MediaPlayer 包含了 Audio 和 Video 的播放功能

B．获得 MediaPlayer 实例的方法有 new MediaPlayer()和 MediaPlayer.create()

C．播放歌曲时，由于系统产生异常，会造成服务中止、数据丢失，再次启动 MediaPlayer 后，只能从一首歌的开始位置播放

D．指定 MediaPlayer 数据的来源时，必须指定播放文件的路径、资源 ID 和网络路径

7．如果使用 Android 智能手机播放视频（或音频）时，恰好有来电，下列说法正确的是（　）。

A．视频（或音频）不会自动退出

B．视频（或音频）会自动退出

C．有的手机会自动退出视频（或音频），有的手机不会自动退出视频（或音频）

D．以上说法都错误

8．下列选项中，不属于 Service 生命周期方法的是（　　）。

A．onCreate () B．onDestroy() C．onStop() D．onStart()

9．关于 Service 生命周期的说法正确的是（ ）。

A． 如果 Service 已经启动，将先后调用 onCreate()方法和 onStart()方法

B． 第一次启动 Service，将先后调用 onCreate()方法和 onStart()方法

C． 第一次启动 Service，将只调用 onCreate()方法

D． 如果 Service 没有启动，则不能调用 StopService()方法停止服务

10． 有关 startService()方法和 bindService()方法的说法错误的是（ ）。

A．使用 startService()方法启动服务，则调用的生命周期方法为 onCreate()→onStart()→onDestory()

B．使用 startService()方法启动服务，访问者与 Service 不绑定在一起，即访问者退出，Service 还在运行

C．使用 bindService()方法启动服务，则调用的生命周期方法为 onCreate()→onBind()→onDestory()→onUnbind()

D. 使用 bindService()方法启动服务，访问者与 Service 绑定在一起，即访问者退出，Service 终止，并解除绑定

6.10 作业

1．举例说明 Service 的应用。

2．Service 有哪两种启动方式？它们的区别是什么？

3．在 Android 系统中如何播放音乐？如何播放视频？两者有什么区别？

第7章 BroadcastReceiver——短信过滤器项目

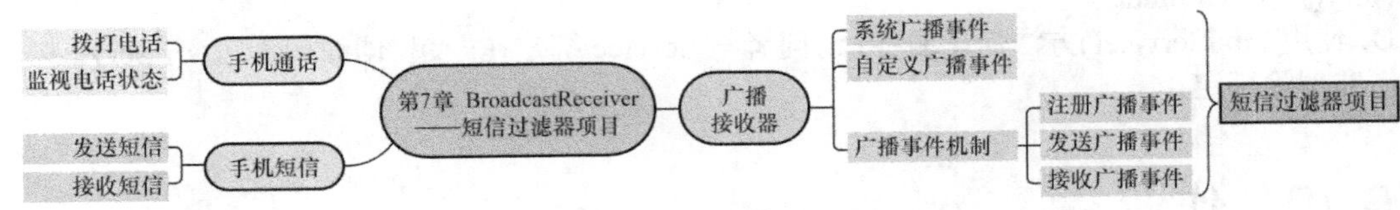

本章知识要点思维导图

Android 系统的四大组件包括 Activity、ContentProvider、Service 和 BroadcastReceiver，本章将通过短信过滤器项目对 BroadcastReceiver（广播接收器）进行重点介绍。此外，还将介绍手机通话和手机短信（SMS，Short Message Service）的相关内容。

7.1 需求分析

开发一个短信过滤器项目，用于屏蔽指定号码（黑名单）发来的垃圾短信，以及管理（包括增加、删除、查询等功能）被屏蔽的电话号码。此外，使用短信过滤器还可以查阅被屏蔽的短信、回复相关短信、给发信者回电、转发相关短信等。

7.2 界面设计

短信过滤器项目只有一个界面，分为上、下两部分。上半部分用于管理黑名单号码；下半部分用于显示被屏蔽的短信，界面如图 7-1 所示。

图 7-1 短信过滤器界面

本项目需要一个用于接收短信的 BroadcastReceiver，其作用为过滤所有收到的短信。如果发送者的号码在黑名单中，则将相关短信记录在数据库的被屏蔽短信表中，并且中止该短信进入手机的短信应用软件，从而实现短信过滤功能。

7.3　数据结构设计

短信过滤器项目的数据结构设计比较复杂，需要两张数据表保存相关数据：第一张表用于保存被屏蔽的电话号码，这些号码可以被管理（添加或删除），详细情况见表7-1；第二张表用于保存被屏蔽的短信，包括短信的电话号码、时间和内容，详细情况见表 7-2。设置第二张表是为了能找到那些被错误屏蔽的短信，避免数据丢失。

综上，需要对两张表中的数据进行增、删、改操作。

本项目的数据库仍采用 SQLite，数据库名称为 sms_filter.db。

表 7-1　被屏蔽的电话号码（phone）

字段名	数据类型	为空性	含义	注释
_id	integer	Not null	ID	主键，自动增长（autoincrement）
phone_number	Text	Not null	电话号码	保存被屏蔽的电话号码

表 7-2　被屏蔽的短信（sms）

字段名	数据类型	为空性	含义	注释
_id	integer	Not null	ID	主键，自动增长（autoincrement）
phone_number	Text	Not null	被屏蔽的电话号码	短信的电话号码，该号码为被屏蔽的电话号码
received_date	Text	Not null	时间	短信的日期和时间，格式 yy-MM-dd hh:mm:ss
sms	Text	Not null	短信内容	完整的短信内容

7.4　实施

7.4.1　创建项目

创建一个名为 SmsFilter 的项目，主界面由 MainActivity 构成，BroadcastReceiver 的子类为 SmsReceiver，SQLiteOpenHelper 的子类为 Helper，对本项目数据库的两张表——被屏蔽的电话号码（phone）和被屏蔽的短信（sms）进行管理。

7.4.2　界面实现

短信过滤器项目的布局文件包括上半部分的 ListView 控件对应的自定义布局文件（phone_layout.xml）、下半部分的 ListView 控件对应的自定义布局文件（sms_layout.xml）、主界面的布局文件（activity_main.xml）。

1. phone_layout.xml

上半部分的 ListView 控件对应的自定义布局文件为 phone_layout.xml，代码如下：

```
<?xml version="1.0" encoding="utf-8"?>
<LinearLayout xmlns:android="http://schemas.android.com/apk/res/android"
    android:layout_width="match_parent"
    android:layout_height="match_parent"
    android:orientation="vertical" >

    <LinearLayout
        android:layout_width="match_parent"
```

```
            android:layout_height="wrap_content" >

            <TextView
                android:id="@+id/tvphone"
                android:layout_width="wrap_content"
                android:layout_height="wrap_content"
                android:layout_gravity="left"
                android:layout_weight="1"
                android:text="电话号码"/>

        </LinearLayout>

</LinearLayout>
```

2. sms_layout.xml

下半部分的 ListView 控件对应的自定义布局文件为 sms_layout.xml，代码如下：

```
<?xml version="1.0" encoding="utf-8"?>
<LinearLayout xmlns:android="http://schemas.android.com/apk/res/android"
    android:layout_width="match_parent"
    android:layout_height="match_parent"
    android:orientation="vertical" >

    <LinearLayout
        android:layout_width="match_parent"
        android:layout_height="wrap_content" >

        <TextView
            android:id="@+id/tvphone_number"
            android:layout_width="wrap_content"
            android:layout_height="wrap_content"
            android:layout_gravity="left"
            android:layout_weight="1"
            android:text="电话号码"
            android:textColor="#FF0000" />

        <TextView
            android:id="@+id/tvreceived_date"
            android:layout_width="wrap_content"
            android:layout_height="wrap_content"
            android:layout_marginLeft="10dp"
            android:layout_weight="1"
            android:text="日期" />

        <TextView
            android:id="@+id/tvsms"
            android:layout_width="wrap_content"
            android:layout_height="wrap_content"
            android:layout_marginLeft="10dp"
            android:layout_weight="1"
            android:text="短信" />
    </LinearLayout>

</LinearLayout>
```

3. activity_main.xml

主界面的布局文件为 activity_main.xml，代码如下：

```
<LinearLayout xmlns:android="http://schemas.android.com/apk/res/android"
    xmlns:tools="http://schemas.android.com/tools"
```

```
    android:id="@+id/LinearLayout1"
    android:layout_width="match_parent"
    android:layout_height="match_parent"
    android:orientation="vertical"
    android:paddingBottom="@dimen/activity_vertical_margin"
    android:paddingLeft="@dimen/activity_horizontal_margin"
    android:paddingRight="@dimen/activity_horizontal_margin"
    android:paddingTop="@dimen/activity_vertical_margin"
    tools:context=".MainActivity" >

    <TextView
        android:id="@+id/textView1"
        android:layout_width="wrap_content"
        android:layout_height="wrap_content"
        android:text="短信过滤器"
        android:layout_gravity="center"
        android:textAppearance="?android:attr/textAppearanceLarge" />

    <TextView
        android:id="@+id/textView2"
        android:layout_width="wrap_content"
        android:layout_height="wrap_content"
        android:text="黑名单号码"
        android:layout_gravity="center"
         android:layout_marginTop="10dp"
        android:textAppearance="?android:attr/textAppearanceMedium" />

    <LinearLayout
        android:layout_width="wrap_content"
        android:layout_height="wrap_content" >

        <Button
            android:id="@+id/btnAdd"
            android:layout_width="wrap_content"
            android:layout_height="wrap_content"
                android:layout_marginLeft="10dp"
            android:text="增加" />
        <Button
            android:id="@+id/btnDelete"
            android:layout_width="wrap_content"
            android:layout_height="wrap_content"
            android:layout_marginLeft="30dp"
            android:text="删除" />
        <Button
            android:id="@+id/btnFresh"
            android:layout_width="wrap_content"
            android:layout_height="wrap_content"
            android:layout_marginLeft="30dp"
            android:text="刷新" />

    </LinearLayout>

    <LinearLayout
        android:layout_width="match_parent"
        android:layout_height="wrap_content"
         android:layout_marginTop="10dp" >

        <TextView
```

```
                imageView.clearAnimation(); // 停止动画
                Toast.makeText(MainActivity.this, "播放停止", 0).show();
                break;
            }
            startService(it);
        }

        /**
         * 控制歌曲名和歌手的 TextView 的方法
         */
        public void setText() {
            song.setText(list.get(playPosition).song);
            song.setSelected(true); // 若歌曲名较长，则令其滚动显示
            singer.setText(list.get(playPosition).singer);
        }

        @Override
        public boolean onCreateOptionsMenu(Menu menu) {
            getMenuInflater().inflate(R.menu.main, menu);
            return true;
        }

        @Override
        public boolean onOptionsItemSelected(MenuItem item) {
            switch (item.getItemId()) {
            case R.id.item_stop:
                Intent it = new Intent(MainActivity.this, MyService.class);
                stopService(it);
                System.exit(0);
                break;
            }
            return true;
        }
}
```

6．MyService.java

定义服务类 MyService，用于后台控制音乐播放程序。MyService.java 的代码如下：

```
package com.example.administrator.musicplayer;

import java.io.IOException;

import android.annotation.SuppressLint;
import android.annotation.TargetApi;
import android.app.Service;
import android.content.Intent;
import android.media.AudioManager;
import android.media.MediaPlayer;
import android.media.MediaPlayer.OnCompletionListener;
import android.os.Build;
import android.os.Bundle;
import android.os.IBinder;
import android.view.View;
import android.widget.Toast;

public class MyService extends Service{
    String filePath;
    String fileName;
    public static MediaPlayer player= new MediaPlayer();
```

```
            android:id="@+id/textView3"
            android:layout_width="wrap_content"
            android:layout_height="wrap_content"
            android:text="电话号码："
            android:textAppearance="?android:attr/textAppearanceMedium" />

        <EditText
            android:id="@+id/etPhone"
            android:layout_width="wrap_content"
            android:layout_height="wrap_content"
            android:layout_weight="1"
            android:ems="10"
            android:inputType="phone" >

            <requestFocus />
        </EditText>

    </LinearLayout>

    <ListView
        android:id="@+id/lstPhone"
        android:layout_width="match_parent"
        android:layout_height="93dp" >
    </ListView>

    <TextView
        android:id="@+id/textView4"
        android:layout_width="wrap_content"
        android:layout_height="wrap_content"
        android:text="被屏蔽短信"
        android:layout_marginTop="40dp"
         android:layout_gravity="center"
        android:textAppearance="?android:attr/textAppearanceMedium" />

    <ListView
        android:id="@+id/lstSms"
        android:layout_width="match_parent"
        android:layout_height="wrap_content" >
    </ListView>

</LinearLayout>
```

7.4.3 Java 代码

1．数据库管理

（1）号码类 Phone，用于存储手机编号、手机号码，代码如下：

```
package com.example.administrator.smsfilter;

public class Phone {
    int _id;
    String phone_number;

    public Phone() {
        super();
    }

    public Phone(int _id, String phone_number) {
```

```
            super();
            this._id = _id;
            this.phone_number = phone_number;
        }
}
```

（2）短信类 Sms，用于存储短信编号、手机号码、接收日期、短信内容，代码如下：

```
package com.example.administrator.smsfilter;

public class Sms {
        int _id;
        String phone_number;
        String received_date;
        String sms;

        public Sms() {
                super();
        }

        public Sms(int _id, String phone_number, String received_date, String sms) {
                super();
                this._id = _id;
                this.phone_number = phone_number;
                this.received_date = received_date;
                this.sms = sms;
        }
}
```

（3）数据库管理类 Helper，Helper 包含八个方法，详细情况如下。

Helper(Context context)方法用于创建数据库；onCreate (SQLiteDatabase db) 方法用于创建两张数据表，并各添加一条初始信息；queryPhone()方法用于查询号码；insertPhone(Phone phone)方法用于增加号码；deletePhone(String phone_number) 方法用于删除号码；querySms()方法用于查询短信；insertSms(Sms sms)方法用于增加短信，onUpgrade(SQLiteDatabase arg0, int arg1, int arg2) 方法用于数据库升级。

数据库管理类 Helper 的代码如下：

```
package com.example.administrator.smsfilter;

import java.text.SimpleDateFormat;
import android.content.ContentValues;
import android.content.Context;
import android.database.Cursor;
import android.database.sqlite.SQLiteDatabase;
import android.database.sqlite.SQLiteOpenHelper;

public class Helper extends SQLiteOpenHelper{
public final static SimpleDateFormat df = new SimpleDateFormat("yy-MM-dd");
SQLiteDatabase db;

        public Helper(Context context) {
                super(context, "sms_filter.db", null, 1);
        }

        @Override
        public void onCreate(SQLiteDatabase db) {
                db.execSQL("create table phone(_id integer not null primary key autoincrement,
phone_number text not null)");
                db.execSQL("create table sms(_id integer not null primary key autoincrement,
phone_number text not null," +"received_date text not null,sms text not null)");
                db.execSQL("insert into phone(phone_number) values('10086')");
```

```
            db.execSQL("insert into sms(phone_number,received_date,sms)
values('10086','2018-6-12','nihao')");
        }

    public Cursor queryPhone()
    {
        db=this.getReadableDatabase();
        Cursor cursor=db.query("phone", null, null, null, null, null, null);
        return cursor;
    }

    public long insertPhone(Phone phone)
    {
        db=this.getWritableDatabase();
        ContentValues cv=new ContentValues();
        cv.put("phone_number", phone.phone_number);
        return db.insert("phone", null, cv);
    }

    public int deletePhone(String phone_number)
    {
        db=this.getWritableDatabase();
        return db.delete("phone", "phone_number='"+phone_number+"'", null);
    }

    public Cursor querySms()
    {
        db=this.getReadableDatabase();
        Cursor cursor=db.query("sms", null, null, null, null, null, null);
        return cursor;
    }

    public long insertSms(Sms sms)
    {
        db=this.getWritableDatabase();
        ContentValues cv=new ContentValues();
        cv.put("phone_number", sms.phone_number);
        cv.put("received_date", sms.received_date);
        cv.put("sms", sms.sms);
        return db.insert("sms", null, cv);
    }

    @Override
    public void onUpgrade(SQLiteDatabase arg0, int arg1, int arg2) {
        // TODO Auto-generated method stub

    }
}
```

2. MainActivity.java

主界面的 MainActivity.java 代码如下：

```
package com.example.administrator.smsfilter;

import android.support.v7.app.AppCompatActivity;
import android.database.Cursor;
import android.database.sqlite.SQLiteCursor;
import android.os.Bundle;
import android.support.v4.widget.SimpleCursorAdapter;
```

```
import android.view.Menu;
import android.view.View;
import android.view.View.OnClickListener;
import android.widget.AdapterView;
import android.widget.AdapterView.OnItemClickListener;
import android.widget.Button;
import android.widget.EditText;
import android.widget.ListView;

public class MainActivity extends AppCompatActivity implements OnClickListener{
	Cursor cursor,cursor_sms;
	Helper helper;
	ListView lstPhone,lstSms;
	Button btnAdd,btnDelete,btnFresh;
	EditText etPhone;

	@Override
	protected void onCreate(Bundle savedInstanceState) {
		super.onCreate(savedInstanceState);
		setContentView(R.layout.activity_main);

		lstPhone=(ListView)this.findViewById(R.id.lstPhone);
		lstSms=(ListView)this.findViewById(R.id.lstSms);

		btnAdd=(Button)this.findViewById(R.id.btnAdd);
		btnAdd.setOnClickListener(this);
		btnDelete=(Button)this.findViewById(R.id.btnDelete);
		btnDelete.setOnClickListener(this);
		btnFresh=(Button)this.findViewById(R.id.btnFresh);
		btnFresh.setOnClickListener(this);

		etPhone=(EditText)this.findViewById(R.id.etPhone);

		helper=new Helper(this);
		DataRefresh();

		lstPhone.setOnItemClickListener(new OnItemClickListener(){
			@Override
			public void onItemClick(AdapterView<?> av, View arg1, int position,
					long arg3) {
				SQLiteCursor sc=(SQLiteCursor) av.getItemAtPosition(position);
				etPhone.setText(sc.getString(1));
			}});
	}

	private void DataRefresh() {
		cursor=helper.queryPhone();
		SimpleCursorAdapter adapter=new    SimpleCursorAdapter(this,R.layout.phone_layout,
				cursor,new String[]{"phone_number"},new int[]{R.id.tvphone});
//Context,int,Cursor,String[],int[]
		lstPhone.setAdapter(adapter);

		cursor_sms=helper.querySms();
		SimpleCursorAdapter adapter_sms=new    SimpleCursorAdapter(this,R.layout.sms_layout,
				cursor_sms,new String[]{"phone_number","received_date","sms"},
						new int[]{R.id.tvphone_number,R.id.tvreceived_date,R.id.tvsms});
//Context,int,Cursor,String[],int[]
		lstSms.setAdapter(adapter_sms);
	}
```

```
        @Override
        public void onClick(View v) {
            switch(v.getId()){
            case R.id.btnAdd:
                Phone p=new Phone();
                p.phone_number=etPhone.getText().toString();
                helper.insertPhone(p);
                DataRefresh();
                etPhone.setText("");
                break;
            case R.id.btnDelete:
                helper.deletePhone(etPhone.getText().toString());
                DataRefresh();
                etPhone.setText("");
                break;
            case R.id.btnFresh:
                DataRefresh();
                break;
            }
        }
}
```

3. SmsReceiver.java

BroadcastReceiver 的子类为 SmsReceiver，用于短信接收和过滤，SmsReceiver.java 的代码如下：

```
package com.example.administrator.smsfilter;

import java.util.Date;
import android.content.BroadcastReceiver;
import android.content.Context;
import android.content.Intent;
import android.database.Cursor;
import android.os.Bundle;
import android.telephony.SmsMessage;
import android.util.Log;

public class SmsReceiver extends BroadcastReceiver {

    @Override
    public void onReceive(Context context, Intent intent) {
        // 短信通过 Intent 中的 Bundle 传递
                Bundle bundle = intent.getExtras();

                // 短信被保存在名为 pdus 的数据中
                //一条短信可能被分为多条信息片段，因此要建立数组，将多条信息片段合并
                Object[] pdus = (Object[]) bundle.get("pdus");
                SmsMessage[] msgs = new SmsMessage[pdus.length];
                for (int i = 0; i < pdus.length; i++) {
                    msgs[i] = SmsMessage.createFromPdu((byte[]) pdus[i]);
                }

                // 获取来信号码
                String phone = msgs[0].getOriginatingAddress();
                if (phone == null) {
                    return;
                }
```

```
                Log.v("SmsFilter", "收到来信：" + phone);

                // 判断是否进行屏蔽
                boolean isBlock = false;
                Helper helper = new Helper(context);
                Cursor cursor = helper.queryPhone();
                while (cursor.moveToNext()) {
                    if (phone.equals(cursor.getString(1))) {
                        isBlock = true;
                        break;
                    }
                }

                if (isBlock) {
                    // 需要屏蔽
                    String body = "";
                    for (SmsMessage msg : msgs) {
                        body += msg.getMessageBody();
                    }

                    String date = Helper.df.format(new Date());

                    // 将短信存入数据库的被屏蔽短信表中
                    //SqlRowSms row = new SqlRowSms(0, phone, date, body);
                    helper.insertSms(new Sms(0, phone, date, body));

                    Log.v("SmsFilter", "来信需要屏蔽：" + phone + ":" + body);
                    // 不再进入手机的短信中
                    abortBroadcast();
                }
        }
}
```

7.4.4 注册

因项目需要，在 AndroidManifest.xml 中手动注册 BroadCastReceiver 的子类，代码如下：

```
<!-- 注册短信广播接收器，优先级 priority 设置为最高级别，取值为 2147483647 -->
<receiver android:name=".SmsReceiver" >
    <intent-filter android:priority="2147483647" >
        <action android:name="android.provider.Telephony.SMS_RECEIVED" />
    </intent-filter>
</receiver>
```

7.4.5 开启接收短信的权限

因项目需要，在 AndroidManifest.xml 中开启接收短信的权限，代码如下：

```
<!-- 需要开启接收短信的权限 -->
    <uses-permission android:name="android.permission.RECEIVE_SMS" />
```

7.4.6 运行测试

测试本项目前，读者需要掌握使用 AVD 发送短信的方法，Android SDK 提供了两种发送短信的方法，即使用 telnet 命令和模拟器界面。

1．使用 telnet 命令

使用 telnet 命令发送短信，只需开启一个 AVD。

在 Windows 的命令行模式下，使用 telnet 命令进入 AVD，然后发送短信。

注意：如果执行 telnet 命令后，出现如图 7-2 所示的问题，则应单击“开始”菜单，选择“控制面板”选项，在弹出的“控制面板”界面中，执行“程序”→“程序和功能”→“打开或关闭 Windows 功能”命令，弹出如图 7-3 所示的“Windows 功能”对话框，在列表中找到“Telnet 客户端”选项，并选中该选项前面的复选框，最后，单击“确定”按钮。

'telnet' 不是内部或外部命令，也不是可运行的程序
或批处理文件。

图 7-2　执行 telnet 命令后遇到问题

图 7-3　“Windows 功能”对话框

问题处理后，即可发送短信，步骤如下。

（1）进入 telnet 命令模式，其语法格式为：

```
C:\Documents and Settings\Administrator>telnet localhost 5554
```

其中，localhost 表示本机，5554 为端口号，即 AVD 运行时在标题栏显示的数字。

（2）进入 telnet 命令模式后，输入命令发送短信，代码如下：

```
sms send 123 hello
```

其中，123 是发送方的号码，hello 是发送的短信内容。

如图 7-4 所示，窗口中的命令行显示“OK”，表示发送短信成功。再输入以下代码：

```
sms send 12345 hello
```

窗口中的命令行再次显示“OK”，输入“exit”退出 telnet 命令模式。

注意：在 telnet 命令模式下，如果按“Backspace”键修改文字，命令将无法执行。

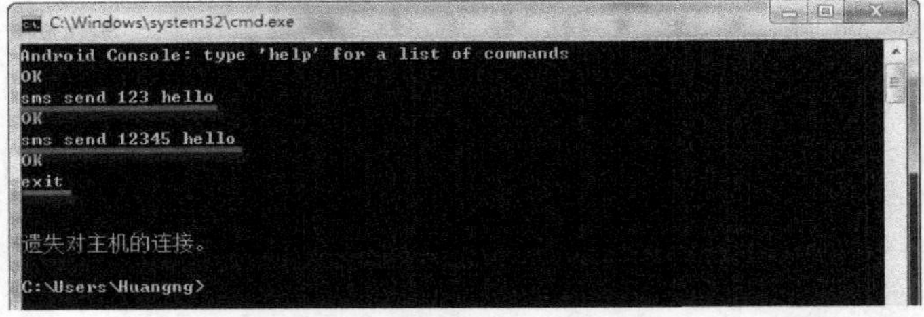

图 7-4　通过 telnet 发送短信

2．模拟器界面

使用模拟器界面发送短信，其流程与使用普通手机一样。但是，该方法需要开启两个 AVD，模拟的手机号码就是端口号，通常为 5554 或 5556。

7.5　BroadcastReceiver

BroadcastReceiver 是一种专注于接收广播通知信息，并能做出相应处理的组件。通常，多数广播是由系统发出的，如地域变换、电量不足、来电与来信等；此外，有的应用程序也可以发出广播，例如，程序 A 通知其他应用程序，当前的数据已从互联网下载完毕，且数据处于可用状态。

一个应用程序可以拥有任意数量的 BroadcastReceiver，从而对所有它感兴趣的通知信息予以响应。BroadcastReceiver 没有用户界面，所有的接收器均继承 BroadcastReceiver 基类。

BroadcastReceiver 可以通过多种方式通知用户，比如启动 Activity、使用 NotificationManager、开启背景灯、振动设备、播放声音等，最常见的通知方式为在状态栏显示图标，用户可以单击该图标，从而查看通知内容，这种通知方式的典型应用为短信通知。

Android 系统的广播事件有两种：第一种，系统广播事件，由 Android 系统或内置应用程序发出；第二种，自定义广播事件，由用户开发的应用程序发出。

7.5.1　系统广播事件

系统广播是由系统或内置应用程序发出的，如 ACTION_BOOT_COMPLETED（系统启动完成后触发），ACTION_TIME_CHANGED（系统时间改变时触发），ACTION_BATTERY_LOW（电量低时触发）等。其他常见的系统广播 ACTION 常量见表 7-3。

表 7-3　系统广播 ACTION 常量

常 量 名 称	常　量　值	意　　义
ACTION_BOOT_COMPLETED	android.intent.action.BOOT_COMPLETED	系统启动完成
ACTION_TIME_CHANGED	android.intent.action.ACTION_TIME_CHANGED	时间改变
ACITON_DATE_CHANGED	android.intent.action.ACTION_DATE_CHANGED	日期改变
ACTION_TIMEZONE_CHANGED	android.intent.action.ACTION_TIMEZONE_CHANGED	时区改变
ACTION_BATTERY_LOW	android.intent.action.ACTION_BATTERY_LOW	电量低
ACTION_MEDIA_EJECT	android.intent.action.ACTION_MEDIA_EJECT	插入或拔出外部媒体
ACTION_MEDIA_BUTTON	android.intent.action.ACTION_MEDIA_BUTTON	按下媒体按钮
ACTION_PACKAGE_ADDED	android.intent.action.ACTION_PACKAGE_ADDED	添加包
ACTION_PACKAGE_REMOVED	android.intent.action.ACTION_PACKAGE_REMOVED	删除包

【例 7.1】编写一个能接收并修改日期的 BroadcastReceiver。

（1）编写 BroadcastReceiver 类的子类，代码如下：

```
public class MyBroadcastReceiver extends BroadcastReceiver {
    @Override
    public void onReceive(Context context, Intent intent) {
        Log.v("BroadcastReceiver","你修改了日期。");
    }
}
```

（2）在 AndroidManifest.xml 中手动注册，代码如下：

```
<receiver android:name=".MyBroadcastReceiver">
    <intent-filter>
        <action android:name="android.intent.action.DATE_CHANGED">
        </action>
    </intent-filter>
</receiver>
```

或者采用动态注册的方式，代码如下：

```
IntentFilter filter = new IntentFilter();
```

```
filter.addAction("org.ngweb.broadcast.Hello");
getApplicationContext().registerReceiver(new MyBroadcastReceiver(),
        filter);
```

该应用程序会在用户修改系统日期时收到广播，并对此做出响应。

7.5.2 自定义广播事件

自定义广播由应用程序发出，一般涉及两个应用程序，其中一个应用程序发出广播，另一个应用程序接收广播。

【例 7.2】应用程序发出广播。

使用 sendBroadcast()方法发出广播，代码如下：

```
public class BroadcastActivity extends AppCompatActivity{
    @Override
    public void onCreate(Bundle savedInstanceState) {
        super.onCreate(savedInstanceState);
        setContentView(R.layout.main);
        Button b1 = (Button) findViewById(R.id.button1);

        b1.setOnClickListener(new View.OnClickListener() {

            @Override
            public void onClick(View v) {
                // 定义一个 intent
                Log.v("=========", "broadcast it!");
                Intent intent = new Intent("org.ngweb.broadcast.Hello");
                intent.putExtra("data", "广播开始了！ ");
                // 广播出去
                sendBroadcast(intent);
            }
        });
    }
}
```

【例 7.3】应用程序接收另一个应用程序发出的广播。

应用程序接收另一个应用程序发出的广播与接收系统广播有很多相似之处，本例中，需要完成以下两项内容。

（1）因为在一个应用程序中，可以接收多个广播，所以在 AndroidManifest.xml 中手动注册准备接收的广播，代码如下：

```
<receiver android:name=".MyBroadcastReceiver">
    <intent-filter>
        <action android:name="android.intent.action.DATE_CHANGED">
        </action>
    </intent-filter>
    <intent-filter>
        <action android:name="org.ngweb.broadcast.Hello">
        </action>
    </intent-filter>
</receiver>
```

（2）编写 BroadcastReceiver 的代码：

```
public class MyBroadcastReceiver extends BroadcastReceiver {
    @Override
    public void onReceive(Context context, Intent intent) {
        Log.v("BroadcastReceiver","收到广播了。");

          //对比 Action 后，决定输出的信息
        if(intent.getAction().equals("android.intent.action.DATE_CHANGED")){
```

```
            Log.e("状态是：", "日期被修改");
        }

        if(intent.getAction().equals("org.ngweb.broadcast.Hello")){
            Log.e("收到的数据是：", intent.getStringExtra("data"));
        }
    }
}
```

这样，将例 7.2 和例 7.3 结合，就能通过广播的方式实现应用程序间的通信。注意，上述例子中的广播名字为 org.ngweb.broadcast.Hello，双方应该协商好。

7.5.3　广播事件机制

广播事件机制和事件处理机制类似，两种机制的流程也比较相似。只不过，事件处理机制是程序组件级别的，而广播事件机制是系统级别的，有关广播事件机制的流程如下。

- 注册广播事件：广播事件的注册方式有两种，一种是静态注册，即在 AndroidManifest.xml 中定义，注册的广播接收器必须继承 BroadcastReceiver；另一种是动态注册，即在程序中使用 Context.registerReceiver 注册，注册的广播接收器相当于一个匿名类。两种注册方式都需要 IntentFilter。
- 发送广播事件：通过 Context.sendBroadcast 发送广播事件，由 Intent 传递注册广播事件时用到的 Action。
- 接收广播事件：当接收器监听到发送的广播时，会调用 onReceive()方法，并将包含消息的 Intent 传给广播接收者。onReceive()方法中的代码执行时间不要超过 5 秒，否则 Android 系统会弹出超时 Dialog。

7.6　手机通话

本节将通过两个小项目介绍实现拨打电话和监视电话状态等功能的方法。

7.6.1　拨打电话

拨打电话是通过隐式 Intent 实现的，下面举例说明。

【例 7.4】制作拨打电话项目（CallActivity），界面如图 7-5 所示。

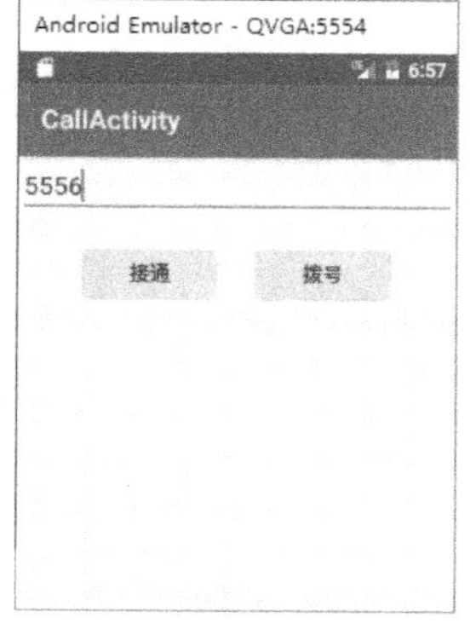

图 7-5　拨打电话项目界面

（1）布局文件的代码如下：

```
<?xml version="1.0" encoding="utf-8"?>
<LinearLayout xmlns:android="http://schemas.android.com/apk/res/android"
    android:orientation="vertical"
    android:layout_width="fill_parent"
    android:layout_height="fill_parent">
    <EditText android:id="@+id/phone"
        android:layout_height="wrap_content"
        android:layout_width="match_parent"
        android:text="5556">
        <requestFocus>

        </requestFocus>
    </EditText>

    <LinearLayout
        android:orientation="horizontal"
        android:layout_gravity="center"
        android:layout_marginTop="20dp"
        android:layout_width="wrap_content"
        android:layout_height="wrap_content">

        <Button android:id="@+id/call"
            android:text="接通"
            android:layout_height="wrap_content"
            android:layout_width="106dp">
        </Button>
        <Button android:id="@+id/dial"
            android:text="拨号"
            android:layout_marginLeft="20dp"
            android:layout_height="wrap_content"
            android:layout_width="106dp">
        </Button>
    </LinearLayout>
</LinearLayout>
```

（2）Activity 类的代码如下：

```
package com.example.administrator.callactivity;

import android.Manifest;
import android.content.Intent;
import android.content.pm.PackageManager;
import android.net.Uri;
import android.support.v4.app.ActivityCompat;
import android.support.v7.app.AppCompatActivity;
import android.os.Bundle;
import android.view.View;
import android.widget.Button;
import android.widget.EditText;

public class MainActivity extends AppCompatActivity implements View.OnClickListener {
    Button btnCall;
    Button btnDial;
    EditText etPhone;

    @Override
    public void onCreate(Bundle savedInstanceState) {
        super.onCreate(savedInstanceState);
        setContentView(R.layout.activity_main);
```

```
        btnCall = (Button) this.findViewById(R.id.call);
        btnCall.setOnClickListener(this);

        btnDial = (Button) this.findViewById(R.id.dial);
        btnDial.setOnClickListener(this);

        etPhone = (EditText) this.findViewById(R.id.phone);
    }

    @Override
    public void onClick(View v) {
        Intent intent;
        switch (v.getId()) {
            case R.id.call:
                intent = new Intent(Intent.ACTION_CALL,
                        Uri.parse("tel:" + etPhone.getText().toString()));
                if                          (ActivityCompat.checkSelfPermission(this,
Manifest.permission.CALL_PHONE) != PackageManager.PERMISSION_GRANTED) {
                    // TODO: Consider calling
                    //    ActivityCompat#requestPermissions
                    // here to request the missing permissions, and then overriding
                    //      public   void   onRequestPermissionsResult(int   requestCode,   String[]
permissions,
                    //                                          int[] grantResults)
                    // to handle the case where the user grants the permission. See the documentation
                    // for ActivityCompat#requestPermissions for more details.
                    return;
                }
                startActivity(intent);
                break;
            case R.id.dial:
                intent = new Intent(Intent.ACTION_DIAL,
                        Uri.parse("tel:"+etPhone.getText().toString()));
                startActivity(intent);
                break;
        }
    }
}
```

（3）在配置文件中设置拨打电话的相关权限，代码如下：

```
<?xml version="1.0" encoding="utf-8"?>
<manifest xmlns:android="http://schemas.android.com/apk/res/android"
    package="com.example.administrator.callactivity">

    <uses-permission android:name="android.permission.CALL_PHONE" />

    <application
        android:allowBackup="true"
        android:icon="@mipmap/ic_launcher"
        android:label="@string/app_name"
        android:supportsRtl="true"
        android:theme="@style/AppTheme">
        <activity android:name=".MainActivity">
            <intent-filter>
                <action android:name="android.intent.action.MAIN" />

                <category android:name="android.intent.category.LAUNCHER" />
            </intent-filter>
        </activity>
```

```
    </application>

</manifest>
```

7.6.2 监视电话状态

制作一个应用程序，实现监视电话状态（无通话、呼入、拨出）、检索来电号码、管理通话、查询移动服务商的信息等功能。

【例 7.5】制作监视电话状态项目（CalledActivity）。

该项目无须界面设计，可以通过 LogCat 监视电话状态。

（1）Activity 类的代码如下：

```
import android.support.v7.app.AppCompatActivity;
import android.content.Context;
import android.os.Bundle;
import android.telephony.PhoneStateListener;
import android.telephony.TelephonyManager;
import android.util.Log;

public class CalledActivity extends AppCompatActivity {
    /** Called when the activity is first created. */
    @Override
    public void onCreate(Bundle savedInstanceState) {
        super.onCreate(savedInstanceState);
        setContentView(R.layout.main);

        TelephonyManager tm = (TelephonyManager) getSystemService
(Context.TELEPHONY_SERVICE);
        tm.listen(callListener, PhoneStateListener.LISTEN_CALL_STATE);

        Log.v("来电管理 getNetworkCountryIso",tm.getNetworkCountryIso());
        Log.v("来电管理 getNetworkOperator",tm.getNetworkOperator());
        Log.v("来电管理 getNetworkOperatorName",tm.getNetworkOperatorName());
        Log.v("来电管理 getSimSerialNumber",tm.getSimSerialNumber());

        Log.v("来电管理 getLine1Number",tm.getLine1Number());
    }

    PhoneStateListener callListener = new PhoneStateListener(){
     @Override
     public void onCallStateChanged(int state, String incomingNumber){
          switch(state){
          case TelephonyManager.CALL_STATE_IDLE:
               Log.v("来电管理","休眠……："+incomingNumber);
               break;
          case TelephonyManager.CALL_STATE_RINGING:
               Log.v("来电管理","你有电话……："+incomingNumber);
               break;
          case TelephonyManager.CALL_STATE_OFFHOOK:
               Log.v("来电管理","通话中……："+incomingNumber);
               break;
          }
     }
    };

}
```

（2）在配置文件中设置相关的权限，代码如下：

```
<?xml version="1.0" encoding="utf-8"?>
<manifest xmlns:android="http://schemas.android.com/apk/res/android"
      package="org.ngweb.called" android:versionCode="1" android:versionName="1.0">
      <uses-sdk android:minSdkVersion="8" />
      <uses-permission android:name="android.permission.READ_PHONE_STATE" />
      <application android:icon="@drawable/icon" android:label="@string/app_name">
            <activity android:name=".CalledActivity" android:label="@string/app_name">
                  <intent-filter>
                        <action android:name="android.intent.action.MAIN" />
                        <category android:name="android.intent.category.LAUNCHER" />
                  </intent-filter>
            </activity>
      </application>
</manifest>
```

7.7 手机短信

若想使用手机短信功能，则需要在配置文件中设置权限，方法如下。

设置接收短信的权限，代码如下：

```
<uses-permission android:name="android.permission.READ_SMS" />
<uses-permission android:name="android.permission.RECEIVE_SMS" />
```

设置发送短信的权限，代码如下：

```
<uses-permission android:name="android.permission.SEND_SMS" />
```

7.7.1 发送短信

发送短信的实现方法有两种：第一种，通过调用系统的短信功能发送短信；第二种，直接调用短信类发送短信。

1．调用系统的短信功能

通过调用系统的短信功能发送短信，即通过隐式 Intent 实现目标，步骤如下。

（1）调用短信程序，代码如下：

```
Intent it = new Intent(Intent.ACTION_VIEW, uri);
it.putExtra("sms_body", "The SMS text");
it.setType("vnd.android-dir/mms-sms");
startActivity(it);
```

（2）直接发送短信，代码如下：

```
Uri uri = Uri.parse("smsto://0800000123");
Intent it = new Intent(Intent.ACTION_SENDTO, uri);
it.putExtra("sms_body", "The SMS text");
startActivity(it);
```

2．调用短信类

（1）根据类型区分，手机信息可以分为短信类和彩信类。

发送短信类的方法为 sendTextMessage()，代码如下：

```
sendTextMessage(String destinationAddress, String scAddress, String text, PendingIntent sentIntent,
PendingIntent deliveryIntent)
```

发送彩信类的方法为 SendDataMessage()，代码如下：

```
void SendDataMessage(String destinationAddress, String scAddress, short destinationPort, byte[] data,
PendingIntent sentIntent, PendingIntent deliveryIntent)
```

其中的参数含义如下。

- destinationAddress：消息的目标地址。
- scAddress：服务中心的地址，当地址为空时，使用默认的 SMSC。

- destinationPort：消息的目标端口号码。
- data：消息的主体，即要发送的数据。
- sentIntent：如果 sentIntent 不为空，无论消息发送成功或失败，都广播该意图。
- deliveryIntent：如果 deliveryIntent 不为空，当消息发送成功时，广播该意图。

（2）确认短信是否发送成功，可以通过 BroadcastReceiver 接收 sentIntent 中包含的结果进行判定。短信类对应的系统广播 ACTION 常量见表 7-4。

表 7-4 短信类对应的系统广播 ACTION 常量

常量名称	意义
Activity.RESULT_OK	成功
RESULT_ERROR_GENERIC_FAILURE	一般性失败
RESULT_ERROR_RADIO_OFF	无线服务被关闭
RESULT_ERROR_NULL_PDU	无 PDU
RESULT_ERROR_NO_SERVICE	无可用服务

其中，PDU（Protocol Data Unit）是协议数据单元，是指对等层次之间传递的数据单位。

【例 7.6】 制作发送短信项目（SmsActivity）。

Activity 类的代码如下：

```
import android.support.v7.app.AppCompatActivity;
import android.os.Bundle;
import android.telephony.SmsManager;
import android.util.Log;
import android.view.View;
import android.view.View.OnClickListener;
import android.widget.Button;

public class SmsActivity extends AppCompatActivity implements OnClickListener {
    /** Called when the activity is first created. */
    @Override
    public void onCreate(Bundle savedInstanceState) {
        super.onCreate(savedInstanceState);
        setContentView(R.layout.main);

        Button b = (Button) this.findViewById(R.id.button1);
        b.setOnClickListener(this);

    }

    @Override
    public void onClick(View arg0) {
        Log.v("发送短信中....", "");
        sendSms("5554", "Hello,");
        Log.v("发送结束", "");
    }

    public void sendSms(String address, String text) {
        SmsManager sm = SmsManager.getDefault();

        sm.sendTextMessage(address, null, text, null, null);
    }
}
```

7.7.2 接收短信

接收短信必须通过 BroadcastReceiver 实现，读者需要先掌握 BroadcastReceiver 的相关知识，然后再理解接收短信的实现过程，这样就比较简单了。

手机收到一条短信后，将广播一条“android.provider.Telephony.SMS_RECEIVED”的广播，短信内容被放在 Intent 里名为 pdu 的附加数据中，每条短信是一个字节数组，需要转换为 SmsMessage 类型，而一条 pdu 数据可能包含多条短信，因此 pdu 是一个二维字节数组。

详细应用请参照本章的短信过滤器项目。

7.8　实训：完善短信过滤器项目

围绕本章的短信过滤器项目，增加以下需求。

- 增加屏蔽骚扰电话功能。为了简单起见，所有被屏蔽短信的电话号码会被屏蔽。
- 增加转发垃圾短信功能，即某条垃圾短信对当前用户的广告意义很小，但对该用户的朋友有一定参考价值，则当前用户可以将该垃圾短信转发给他的朋友。

7.9　练习题

1．下列关于广播作用的说法，正确的是（　　）。

A．广播用于接收系统发布的一些消息　　B．广播可以帮助 Service 修改用户界面

C．广播无法启动一个 Service　　D．广播可以启动一个 Activity

2．下列关于广播的说法，错误的是（　　）。

A．BroadcastReceiver 不是 Android 四大组件之一

B．广播事件有两种注册方式：静态注册和动态注册

C．广播事件的静态注册方式需要在 AndroidManifest.xml 中进行配置

D. 如果选择广播事件的动态注册方式，则只有处于注册后且取消注册前的这段有效期内，才能接收广播

3．下列关于 BroadcastReceiver 的说法，不正确的是（　　）。

A．BroadcastReceiver 用于接收广播的 Intent

B．一个广播的 Intent 只能被一个订阅了此广播的 BroadcastReceiver 所接收

C．对于有序广播，系统会根据接收者声明的优先级别，按顺序对接收者执行处理流程

D．接收者的优先级别在 android:priority 属性中声明，数值越大，优先级别越高

4．下面关于广播的说法，错误的是（　　）。

A．广播分为有序广播和无序广播

B．使用 abortBroadcast()方法可以中断所有广播的传递

C．广播事件的注册方式有两种：静态注册和动态注册

D．sendOrderBroadcast()方法用于向系统广播有序事件，sendBroadcast()方法用于广播无序事件

5．要创建广播接收器，需重写（　　）方法。

A．onReceive()　　B．sendBroadcast ()　　C．onCreate()　　D．onStartCommand()

6．下列关于 Intent 启动组件的方法，错误的是（　　）。

A．startActivity()　　B．startService()

C．startBroadcastReceiver()　　D．startActivityForResult()

7．发送广播时，Intent 的启动方式为（　　）。

A．显式启动　　B．隐式启动　　C．A 和 B 都可以　D．以上说法都不正确

8．下列关于广播的说法，正确的是（　　）。

A．BroadcastReceiver 只能在配置文件中进行注册

B．BroadcastReceiver 注册后不能注销

C．BroadcastReceiver 只能接收自定义广播消息
D．BroadcastReceiver 可以在 Activity 中单独注册及注销
9．若使用 Android 应用程序发送短信，则应在 AndroidManifest.xml 中增加（　　）权限。
A．发送短信，无须配置权限　　B．permission.SMS
C．android.permission.RECEIVE_SMS　　D．android.permission.SEND_SMS
10．下面在 AndroidManifest.xml 中注册 BroadcastReceiver 的方式，正确的是（　　）。

A．
```
<receiver android:name="NewBroad">
        <intent-filter>
            <action
                android:name="android.provider.action.NewBroad"/>
            <action>
        </intent-filter>
</receiver>
```

B．
```
<receiver android:name="NewBroad">
            <intent-filter>
                android:name="android.provider.action.NewBroad"/>
            </intent-filter>
</receiver>
```

C．
```
<receiver android:name="NewBroad">
            <action
                android:name="android.provider.action.NewBroad"/>
            </action>
</receiver>
```

D．
```
<intent-filter>
        <receiverandroid:name="NewBroad">
            <action>
                android:name="android.provider.action.NewBroad"/>
            </action>
        </receiver>
</intent-filter>
```

7.10 作业

1．在 Android 系统中，如何接收短信？请写出关键代码。

2．请说明 Intent.ACTION_CALL 和 Intent.ACTION_DIAL 两种隐式意图的区别。

3．什么是 BroadcastReceiver？广播事件有哪两种？

4．注册广播事件有几种方式？这些方式有何优、缺点？请说明 Android 系统引入广播事件机制的用意。

5．编写并配置一个 BroadcastReceiver，当项目启动时，在 LogCat 中输出信息“系统已启动成功”。

第8章

多线程——射击游戏项目

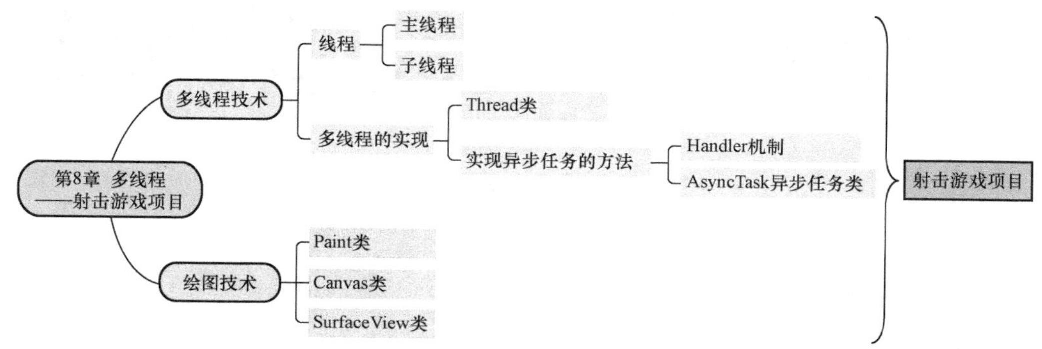

本章知识要点思维导图

多线程是 Android 应用程序开发，特别是游戏开发、网络编程和串口编程中必不可少的一项技术。本章将通过射击游戏项目介绍多线程。

8.1 需求分析

开发一个简单的 Android 游戏项目——射击游戏，满足如下需求。

- 射击目标为一个可左右移动的圆形靶子。
- 射击所使用的枪可以瞄准目标，灵活转动。瞄准目标的操作是通过手指触摸屏幕进行控制的，即调整枪管的发射角度；当手指离开屏幕时，枪管发射子弹。
- 子弹按照枪管瞄准的方向，发射后做直线运动。
- 子弹在前进过程中若接触靶子，表示击中目标，并在屏幕上方的“命中次数”TextView 中增加 1。
- 不论子弹是否击中靶子，都要在屏幕上方的“发射子弹”TextView 中增加 1。

绘制本项目界面中的各种线段、圆弧线、图形等物体需要使用绘图技术，完成圆形靶子和子弹的动作需要使用多线程技术，以上内容将在本章详细介绍。

8.2 界面设计

射击游戏项目只有一个界面，如图 8-1 所示，整体采用垂直线性布局，在垂直线性布局中嵌套一个水平线性布局（内含两个 TextView 控件，分别用于显示“命中次数”和“发射子弹”）和一个相对布局，相对布局的内容由 GameSurfaceView 类实现。

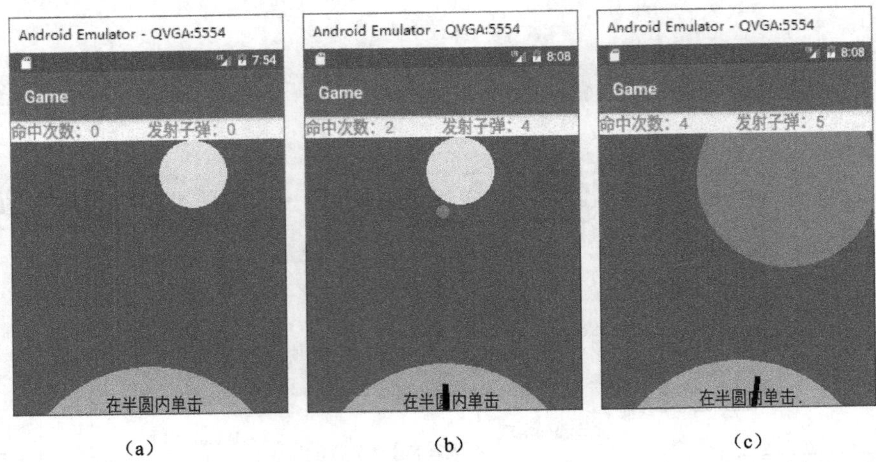

(a) (b) (c)

图 8-1 射击游戏项目界面

8.3 实施

8.3.1 创建项目

创建一个名为 Game 的项目。

8.3.2 界面实现

射击游戏项目的布局比较简单，界面顶端是两个 TextView 控件，用于显示“命中次数”和“发射子弹”的数值；TextView 控件下方采用 RelativeLayout 布局结构，它占据了剩余的空间，用于显示射击场景，该 RelativeLayout 内没有任何控件，在 MainActivity.java 代码中用 GameSurfaceView 类（SurfaceView 的子类）进行填充。

主界面的布局文件为 activity_main.xml，代码如下：

```
<?xml version="1.0" encoding="utf-8"?>
<LinearLayout xmlns:android="http://schemas.android.com/apk/res/android"
    xmlns:tools="http://schemas.android.com/tools"
    android:id="@+id/LinearLayout1"
    android:layout_width="match_parent"
    android:layout_height="match_parent"
    android:orientation="vertical"
    android:paddingBottom="0dp"
    android:paddingLeft="0dp"
    android:paddingRight="0dp"
    android:paddingTop="0dp"
    tools:context=".MainActivity" >

    <LinearLayout
        android:layout_width="match_parent"
        android:layout_height="wrap_content" >

        <TextView
            android:id="@+id/score"
            android:layout_width="wrap_content"
            android:layout_height="wrap_content"
            android:layout_weight="1"
```

```
                android:text="命中次数：0"
                android:textAppearance="?android:attr/textAppearanceMedium" />

            <TextView
                android:id="@+id/bullet"
                android:layout_width="wrap_content"
                android:layout_height="wrap_content"
                android:layout_weight="1"
                android:text="发射子弹：0"
                android:textAppearance="?android:attr/textAppearanceMedium" />
        </LinearLayout>

        <RelativeLayout
            android:id="@+id/drawCanvas"
            android:layout_width="match_parent"
            android:layout_height="match_parent" >
        </RelativeLayout>

    </LinearLayout>
```

8.3.3　Java 代码

射击游戏项目采用面向对象的设计思路，子弹和圆形靶子都是对象，由于子弹和圆形靶子都是绘制的图形，所以抽象出一个可绘制接口，作为子弹和圆形靶子必须实现的接口。

1．Activity 类

在 Activity 类的代码中，需要注意名为 updateScoreHandler 的 Handler 机制，它负责处理子线程与主线程的通信流程，即更新“命中次数”和“发射子弹”的数值，并播放游戏的音效。

Activity 类的代码如下：

```
package com.example.administrator.game;

import android.support.v7.app.AppCompatActivity;
import android.media.MediaPlayer;
import android.os.Bundle;
import android.os.Handler;
import android.os.Message;
import android.util.DisplayMetrics;
import android.widget.RelativeLayout;
import android.widget.TextView;

public class MainActivity extends AppCompatActivity{
    // 以下静态属性用于从其他类进行访问
    static MainActivity meself; // 保存自身的实例，用于访问 updateScoreHandler 属性（非静态）
    static int scrWidth; // 屏幕的宽度
    static int scrHeight; // 屏幕的高度
    static int score; // 命中次数
    static int bullet; // 发射子弹

    TextView tvScore;
    TextView tvBullet;

    @Override
    protected void onCreate(Bundle savedInstanceState) {
        super.onCreate(savedInstanceState);
        setContentView(R.layout.activity_main);
```

```
            // 设置自身实例，获取屏幕的宽度和高度（被其他类访问）
            meself = this;
            DisplayMetrics metric = new DisplayMetrics();
            getWindowManager().getDefaultDisplay()
                    .getMetrics(metric);
            scrWidth = metric.widthPixels;
            scrHeight = metric.heightPixels;

            tvScore = (TextView) findViewById(R.id.score);
            tvBullet = (TextView) findViewById(R.id.bullet);

            RelativeLayout layout = (RelativeLayout) findViewById(R.id.drawCanvas);
            layout.addView(new GameSurfaceView(this)); // 将相对布局的内容设为 GameSurfaceView
        }

        public Handler updateScoreHandler = new Handler(){ // 处理子线程与主线程的通信流程
            @Override
            public void handleMessage(Message msg){
                if(msg.what==1){
                    tvBullet.setText("发射子弹："+bullet);
                    MediaPlayer player = MediaPlayer.create(meself, R.raw.gun);
                    player.start(); // 播放音效
                }else{
                    tvScore.setText("命中次数："+score);
                    MediaPlayer player = MediaPlayer.create(meself, R.raw.end);
                    player.start(); // 播放音效
                }
            }
        };
}
```

2．可绘制接口

Drawable 的含义是可绘制接口，它的实现类有能力在屏幕中完成自行绘制操作，即拥有draw()方法，代码如下：

```
package com.example.administrator.game;

import android.graphics.Canvas;
import android.graphics.Paint;

public interface Drawable{
        public boolean isFinish();
        public int getX();
        public int getY();
        public int getSize();

        public void draw(Canvas canvas, Paint paint);
}
```

3．靶子类

靶子类 Target 的核心是 draw()方法，该方法用于自行绘制操作，即在当前的位置画一个以 size 为半径的圆，并且每次绘制圆形时随机向左或向右移动一定的距离，代码如下：

```
package com.example.administrator.game;

import android.graphics.Canvas;
import android.graphics.Color;
import android.graphics.Paint;
```

```
public class Target implements Drawable {
    boolean finish;
    int size = 30; // 圆形靶子的大小
    int x = MainActivity.scrWidth / 2; // 圆形靶子的初始位置
    int y = 30; // 该值不变，靶子只能横向移动

    int step = 4; // 每次移动的步长

    @Override
    public boolean isFinish() {
        return finish;
    }

    @Override
    public int getX() {
        return x;
    }

    @Override
    public int getY() {
        return y;
    }

    @Override
    public int getSize() {
        return size;
    }

    @Override
    public void draw(Canvas canvas, Paint paint) {
        // 在当前位置完成自行绘制操作
        paint.setColor(Color.YELLOW);
        paint.setStyle(Paint.Style.FILL);
        paint.setStrokeWidth(1);
        canvas.drawCircle(x, y, size, paint);
        // 随机移动
        x = Math.random() > 0.5 ? x + step : x - step;

        // 防止圆形靶子超出边界
        x = x < 0 ? 0 : x;
        x = x > MainActivity.scrWidth ? MainActivity.scrWidth : x;
    }
}
```

4．子弹类

子弹类 Bullet 与靶子类 Target 相比，两者的 draw()方法有许多相似之处。但是，需要注意其中的三个细节。

（1）移动方式不同。子弹按照用户控制的枪管方向沿直线弹道移动，所以要指定子弹发出的位置和弹道的斜率。当子弹碰到屏幕两侧的边界时，将被反弹回来（但只能反弹一次）。

（2）子弹移动时，系统要随时判断是否击中靶子，如果子弹击中圆形靶子，则通过 Handle 机制通知主界面更新“命中次数”的数值，并发出击中的音效。

（3）当子弹击中圆形靶子后，或射出屏幕有效范围，子弹将失效并被删除。

子弹类 Bullet 的代码如下：

```
package com.example.administrator.game;

import android.graphics.Canvas;
```

```
import android.graphics.Color;
import android.graphics.Paint;
import android.os.Message;

public class Bullet implements Drawable {
    boolean finish; // 击中圆形靶子或失效（射出屏幕有效范围）
    int size = 6; // 子弹的大小
    int x = 0; // 子弹的位置，子弹的出发位置为枪管延长线与半圆的交点
    int y = 0;

    int x0 = 0; // 枪管的位置，用于计算弹道斜率，枪管下部的顶点与界面底部的中点重合
    int y0 = 0; // 底部

    double slope = 1; // 子弹发射时的弹道斜率
    int step = 8; //子弹的速度（子弹每次移动的步长）

    public Bullet(int x, int y, double slope) {
        super();
        // 初始化子弹的参数（位置和斜率）
        this.x = x;
        this.y = y;
        this.slope = slope;
        x0 = MainActivity.scrWidth / 2;
        y0 = MainActivity.scrHeight;
    }

    @Override
    public boolean isFinish() {
        return finish;
    }

    @Override
    public int getX() {
        return x;
    }

    @Override
    public int getY() {
        return y;
    }

    @Override
    public int getSize() {
        return size;
    }

    @Override
    public void draw(Canvas canvas, Paint paint) {
        if (!finish) {
            // 每次向上移动，即 y 的值减少 step
            y -= step;
            // 根据公式 x = slope * y，计算 x 的值
            x = (int) ((MainActivity.scrHeight - y) * slope) + x0;

            checkShutted();//计算并判断子弹是否击中圆形靶子

            // 在当前位置完成自行绘制操作
```

```
                paint.setColor(Color.RED);
                paint.setStyle(Paint.Style.FILL);
                paint.setStrokeWidth(1);
                canvas.drawCircle(x, y, size, paint);

                if (y < 0) {
                    // 如果子弹碰到屏幕上边界，则超出有效范围，子弹失效并被删除
                    finish = true;
                }

                if (x < 0) {
                    // 如果子弹碰到屏幕左边界，子弹反弹一次
                    slope = -slope;
                    x0 = x0 - MainActivity.scrWidth;
                }
                if (x > MainActivity.scrWidth) {
                    //如果子弹碰到屏幕右边界，子弹反弹一次
                    slope = -slope;
                    x0 = x0 + MainActivity.scrWidth;
                }
            }
        }

        void checkShutted() {
            //计算并判断子弹是否击中圆形靶子，若击中则更新界面（包括播放音效）
            if (finish) {
                return;
            }
            for (Drawable d : GameSurfaceView.drawableList) {
                if (d instanceof Target) {
                    int diffX = d.getX() - x; // 直角边的边长
                    int diffY = d.getY() - y; // 另一直角边的边长
                    int space = d.getSize() + size; // 两个圆的半径之和
                    // 用勾股定理判断两者的距离是否小于两个圆的半径之和
                    if (diffX * diffX + diffY * diffY < space * space) {
                        size = 80; // 若子弹击中圆形靶子，则用一个红色大圆表示，并播放音效
                        finish = true;
                        MainActivity.score++; // “命中次数”加 1

                        Message msg = new Message();
                        msg.what = 2;// 参数值为 2 表示命中靶子，并通知 Handle 机制处理消息
                        MainActivity.meself.updateScoreHandler.sendMessage(msg);
                    }
                }
            }
        }
    }
```

5．绘图类

绘图类是实现绘图功能的核心，它继承 SurfaceView 类，其功能如下。

开始游戏时，创建靶子类 Target 的实例，启动线程并添加到 drawableList 列表中，该列表保存了所有可自行绘制的类（接口 Drawable 的实现类）。

线程体 run()方法的功能是每隔一段时间重绘屏幕，重绘屏幕的代码在 reDraw()方法中，reDraw()方法的功能有三项：第一，绘制发射装置（半圆形底座、枪管、提示文字）；第二，重绘 drawableList 列表中的所有活跃的对象，删除不活跃的对象（由 isFinish()方法判断）；第三，完成发射子弹动作，从本质上讲，发射子弹的过程就是创建一个子弹类 Bullet 的实例，

Bullet 类的构造方法中的三个参数用于定位发射的初始位置和斜率，参数由 onTouch()事件处理程序设置，然后将该实例添加到 drawableList 列表中，还要通过 Handler 机制通知主界面更新“发射子弹”的数值，并发出发射子弹的音效。

onTouch()事件处理程序记录用户手指离开屏幕时的位置，该位置将作为子弹发射的位置，此外，系统将计算子弹的发射角度（弹道斜率）。

绘图类的代码如下：

```
package com.example.administrator.game;

import android.content.Context;
import android.graphics.Canvas;
import android.graphics.Color;
import android.graphics.Paint;
import android.graphics.Point;
import android.os.Message;
import android.view.MotionEvent;
import android.view.SurfaceHolder;
import android.view.SurfaceView;
import android.view.View;
import java.util.LinkedList;
import java.util.List;

public class GameSurfaceView extends SurfaceView implements SurfaceHolder.Callback,
        View.OnTouchListener {
    DrawThread drawThread; // 刷新屏幕的线程
    static List<Drawable> drawableList = new LinkedList<Drawable>(); // 保存所有可绘制的物体

    Point trigger = new Point(); // 记录用户单击屏幕的位置（枪管瞄准的位置，由用户控制）
    double slope = 1; // 斜率（射击时枪管的斜率，枪管默认是竖直状态，与屏幕底边垂直）
    boolean isRelease = false; // 用户手指离开屏幕

    public GameSurfaceView(Context context) {
        super(context);
        SurfaceHolder holder = getHolder();
        holder.addCallback(this);
        // 以上三行代码是必须编写的

        // 设置用户单击屏幕的触发器
        this.setOnTouchListener(this);
        // 创建线程
        drawThread = new DrawThread(holder, getContext());
    }

    @Override
    public void surfaceChanged(SurfaceHolder holder, int format, int width,
            int height) {
        // do nothing
    }

    @Override
    public void surfaceCreated(SurfaceHolder holder) {
        // 启动线程
        drawThread.start();
        startGame();
    }

    @Override
```

```
public void surfaceDestroyed(SurfaceHolder holder) {
    // 关闭线程
    drawThread.isRunning = false;
}

void startGame() {
    // 游戏开始，添加一个圆形靶子
    Target target = new Target();
    drawableList.add(target);
}

class DrawThread extends Thread {
    // 在 SurfaceView 内的绘图线程
    boolean isRunning;
    SurfaceHolder surfaceHolder;
    Context context;
    Paint paint;

    public DrawThread(SurfaceHolder surfaceHolder, Context context) {
        this.surfaceHolder = surfaceHolder;
        this.context = context;
        paint = new Paint();
        paint.setDither(true);
    }

    @Override
    public void run() {
        isRunning = true;
        Canvas canvas = null;
        while (isRunning) {
            try {
                synchronized (surfaceHolder) {
                    canvas = surfaceHolder.lockCanvas(null);
                    reDraw(canvas);
                    Thread.sleep(20); // 每 20 毫秒重新绘制一次
                }
            } catch (InterruptedException e) {
                e.printStackTrace();
            } finally {
                surfaceHolder.unlockCanvasAndPost(canvas);
            }
        }
    }

    // 重绘屏幕的线程体
    public void reDraw(Canvas canvas) {
        canvas.drawColor(Color.BLUE); // 清屏操作，同时设置背景色
        try {
            for (Drawable d : drawableList) {
                if (d.isFinish()) {
                    drawableList.remove(d);// 如果对象不再有效时则将它移除
                } else {
                    d.draw(canvas, paint);// 绘制每个对象
                }
            }
        } catch (Exception e) {
            e.printStackTrace();
```

```
            }

            // 绘制半圆形底座
            paint.setStyle(Paint.Style.FILL);
            paint.setStrokeWidth(1);
            paint.setColor(Color.GREEN);
            canvas.drawCircle(MainActivity.scrWidth / 2,
                    MainActivity.scrHeight, MainActivity.scrWidth / 2,
                    paint);

            // 输入提示文字
            paint.setStrokeWidth(1);
            paint.setTextSize(14);
            paint.setColor(Color.RED);
            canvas.drawText("在半圆内单击", MainActivity.scrWidth / 2 - 35,
                    MainActivity.scrHeight - 80, paint);

            //用户单击屏幕的位置在半圆形底座内才绘制枪管
            int diffX = MainActivity.scrWidth / 2 - trigger.x;
            int diffY = (int) (MainActivity.scrHeight) - trigger.y;
            if (diffX * diffX + diffY * diffY < (MainActivity.scrWidth / 2)
                    * (MainActivity.scrWidth / 2)) {
                // 如果用户单击屏幕的位置在半圆形底座内，则绘制枪管
                paint.setStrokeWidth(5);
                paint.setColor(Color.BLACK);
                canvas.drawLine(MainActivity.scrWidth / 2,
                        (int) (MainActivity.scrHeight), trigger.x,
                        trigger.y, paint);

                if (isRelease) {
                    isRelease = false;
                    // 如果手指离开屏幕，则发射子弹
                    Bullet bullet = new Bullet(trigger.x, trigger.y, slope); // 创建子弹
                    drawableList.add(bullet);
                    MainActivity.bullet++;// “发射子弹”数值加 1

                    Message msg = new Message();
                    msg.what = 1;
                    MainActivity.meself.updateScoreHandler.sendMessage(msg);
                }
            }
        }
    }

    @Override
    public boolean onTouch(View v, MotionEvent event) {
        if (event.getAction() == MotionEvent.ACTION_UP) {
            // 只有手指离开屏幕时，系统才开始记录
            isRelease = true;
            // 记录用户单击屏幕的位置
            trigger.x = (int) event.getX();
            trigger.y = (int) event.getY();
            //以枪管下部的顶点为原点计算斜率，公式为 slope = x / y; 因此有 x = slope * y;
            slope = (trigger.x - MainActivity.scrWidth / 2)
                    / (double)(MainActivity.scrHeight - trigger.y);
        }
```

```
            return true;
        }
    }
```

射击游戏项目综合了接口、实现类、多线程、子线程与主线程之间的通信等技术，演示了面向对象的程序设计方法，并充分展示了多线程技术的应用。

8.3.4　运行测试

采用黑盒测试方法对射击游戏项目进行测试，验证其是否达到设计目标。

8.4　多线程技术

多线程技术在 Android 系统中被广泛应用，比如更新 UI 界面、处理网络通信和串口通信等。其实，Java 语言已有成熟的多线程技术，这些技术可以在 Android 系统中直接使用。另外，Android 系统还对 Java 语言的多线程技术进行了封装，提供了更方便、更安全的多线程类——AsyncTask。除此之外，Android 系统还有很多与多线程处理相关的类，如 Thread、Handler、Message、MessageQueue、Looper、HandlerThread 等。

8.4.1　理解 Android 多线程

1．使用多线程技术的原因

使用多线程技术的原因主要有以下三点。

（1）避免出现 ANR（Application Not Responding）现象：有些事件处理代码包含了运行时间较长的代码，这时需要使用多线程技术，否则会出现 ANR 现象，造成应用程序崩溃。

（2）提升用户体验：应用程序响应较慢会降低用户体验。使用多线程技术可以提高效率，从而改善应用程序的用户体验。

（3）异步通信：在某些情况下，不必使用同步阻塞等待返回结果，可以通过多线程技术实现异步处理。

2．使用多线程技术的场合

使用多线程技术的场合主要有以下四种。

（1）运行速度慢的代码：如果某段代码的运行时间可能超过 5 秒，就应该为其创建一个线程，放在子线程中运行。

（2）访问网络的代码：运行访问网络的代码，其响应速度有时很快，有时却很慢。例如，遇到网络故障时，可能阻塞较长时间。因此，必须将所有访问网络的代码放在子线程中运行。

（3）访问串口的代码：访问串口的代码与访问网络的代码有相似之处，因此也需要放在子线程中运行。

（4）需要高效运行的代码：多线程技术可以提高运行效率，改善用户体验，因此将一些需要高效运行的代码放在子线程中运行，这是游戏类应用程序常用的做法。

8.4.2　主线程和子线程

Android 应用程序运行时，会单独启动一个进程，这个进程就是主线程。默认情况下，这个应用程序中的所有 Activity（包括其他组件，如 Service 等）都运行在主线程中。因为主线程通常包含 Activity 所创建的 UI 控件，因此主线程有时也被称为 UI 线程，即负责控制 UI 界面的显示、更新和控件的交互操作。所以在 Android 应用程序中，主线程和 UI 线程的意义相同。

在这种环境下，所有的任务都在一个线程中运行。因此，线程中所有正在执行的任务，其花费的时间应当越短越好；而其他比较费时的任务（如访问网络、下载数据、查询数据库等），都应该交由子线程去执行，以免阻塞主线程。应用程序可以在主线程外创建新的线程，即子线程。

8.4.3 Thread 类

【例 8.1】使用多线程技术，创建一个名为 ThreadTest 的项目，用于实时显示系统时间。

题目要求：

- 以秒为单位进行更新，显示系统时间；
- 为了简化题目，系统时间更新 10 次后，项目自动结束。

实现功能：

- 用户单击“开始”按钮，在第一个 TextView 控件中显示系统时间（每秒更新一次），在第二个 TextView 控件中显示“开始”状态提示，与此同时，进度条显示更新的进度。当系统时间更新 10 次后，项目自动结束，并在第二个 TextView 控件中显示“结束”状态提示。

1. 界面设计

本项目的界面包括两个 TextView 控件、一个进度条和一个“开始”按钮，如图 8-2 所示。

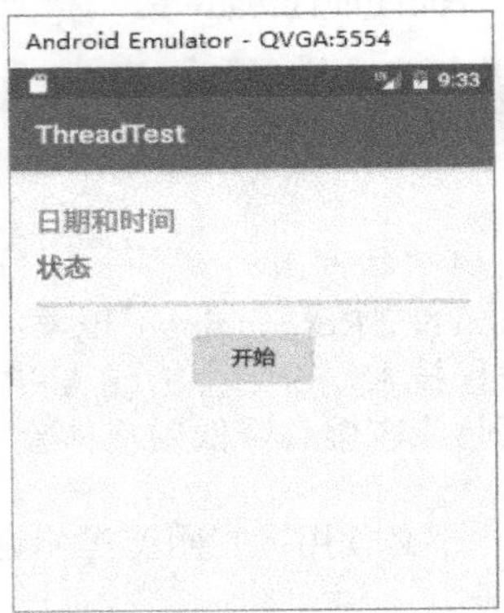

图 8-2 【例 8.1】界面

界面布局文件的代码如下：

```
<?xml version="1.0" encoding="utf-8"?>
<LinearLayout xmlns:android="http://schemas.android.com/apk/res/android"
    xmlns:tools="http://schemas.android.com/tools"
    android:id="@+id/LinearLayout1"
    android:layout_width="match_parent"
    android:layout_height="match_parent"
    android:orientation="vertical"
    android:paddingBottom="@dimen/activity_vertical_margin"
    android:paddingLeft="@dimen/activity_horizontal_margin"
    android:paddingRight="@dimen/activity_horizontal_margin"
    android:paddingTop="@dimen/activity_vertical_margin"
    tools:context=".MainActivity" >

<TextView
    android:id="@+id/tvDateTime"
    android:layout_width="wrap_content"
    android:layout_height="wrap_content"
    android:text="日期和时间"
    android:layout_marginTop="10dp"
    android:textAppearance="?android:attr/textAppearanceMedium" />
```

```
<TextView
    android:id="@+id/tvStatus"
    android:layout_width="wrap_content"
    android:layout_height="wrap_content"
    android:text="状态"
    android:textColor="#FF0000"
    android:layout_marginTop="10dp"
    android:textAppearance="?android:attr/textAppearanceMedium" />

<ProgressBar
    android:id="@+id/progressBar1"
    style="?android:attr/progressBarStyleHorizontal"
    android:layout_marginTop="10dp"
    android:layout_width="match_parent"
    android:layout_height="wrap_content" />

<LinearLayout
    android:layout_width="wrap_content"
    android:layout_height="wrap_content"
    android:layout_gravity="center"
    android:layout_marginTop="10dp">

    <Button
        android:id="@+id/btnStart"
        android:layout_width="wrap_content"
        android:layout_height="wrap_content"
        android:text="开始" />

</LinearLayout>

</LinearLayout>
```

2．Activity 类

先不使用多线程技术，编写如下代码：

```
package com.example.administrator.threadtest;

import android.support.v7.app.AppCompatActivity;
import android.os.Bundle;
import android.util.Log;
import android.view.View;
import android.widget.Button;
import android.widget.ProgressBar;
import android.widget.TextView;
import java.text.SimpleDateFormat;
import java.util.Date;

public class MainActivity extends AppCompatActivity implements View.OnClickListener {
    TextView tvDateTime;// 用于显示日期和时间（格式为年-月-日 时:分:秒）
    TextView tvStatus;// 用于显示运行状态
    ProgressBar progressBar1; // 进度条显示更新进度
    Button btnStart;
    public final static SimpleDateFormat df = new SimpleDateFormat("yyyy-MM-dd hh:mm:ss"); // 格式化时间

    @Override
    protected void onCreate(Bundle savedInstanceState) {
        super.onCreate(savedInstanceState);
        setContentView(R.layout.activity_main);
```

```
        tvDateTime = (TextView) this.findViewById(R.id.tvDateTime);
        tvStatus = (TextView) this.findViewById(R.id.tvStatus);
        progressBar1 = (ProgressBar) this.findViewById(R.id.progressBar1);

        btnStart = (Button) this.findViewById(R.id.btnStart);
        btnStart.setOnClickListener(this);
    }

    @Override
    public void onClick(View v) {
        showTime();
    }

    private void showTime(){ // 每秒更新一次时间
        // 线程开始时
        tvStatus.setText("开始");
        for (int i = 0; i < 10; i++) {
            String date=df.format(new Date());
            Log.v("== date & time ==", date); // 在 LogCat 中输出当前时间
            tvDateTime.setText(date); //在 TextView 控件中显示时间
            progressBar1.setProgress((i+1) / 10); // 显示进度
            try {
                Thread.sleep(1000); // 暂停 1000 毫秒
            } catch (InterruptedException e) {
                e.printStackTrace();
            }
        }
        // 线程结束时
        tvStatus.setText("结束");
    }
}
```

运行上述代码，在 LogCat 中会看到每秒更新的时间，在最后会出现一条运行提示“The application may be doing too much work on its main thread.”

此外，在前端看不到每次更新的进度，只能显示最后一次的结果。种种现象都表明，如果程序不使用 Thread 类，势必会出现运行异常。运行结果如图 8-3 所示。

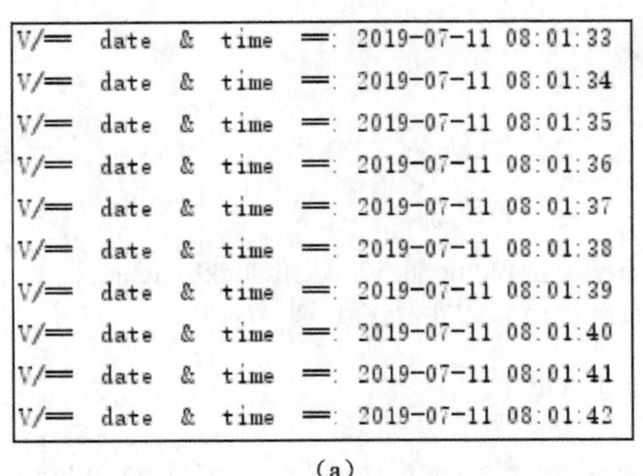

```
V/== date & time ==: 2019-07-11 08:01:33
V/== date & time ==: 2019-07-11 08:01:34
V/== date & time ==: 2019-07-11 08:01:35
V/== date & time ==: 2019-07-11 08:01:36
V/== date & time ==: 2019-07-11 08:01:37
V/== date & time ==: 2019-07-11 08:01:38
V/== date & time ==: 2019-07-11 08:01:39
V/== date & time ==: 2019-07-11 08:01:40
V/== date & time ==: 2019-07-11 08:01:41
V/== date & time ==: 2019-07-11 08:01:42
```

（a）

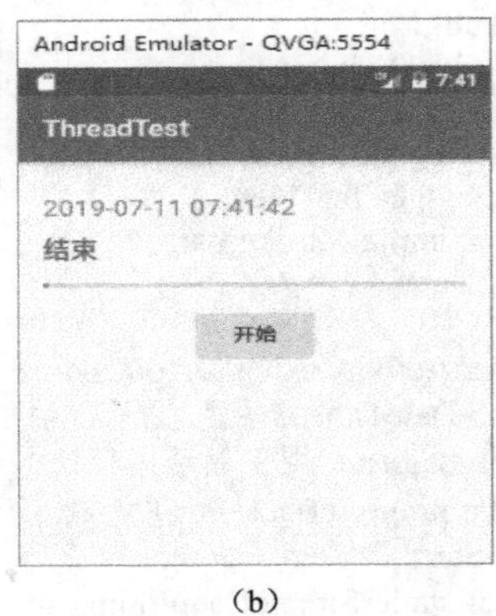

（b）

图 8-3　没有采用多线程技术时的运行结果

本项目只有采用多线程技术，才能实现题目的要求与功能。因此，对代码进行修改，采用多线程技术编写 Activity 类代码：

```
package com.example.administrator.threadtest;
```

```
// 省略导包部分
public class MainActivity extends AppCompatActivity implements OnClickListener {
    // 省略前面重复的代码

    @Override
    public void onClick(View arg0) {
        //showTime();
        new Thread(new ShowTimeThread()).start();
    }

    // 多线程，在线程体中执行 showTime()方法
    private class ShowTimeThread implements Runnable{
        @Override
        public void run() {
            showTime();
        }
    }
}
```

上述代码将占用时长的 showTime()方法放在一个线程中运行，减少了主线程的工作，但是这种方式也导致出现另外一个问题，即 tvDateTime.setText(date)在主线程中执行时可以得到正确的结果，但是 tvDateTime.setText(date)在子线程中执行时会出现下面的异常现象：

```
android.view.ViewRootImpl$CalledFromWrongThreadException: Only the original thread that created a
view hierarchy can touch its views.
```

上述异常现象表示只有创建这个 View 的线程才能访问（touch）它，即 tvDateTime 是在主线程中被创建的，子线程是不能访问 tvDateTime 的。在 Android 系统中，所有可视化组件（UI 控件）都是由主线程创建的，因此，所有子线程都不允许访问可视化组件（UI 控件）。

因此，截至目前，本例无法动态更新时间和进度，甚至无法在界面中显示状态提示。为了避免子线程对主线程造成干扰，Android 系统规定，不允许子线程访问主线程中的 UI 控件。那么，如何解决上述问题？

Android 系统提供了两种用于实现异步任务的方法——Handler 机制和 AsyncTask 异步任务类。我们将在 8.4.4 节进行介绍。

8.4.4 Handler 机制和 AsyncTask 异步任务类

1. Handler 机制

Android 系统提供了 Handler 机制来实现子线程对主线程 UI 控件的访问。

新建名为 ThreadAndHandler 的项目，采用 Handler 机制改进例 8.1 的代码，详情如下：

```
package com.example.administrator.threadandhandler;

import android.os.Handler;
import android.os.Message;
import android.support.v7.app.AppCompatActivity;
import android.os.Bundle;
import android.util.Log;
import android.view.View;
import android.widget.Button;
import android.widget.ProgressBar;
import android.widget.TextView;
import android.widget.Toast;
import java.text.SimpleDateFormat;
import java.util.Date;

public class MainActivity extends AppCompatActivity implements View.OnClickListener {
```

```
static TextView tvDateTime;
static TextView tvStatus;
static ProgressBar progressBar1;
Button btnStart;
public final static SimpleDateFormat df = new SimpleDateFormat("yyyy-MM-dd hh:mm:ss");
public static MyThread r = new MyThread();
public static Thread t = new Thread(r);

@Override
protected void onCreate(Bundle savedInstanceState) {
    super.onCreate(savedInstanceState);
    setContentView(R.layout.activity_main);

    tvDateTime = (TextView) this.findViewById(R.id.tvDateTime);
    tvStatus = (TextView) this.findViewById(R.id.tvStatus);
    progressBar1 = (ProgressBar) this.findViewById(R.id.progressBar1);

    btnStart = (Button) this.findViewById(R.id.btnStart);
    btnStart.setOnClickListener(this);
}

@Override
public void onClick(View v) {
    t.start();
    Toast.makeText(MainActivity.this, "线程开始", Toast.LENGTH_LONG).show();
    System.out.println("线程开始");
}

static void showTime() {
    Message msg = new Message();
    Bundle bundle = new Bundle();
    msg.setData(bundle);
    msg.what = 1; // 开始
    showTimeHandler.sendMessage(msg);
    for (int i = 0; i < 10; i++) {
        String date = df.format(new Date());
        Log.v("Thread" + i, MainActivity.df.format(new Date()));
        msg = new Message();
        bundle = new Bundle();
        msg.setData(bundle);
        msg.what = 2; // 更新
        bundle.putString("date", date);
        bundle.putInt("progress", i + 1);
        showTimeHandler.sendMessage(msg);
        try {
            Thread.sleep(1000); // 暂停 1000 毫秒
        } catch (InterruptedException e) {
            e.printStackTrace();
        }
    }
    msg = new Message();
    bundle = new Bundle();
    msg.setData(bundle);
    msg.what = 3; // 结束
    bundle.putString("result", "ok");
    showTimeHandler.sendMessage(msg);
    System.out.println("线程结束");
}
```

```
    private static Handler showTimeHandler = new Handler() {

        @Override
        public void handleMessage(Message msg) {
            super.handleMessage(msg);
            Bundle bundle = msg.getData();
            if (msg.what == 1) {
                tvStatus.setText("开始执行线程");
            } else if (msg.what == 2) {
                String dateTime = bundle.getString("date");
                tvDateTime.setText(dateTime); // 显示时间

                int progress = bundle.getInt("progress");
                progressBar1.setProgress((progress * 100) / 10); // 显示进度
            } else if (msg.what == 3) {
                String result = bundle.getString("result");
                tvStatus.setText("执行线程结束" + result);
            }
        }
    };
}

class MyThread implements Runnable {

    @Override
    public void run() {
        MainActivity.showTime();
    }
}
```

运行程序，结果如图 8-4 所示，符合预期效果。

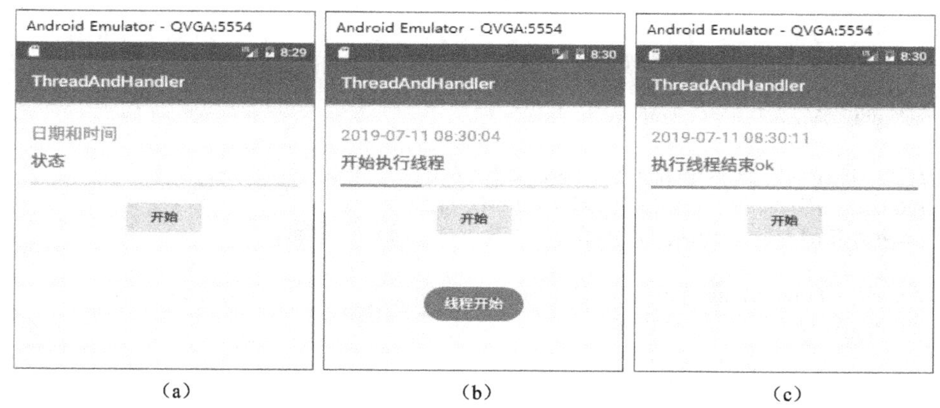

图 8-4　采用 Handler 机制的运行结果

在上述代码中，Handler 的子类 showTimeHandler 是在主线程中被创建的，它能访问主线程的 UI 控件。Android 系统提供了一种机制，允许子线程向这个 Handler 的子类发送消息，触发并执行 Handler 子类中的 handleMessage()方法。

通过这种机制，较好地解决了子线程与主线程之间的通信问题。请读者注意，子线程中发送消息的方法是 sendMessage()，而主线程中被触发的方法是 handleMessage()。

2. AsyncTask 异步任务类

Handler 机制的代码有两个问题：第一，代码比较复杂，需要通过 Handler 机制解决子线程与主线程之间的通信问题；第二，对线程的安全处理不够完善，需要在代码中解决。

Android 系统为了降低开发难度，提供了 AsyncTask 异步任务类，该类是一个封装过的后台任务类。

下面介绍 AsyncTask 异步任务类的工作过程。

【例 8.2】创建一个名为 AsyncTaskTest 的项目，修改例 8.1 中的代码，增加 AsyncTask 子类，并在 onClick()事件处理程序中使用该类。

修改后的代码如下：

```
package com.example.administrator.asynctasktest;

import android.os.AsyncTask;
import android.os.Bundle;
import android.support.v7.app.AppCompatActivity;
import android.util.Log;
import android.view.View;
import android.widget.Button;
import android.widget.ProgressBar;
import android.widget.TextView;
import java.text.SimpleDateFormat;
import java.util.Date;

public class MainActivity extends AppCompatActivity implements OnClickListener {
    TextView tvDateTime;// 用于显示日期和时间（格式为年-月-日 时:分:秒）
    TextView tvStatus;// 用于显示运行状态
    ProgressBar progressBar1; // 进度条显示更新进度
    Button btnStart;
    public final static SimpleDateFormat df = new SimpleDateFormat("yyyy-MM-dd hh:mm:ss"); // 格式化
时间

    @Override
    protected void onCreate(Bundle savedInstanceState) {
        super.onCreate(savedInstanceState);
        setContentView(R.layout.activity_main);

        tvDateTime = (TextView) this.findViewById(R.id.tvDateTime);
        tvStatus = (TextView) this.findViewById(R.id.tvStatus);
        progressBar1 = (ProgressBar) this.findViewById(R.id.progressBar1);

        btnStart = (Button) this.findViewById(R.id.btnStart);
        btnStart.setOnClickListener(this);
    }

        @Override
        public void onClick(View arg0) {
            new ShowTimeTask().execute(); // 改用 AsyncTask 类
        }

        private class ShowTimeTask extends AsyncTask<Integer, Integer, String> {
            String tag = "TimeTask";
            private String dateTime;  // 保存用于刷新的日期

            // onPreExecute 是前置运行体，运行在主线程中，因此可以访问 UI 控件
            @Override
```

```
        protected void onPreExecute() {
            Log.v(tag, "onPreExecute");
            tvStatus.setText("开始执行异步任务");
        }

        /*
         * doInBackground 相当于线程体，即后台运行的异步子线程，不允许访问 UI 控件
         * 当需要访问 UI 控件时，可调用 publishProgress()方法触发 onProgressUpdate()方法，对
UI 控件进行操作
         *
         * doInBackground 的参数与 AsyncTask 类的第一个参数类型一致
         * doInBackground 的返回值与 AsyncTask 类的第三个参数类型一致，将作为
onPostExecute()方法的传入参数
         */

        @Override
        protected String doInBackground(Integer... params) {
            Log.v(tag, "doInBackground");
            for (int i = 0; i < 10; i++) {
                dateTime = df.format(new Date()); // 只能在 LogCat 中显示
                publishProgress(i+1);
                try {
                    Thread.sleep(1000); // 暂停 1000 毫秒
                } catch (InterruptedException e) {
                    e.printStackTrace();
                }
            }
            return "ok";
        }

        // onProgressUpdate()方法是中间运行体
        // 参数类型与 AsyncTask 类的第二个参数类型一致，接收 publishProgress()方法的参数
        // onProgressUpdate()方法可能多次运行，因为它在主线程中，因此可以访问 UI 控件
        @Override
        protected void onProgressUpdate(Integer... values) {
            Log.v(tag, "onProgressUpdate");
            tvDate.setText(dateTime);
            int progress = values[0];
            progressBar.setProgress((progress*100)/10);
        }

        // onPostExecute()方法是后置运行体
        // onPostExecute()方法接收 doInBackground 的返回值,并作为参数,参数类型与 AsyncTask
类的第三个参数类型一致
        // onPostExecute()方法在主线程中运行，可以访问 UI 控件
        @Override
        protected void onPostExecute(String result) {
            Log.v(tag, "onPostExecute");
            tvStatus.setText("异步任务执行结束" + result);
        }
    }
}
```

运行结果如图 8-5 所示。

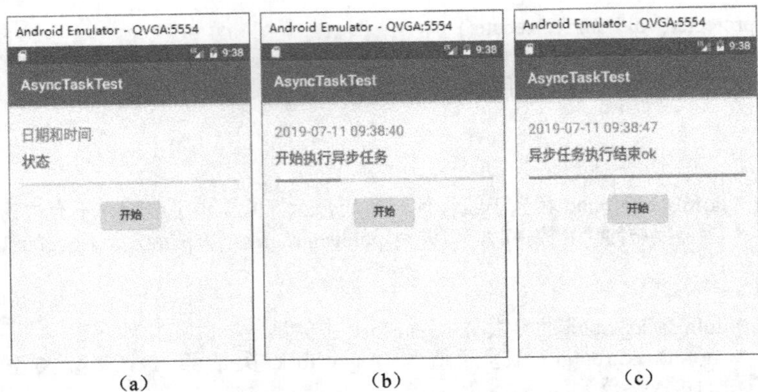

图 8-5 采用 AsyncTask 时的运行结果

AsyncTask 是 Android 系统提供的异步处理辅助类，它可以实现耗时操作在子线程中执行，处理结果在主线程中执行的目的，AsyncTask 类屏蔽了多线程相关的细节处理环节，因此更加便于使用。AsyncTask 抽象类的定义如下：

```
public abstract class AsyncTask<Params, Progress, Result> {
```

Params、Progress 和 Result 为三种泛型类型。其中，Params 指“启动任务执行的输入参数”，如 HTTP 请求的 URL；Progress 指“后台任务执行的进度”；Result 指“后台计算结果的类型”，如 String。

执行异步任务一般包括以下步骤。

（1）execute(Params... params)方法：启动异步任务，需要在主线程中调用本方法，触发并执行异步任务。

（2）onPreExecute()方法：当 execute(Params... params)方法被调用后，立即执行本方法。本方法用于在执行后台任务前对 UI 进行标记。

（3）doInBackground(Params... params)方法：本方法是子线程的线程体，当 onPreExecute()方法完成后，立即执行本方法。本方法用于执行较费时的操作，它可以接收输入参数并返回计算结果。在执行过程中，可以调用 publishProgress(Progress... values)方法更新进度信息。

（4）onProgressUpdate(Progress... values)方法：在 doInBackground(Params... params)方法中调用 publishProgress(Progress... values)方法时，会执行本方法。本方法会将进度信息更新到 UI 控件中。

（5）onPostExecute(Result result)方法：当 doInBackground(Params... params)方法结束时，会调用本方法，计算结果将作为参数传递到本方法中，结果会显示到 UI 控件中。

使用上述方法时，需要注意以下几点事项。

- 异步任务的实例必须在 UI 线程中创建。
- execute(Params... params)方法必须在 UI 线程中调用。
- 不要手动调用 onPreExecute()方法、doInBackground(Params... params)方法、onProgressUpdate(Progress... values)方法和 onPostExecute(Result result)方法。
- 不能在 doInBackground(Params... params)方法中更改 UI 控件的信息。
- 一个任务实例只能执行一次，如果执行第二次将会抛出异常。

8.5 绘图技术

Android 系统采用 OpenGL 图形标准，包括的类有 Paint 类、Canvas 类、View 类和 SurfaceView 类。其中，Paint 类和 Canvas 类可以理解为画家的画笔和画布，而 View 类和 SurfaceView 类可以理解为画家的画室。

8.5.1　Paint 类

Paint 类是画笔，用户可以设置画笔的颜色、粗细等属性。Paint 类的常用方法见表 8-1。

表 8-1　Paint 类的常用方法

方　法	说　明
void setAlpha(int a)	设置 Alpha（透明度），范围是 0~255
void setColor(int color)	设置颜色，Color 类包含了一些常见的颜色定义
void setDither(boolean arg0)	是否使用图像抖动处理功能，该方法可以使图形更加平滑
void setStyle(Style style)	设置线型，在 Paint.Style 中定义了常用线型
void setStrokeWidth(float arg0)	设置线的粗细
void setLinearText(boolean linearText)	设置线性文本
void setTextAlign(Paint.Align align)	设置文本对齐
void setTextSize(float textSize)	设置字体大小
Typeface setTypeface(Typeface typeface)	设置字体属性，如粗细、斜体、颜色等

8.5.2　Canvas 类

Canvas 类是画布，当用户设置好画笔的属性后，可以在画布上绘制任意图形。绘制前，需要设置画布的属性，如画布的颜色、尺寸等属性。Canvas 类的常用方法见表 8-2。

表 8-2　Canvas 类的常用方法

方　法	说　明
Canvas()	构造方法，创建一块空画布
drawColor	设置 Canvas 类的背景颜色，通常用于清除原有图形（清屏）
void setBitmap(Bitmap bitmap)	设置画布上的图像
void rotate(float arg0)	旋转画布
void translate(float arg0, float arg1)	移动画布
void scale(float arg0, float arg1)	缩放画布
void　drawLine(float startX, float startY, float stopX, float stopY, Paint paint)	画线段
void　drawOval(float left, float top, float right, float bottom, Paint paint)	画椭圆
void　drawPicture(Picture picture)	画图像
void　drawRect(float left, float top, float right, float bottom, Paint paint)	画矩形
void　drawText(String text, int start, int end, float x, float y, Paint paint)	输入文字，在画布上输入文本内容

设置好相关属性后，才能用画笔在画布上绘制图形或输入文字等。接下来，通过实例演示绘图功能。

【例 8.3】创建一个名为 DemoPaint 的项目，绘制如图 8-6 所示的效果。

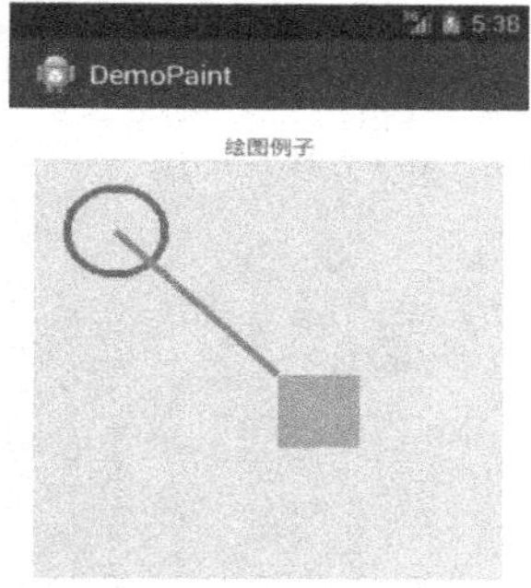

图 8-6　绘制静态的图形

（1）界面布局文件的代码如下：

```
<LinearLayout xmlns:android="http://schemas.android.com/apk/res/android"
    xmlns:tools="http://schemas.android.com/tools"
    android:id="@+id/LinearLayout1"
    android:layout_width="match_parent"
    android:layout_height="match_parent"
    android:orientation="vertical"
    android:paddingBottom="@dimen/activity_vertical_margin"
    android:paddingLeft="@dimen/activity_horizontal_margin"
    android:paddingRight="@dimen/activity_horizontal_margin"
    android:paddingTop="@dimen/activity_vertical_margin"
    tools:context=".MainActivity" >

    <TextView
        android:layout_width="wrap_content"
        android:layout_height="wrap_content"
        android:layout_gravity="center"
        android:text="绘图例子" />

    <LinearLayout
        android:id="@+id/draw"
        android:layout_width="match_parent"
        android:layout_height="match_parent"
        android:orientation="horizontal" >
    </LinearLayout>

</LinearLayout>
```

（2）MainActivity 中用到了 View 类，View 类的子类用于绘制静态的图形，代码如下：

```
package com.example.administrator.demopaint;

import android.support.v7.app.AppCompatActivity;
import android.content.Context;
import android.graphics.Canvas;
import android.graphics.Color;
import android.graphics.Paint;
import android.os.Bundle;
import android.view.View;
import android.widget.LinearLayout;

public class MainActivity extends AppCompatActivity{

    @Override
    protected void onCreate(Bundle savedInstanceState) {
        super.onCreate(savedInstanceState);
        setContentView(R.layout.activity_main);

        LinearLayout linearLayout = (LinearLayout) findViewById(R.id.draw);
        linearLayout.addView(new DrawView(this));
    }

    private class DrawView extends View {
        public DrawView(Context context) {
            super(context);
        }

        @Override
        public void onDraw(Canvas canvas) {
            canvas.drawColor(Color.YELLOW);
```

```
            Paint paint = new Paint();
            paint.setStrokeWidth(5);

            paint.setColor(Color.BLUE);
            paint.setStyle(Paint.Style.STROKE);
            canvas.drawCircle(50, 50, 30, paint);

            paint.setColor(Color.GREEN);
            paint.setStyle(Paint.Style.FILL);
            canvas.drawRect(150, 150, 200, 200, paint);

            paint.setColor(Color.RED);
            canvas.drawLine(50, 50, 150, 150, paint);
        }
    }
}
```

在上述代码中，实现绘画功能的是 DrawView 类，它继承 View 类，从构造方法中获得当前的上下文，全部绘画操作在 onDraw()方法中完成，这是因为系统会在恰当的时机自动调用每个 View 类的子类的 onDraw()方法，从而完成每个子类的绘画操作。

8.5.3　SurfaceView 类

在例 8.3 中，我们绘制的是静态的图形，如果绘制动态的图形，就要用到 SurfaceView 类，并定义它的一个子类，在该子类中定义重绘屏幕的时间，所以要用到多线程技术。

【例 8.4】创建一个名为 DemoPaint 的项目，绘制如图 8-7 所示的效果。

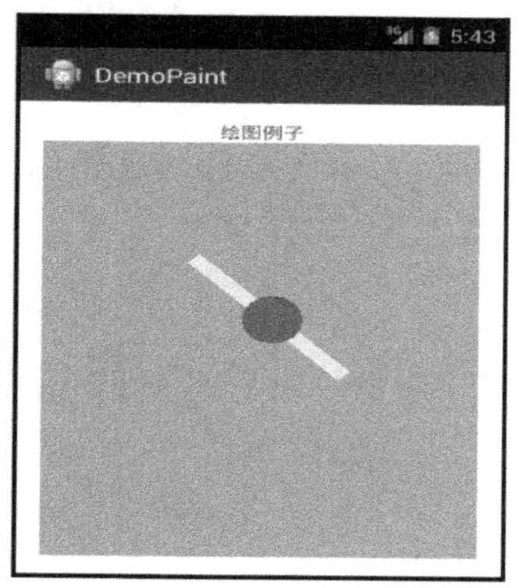

图 8-7　绘制动态的图形（球沿黄线移动）

将例 8.3 的 DrawView 类改为 DrawSurfaceView 类，DrawSurfaceView 类是 SurfaceView 类的子类，代码如下：

```
package com.example.administrator.demopaint;

import android.app.Activity;
import android.content.Context;
import android.graphics.Canvas;
import android.graphics.Color;
import android.graphics.Paint;
import android.os.Bundle;
import android.view.SurfaceHolder;
import android.view.SurfaceHolder.Callback;
import android.view.SurfaceView;
import android.view.View;
```

```
import android.widget.LinearLayout;

public class MainActivity extends Activity {

    @Override
    protected void onCreate(Bundle savedInstanceState) {
        super.onCreate(savedInstanceState);
        setContentView(R.layout.activity_main);

        LinearLayout linearLayout = (LinearLayout) findViewById(R.id.draw);
        linearLayout.addView(new DrawSurfaceView(this));
    }

    private class DrawSurfaceView extends SurfaceView implements Callback {
        // 导入 android.view.SurfaceHolder.Callback 包
        DrawThread drawThread; // 绘画线程实例

        public DrawSurfaceView(Context context) {
            super(context);
            SurfaceHolder holder = getHolder();
            holder.addCallback(this); // 设置 Surface 生命周期回调
            drawThread = new DrawThread(holder, getContext()); // 创建绘画线程
        }

        @Override
        public void surfaceChanged(SurfaceHolder holder, int format, int width,
                int height) {
            // do nothing
        }

        @Override
        public void surfaceCreated(SurfaceHolder holder) {
            drawThread.start(); // 启动绘画线程
        }

        @Override
        public void surfaceDestroyed(SurfaceHolder holder) {
            // 停止绘画线程
            drawThread.isRunning = false;
            try {
                drawThread.join();
            } catch (InterruptedException e) {
                e.printStackTrace();
            }
        }

        // 绘画线程
        class DrawThread extends Thread {
            SurfaceHolder surfaceHolder;
            Context context;
            boolean isRunning;
            Paint paint;

            public DrawThread(SurfaceHolder surfaceHolder, Context context) {
                this.surfaceHolder = surfaceHolder;
                this.context = context;
                paint = new Paint();
            }
```

```
            @Override
            public void run() {
                isRunning = true;
                Canvas canvas = null;
                while (isRunning) {
                    try {
                        synchronized (surfaceHolder) {
                            canvas = surfaceHolder.lockCanvas(null);
                            reDraw(canvas);
                            Thread.sleep(50); // 每 50 毫秒重绘一次
                        }
                    } catch (InterruptedException e) {
                        e.printStackTrace();
                    } finally {
                        surfaceHolder.unlockCanvasAndPost(canvas);
                    }
                }
            }

            int x = 100; // 小球的初始位置
            int y = 100;

            public void reDraw(Canvas canvas) {
                canvas.drawColor(Color.GREEN); // 清屏操作，清楚之前的残留图像

                paint.setColor(Color.YELLOW); // 黄色线段
                paint.setStyle(Paint.Style.STROKE); // 实线
                paint.setStrokeWidth(10); // 线粗为 10
                canvas.drawLine(100, 100, 200, 200, paint); // 一条线段

                paint.setColor(Color.RED); // 红色
                paint.setStyle(Paint.Style.FILL); // 填充
                paint.setStrokeWidth(1); // 线粗为 1
                canvas.drawCircle(x, y, 20, paint);
                // 向右下方位移动
                x = x < 200 ? x + 1 : 100;
                y = y < 200 ? y + 1 : 100;
            }
        }
    }
}
```

分析上述代码，在 DrawSurfaceView 类中有一个绘画线程 DrawThread 类，它是内部类中的内部类。DrawThread 类的作用是每 50 毫秒在屏幕中重绘一次，而每次重绘前都要将屏幕的颜色改为背景色（清屏），然后改变小球的 x 值和 y 值并绘制小球。由于屏幕的刷新频率很快，便形成了小球沿黄线移动的动画效果。DrawThread 类在 DrawSurfaceView 类的构造方法中实例化，在 surfaceCreated()方法中启动线程并运行。

8.6 实训：改进射击游戏项目

围绕本章的射击游戏项目，增加以下需求。

- 增加第二个圆形靶子，应该如何修改代码？

- 当界面中有两个（或三个）圆形靶子时，不允许这两个（或三个）圆形靶子在移动过程中相互重叠，应该如何实现？

8.7 实训：多线程技术的应用——秒表项目

开发一个名为 Timer 的秒表项目，最小计时时间为 0.1 秒，最大计时时间为 59 分。秒表的功能按钮包括“开始”“暂停”“清零”“继续”，秒表项目的界面如图 8-5 所示。

项目要求：采用 AsyncTask 异步任务类实现多线程。

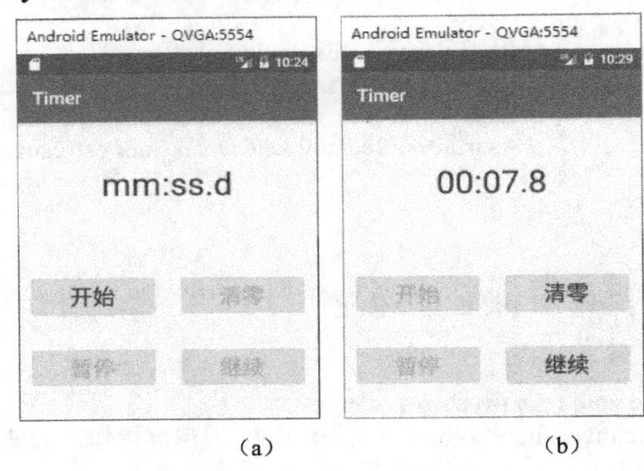

（a）　　（b）

图 8-8　秒表项目界面

activity_main.xml

MainActivity.java

秒表项目的界面包含一个 TextView 控件和四个 Button 控件。

（1）主界面的布局文件为 activity_main.xml，请扫描旁边的二维码进行浏览。

（2）请扫描旁边的二维码，浏览主界面的 MainActivity.java 代码。

8.8 练习题

1．请判断：在 Android 系统中，UI Thread 通常指 Main Thread，启动程序时，系统会为其建立一个 Message Queue。这句话是（　）的。

A．正确　　B．错误

2．Hanlder 类是线程和 Activity 通信的桥梁，如果线程处理不当，装有 Android 系统的设备在运行时就会变得很慢。请问，线程销毁的方法是（　）。

A．onDestroy()　　B．onClear()　　C．onFinish()　　D．onStop()

3．（　）最重要，是最后被销毁的。

A．服务进程　　B．后台进程　　C．可见进程　　D．前台进程

4．下列关于线程的说法，不正确的是（　）。

A．在 Android 系统中，可以在主线程中创建一个新的线程

B．在创建的新线程中可以对 UI 控件进行操作

C．新线程可以和 Handler 机制共同使用

D．Handler 机制隶属于创建它的线程

5．遇到下列（　）情况时，需要把进程移到前台。

A．进程中有一个正在运行的与用户交互的 Activity，它的 onResume()方法被调用

B．进程中有一个正在运行的 BroadcastReceiver，它的 onReceive()方法被调用

C．进程中有一个 Service，该 Service 对应的 Activity 正在与用户进行交互操作

D．上述选项均正确

6．下列说法正确的是（　　）。

A．每个进程都运行在自己的 Java 虚拟机（JVM）中

B．默认情况下，每个应用程序均运行在自己的进程中，而且此进程不会被销毁

C．每个应用程序会被赋予唯一的 Linux 用户 ID，从而使得其他用户也可以访问该应用程序中的文件

D．一个应用程序数据，可以随意被其他应用程序访问

7．关于 Android 进程，下列说法不正确的是（　　）。

A．组件在哪个进程中运行，是由 AndroidManifest.xml 决定的

B．当急需内存时，Android 系统会决定先关闭那些空闲的进程

C．背景进程是不被用户所见的 Activity，但这种进程仍可能被用户看到，故它不能被杀死

D．可视进程一般不被系统杀死

8．在 AsyncTask 中，下列（　　）方法用于执行耗时的后台计算工作。

A．run()　　B．execute()　　C．doInBackground()　　D．onPostExecute()

9．下列关于 Handler 机制的说法，不正确的是（　　）。

A．它是一种用于实现不同进程间通信的机制　　B．它避免了在新线程中刷新 UI 的操作

C．它采用队列的方式存储 Message　　D．它是一种用于实现不同线程间通信的机制

10．下列关于 Handler 机制的说法，正确的是（　　）。

A．它是一种用于实现不同线程间通信的机制　　B．它避免了新线程操作 UI 控件

C．它采用栈的方式组织任务　　D．它可以属于一个新的线程

8.9　作业

1．多线程技术在 Android 应用程序中有什么作用？如果在需要使用多线程技术的场合，却没有使用多线程技术，将会导致什么后果？

2．在 Java 应用程序中，如何实现多线程？试举例说明，并写出代码。

3．在 Android 应用程序中，如何实现多线程？如何在子线程中访问主线程的 UI 界面？

4．在 Android 应用程序中，如何使用 AsyncTask 类？

第9章 嵌入式开发：网络编程——天气预报项目

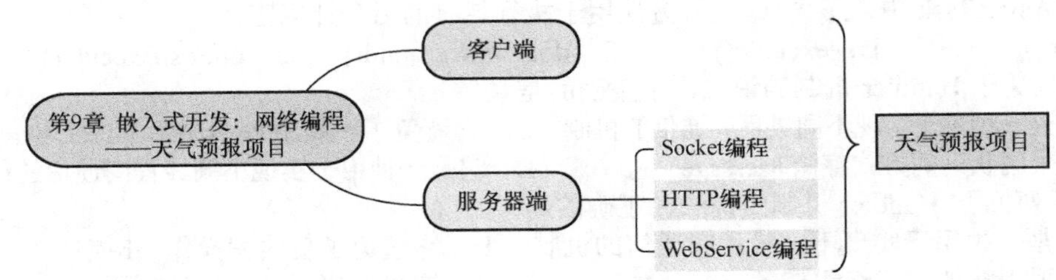

本章知识要点思维导图

Android 系统的嵌入式开发是物联网编程的核心部分，重点内容包括网络编程和串口编程，本章将着重讲解网络编程。

网络编程指基于 TCP/IP 网络的编程，其主要技术包括 Socket 编程、HTTP 编程和 WebService 编程。

本章将通过天气预报项目介绍网络编程。

9.1 需求分析

开发一个天气预报项目，获取中央气象台的天气预报数据，并在界面中显示。

天气预报是一种公共服务，对于这类服务，一般通过 WebService 提供。因此，本项目需要调用 WebService，从而获得天气预报数据。

登录网站 http://www.webxml.com.cn，可以查询到一个提供天气预报的 WebService，数据每 2.5 小时左右自动更新一次，数据来源于中国气象局官方网站，准确可靠。在 WebService 中，包含了众多中国城市和外国城市的天气预报数据。用户可以在 24 小时内免费查询不超过 50 次，并且获取两次数据的时间间隔要大于 600 毫秒。

通过网站链接（http://www.webxml.com.cn/zh_cn/weather_icon.aspx），可以看到提供的天气预报 WebService 的方法及说明。其中 Endpoint 链接（http://www.webxml.com.cn/WebServices/WeatherWS.asmx）提供了详细说明。单击具体的方法名，还可以看到更具体的说明，以及直接调用的例子。

天气预报 WebService 的方法名及说明见表 9-1。

表 9-1　天气预报 WebService 的方法及说明

方 法 名	说 明	参 数
getRegionCountry	获得外国的国家名称和与之对应的 ID	输入参数：无，返回数据：一维字符串数组
getRegionDataset	获得中国的省、直辖市、自治区、特别行政区的名称和与之对应的 ID；获得外国的国家名称和与之对应的 ID	输入参数：无，返回数据：DataSet

续表

方　法　名	说　　明	参　　数
getRegionProvince	获得中国的省、直辖市、自治区、特别行政区的名称和与之对应的 ID	输入参数：无。返回数据：一维字符串数组
getSupportCityDataset	获得支持的城市/地区名称和与之对应的 ID	输入参数：theRegionCode = 城市/地区 ID 或名称。返回数据：DataSet
getSupportCityString	获得支持的城市/地区名称和与之对应的 ID	输入参数：theRegionCode =城市/地区 ID 或名称。返回数据：一维字符串数组
getWeather	获得天气预报数据	输入参数：城市/地区 ID 或名称。返回数据：一维字符串数组

9.2　界面设计

天气预报项目只有一个界面，包括一个标有城市名称的 Button 控件和一个用于显示天气信息的 TextView 控件，如图 9-1 所示。

图 9-1　天气预报项目界面

9.3　实施

在 Android 系统中访问 WebService 需要第三方 jar 包（ksoap2），文件名为 ksoap2-android-assembly-3.6.0-jar-with-dependencies.jar，在项目素材中可以找到。

9.3.1　创建项目

创建天气预报项目，项目名称为 WeatherForecast，将第三方 jar 包复制到项目的 app→libs 目录中，然后右击 jar 包，选择“Add As Library…”命令。

9.3.2　编写 WebServiceCall 类

利用 ksoap2 提供的功能，可以编写一个调用 WebService 的 WebServiceCall 类，代码如下：

```
package com.example.administrator.weatherforecast;

import java.io.IOException;
import java.util.HashMap;
import java.util.Iterator;
import java.util.Map;
```

```
import org.ksoap2.SoapEnvelope;
import org.ksoap2.serialization.SoapObject;
import org.ksoap2.serialization.SoapSerializationEnvelope;
import org.ksoap2.transport.HttpResponseException;
import org.ksoap2.transport.HttpTransportSE;
import org.xmlpull.v1.XmlPullParserException;
import android.os.Handler;
import android.os.Message;

public class WebServiceCall {
    public static void callWebService(String url, final String namespace,
            final String methodName, HashMap<String, String> params,
            final WebServiceCallBack webServiceCallBack, boolean isDotNet) {
        // 初始化 WebSerive 的地址
        final HttpTransportSE httpTransportSE = new HttpTransportSE(url);

        // 设置参数
        SoapObject soapObject = new SoapObject(namespace, methodName);
        if (params != null) {
            for (Iterator<Map.Entry<String, String>> it = params.entrySet()
                    .iterator(); it.hasNext();) {
                Map.Entry<String, String> entry = it.next();
                soapObject.addProperty(entry.getKey(), entry.getValue());
            }
        }

        // 设置信封
        final SoapSerializationEnvelope soapEnvelope = new SoapSerializationEnvelope(
                SoapEnvelope.VER11);
        soapEnvelope.setOutputSoapObject(soapObject);
        soapEnvelope.dotNet = isDotNet; // WebService 不是.Net 技术开发的
        httpTransportSE.debug = true;

        // 设置通信
        final Handler mHandler = new Handler() {
            // 子线程与主线程进行通信
            @Override
            public void handleMessage(Message msg) {
                super.handleMessage(msg);
                // 将返回值回调到 callBack 的参数中
                webServiceCallBack.callBack((SoapObject) msg.obj);
            }

        };

        // 访问 WebService （必须在线程中）
        new Thread(new Runnable() {
            @Override
            public void run() {
                SoapObject resultSoapObject = null;
                try {
                    httpTransportSE.call(namespace + methodName, soapEnvelope);
                    if (soapEnvelope.getResponse() != null) {
                        resultSoapObject = (SoapObject) soapEnvelope.bodyIn;
                    }
                } catch (HttpResponseException e) {
                    e.printStackTrace();
```

```
                    } catch (IOException e) {
                        e.printStackTrace();
                    } catch (XmlPullParserException e) {
                        e.printStackTrace();
                    } finally {
                        // 将结果返回给主线程
                        mHandler.sendMessage(mHandler.obtainMessage(0,
                                resultSoapObject));
                    }
                }
            }).start();
        }
}
```

其中，callWebService()方法包含以下几个参数，含义如下。

- url：指定 WebService WSDL 文档的 URL。
- namespace：指定 WebService 的命名空间。
- methodName：指定 WebService 的方法名。
- params：指定调用 callWebService()方法需要的参数，它是一个映射（HashMap<String,String>）。
- WebServiceCallBack：指定回调方法。

在 callWebService()方法内部，首先创建 HttpTransportSE 对象，并用 WebService WSDL 文档的 URL 初始化。然后，创建 SoapObject 对象和 SoapSerializationEnvelope 对象，这两个对象分别表示 WebService 请求的数据和信封，前者封装了请求的命名空间、方法名以及参数，后者封装了请求的一般性信息，并将前者一并封装进去，就像信封里装入有内容的信一样。

在 Android 系统中，所有与网络访问有关的操作都必须通过线程来完成，因此，实际的调用语句是放在线程体中的，即调用 HttpTransportSE 对象的 call(String soapAction, SoapEnvelope envelope)方法，它的返回信息保存在 soapEnvelope.bodyIn 内。由于这段代码在子线程中，故必须通过 Handler 机制实现子线程与主线程相互通信，最后，在 Handler 机制的 handleMessage(Message msg)方法内将结果回调给 callBack()方法中的参数，而 callBack()方法是 WebServiceCallBack 接口对应的回调方法，代码如下：

```
package com.example.administrator. weatherforecast;

import org.ksoap2.serialization.SoapObject;

public interface WebServiceCallBack {
    public void callBack(SoapObject result);
}
```

9.3.3　Java 代码

在本项目中，用户单击城市名称的按钮后，系统将按钮上的文字作为参数，调用 WebService 的 getWeatherbyCityName()方法，然后，通过 WebServiceCallBack 接口的实现类处理返回的数据，并将结果显示在界面中。

Activity 类的代码如下：

```
package com.example.administrator. weatherforecast;

import java.util.HashMap;
import org.ksoap2.serialization.SoapObject;
import android.support.v7.app.AppCompatActivity;
import android.os.Bundle;
import android.view.Menu;
import android.view.View;
import android.view.View.OnClickListener;
```

```
import android.widget.Button;
import android.widget.TextView;
import android.widget.Toast;

public class MainActivity extends AppCompatActivity implements View.OnClickListener{
    Button button;
    TextView textview;

    @Override
    protected void onCreate(Bundle savedInstanceState) {
        super.onCreate(savedInstanceState);
        setContentView(R.layout.activity_main);

        button = (Button) this.findViewById(R.id.button);
        textview = (TextView) this.findViewById(R.id.textview);

        button.setOnClickListener(this);
    }

    @Override
    public void onClick(View v) {
        String url = "http://www.webxml.com.cn/WebServices/WeatherWS.asmx";
        String namespace = "http://WebXml.com.cn/";
        String methodName = "getWeather";
        HashMap<String, String> params = new HashMap<String, String>();
        Button b = (Button) v;
        params.put("theCityCode",b.getText().toString());
        // 将按钮上的文字作为 theCityCode 的参数值

        WebServiceCall.callWebService(url, namespace, methodName,
                params, webServiceCallBack);
    }

    WebServiceCallBack webServiceCallBack = new WebServiceCallBack() {
        @Override
        public void callBack(SoapObject result) {
            // 处理 WebService 返回的数据
            if (result != null) {
                SoapObject detail = (SoapObject) result
                        .getProperty("getWeatherResult");
                StringBuilder sb = new StringBuilder();
                for (int i = 0; i < detail.getPropertyCount(); i++) {
                    sb.append(detail.getProperty(i)).append("\r\n");
                }
                textview.setText(sb.toString()); //在 TextView 中显示
            } else {
                Toast.makeText(MainActivity.this, "获取 WebService 数据错误",
                        Toast.LENGTH_SHORT).show();
            }
        }
    };
}
```

9.3.4 运行测试

注意：要为本项目设置访问互联网的权限。在 AndroidManifest.xml 中添加以下代码：

```
<uses-permission android:name="android.permission.INTERNET" />
```

天气预报项目的运行结果如图 9-2 所示。

图 9-2　天气预报的运行结果

9.4　网络编程概述

在 Android 网络编程中，Android 设备通常是客户端，通过相应的协议连接到服务器端，服务器端通常是运行 Linux 系统或 Windows 系统的计算机。在某些特殊需求中，Android 设备也可能作为服务器端，为其他客户端提供服务。与其他网络编程一样，Android 网络编程可以分为三大类：Socket 编程、HTTP 编程和 WebService 编程。

- Socket 编程：基于 TCP/IP 协议的编程，可以分为三种，即 TCP、UDP 和 UDP 组播，Socket 编程传输的数据是自定义格式的数据。
- HTTP 编程：基于 HTTP 协议的编程，服务器端是一个 Web 服务器。HTTP 编程传输的数据可以是符合 HTML 协议的超文本数据，也可以是自定义格式的数据。
- WebService 编程：基于 HTTP 协议的编程，服务器端是一个具有 WebService 的 Web 服务器。WebService 编程传输的数据是符合 SOAP 协议的 XML 数据。本章的天气预报项目采用的就是 WebService 编程技术。

上述三种网络编程技术特点分明，需要读者根据项目的需求进行选择，三种网络编程技术的对比情况见表 9-2。

表 9-2　三种网络编程技术的对比情况

类　型	传输协议	数据协议	特　点
Socket 编程	TCP、UDP 和 UDP 组播	自定义格式	简单，没有数据格式标准，数据格式完全自定义，适用于小型系统
HTTP 编程	HTTP	HTML 或自定义格式	服务器端需要 Web 服务器，HTTP 的 80 口可以通过防火墙
WebService 编程	HTTP	SOAP 协议	复杂，数据格式标准化，适合于标准化的应用；服务器端需要 Web 服务器，HTTP 的 80 口可以通过防火墙

9.5　网络编程综合项目

本节将通过案例综合应用 Socket 编程、HTTP 编程和 WebService 编程三种网络编程技术。

创建一个名为 NetClient 的项目，作为三种协议的客户端，将一组温度和湿度传感器的实时测量数据上传到服务器，服务器将温度和湿度数据保存在文本文件或数据库中。

客户端开发工具为 Android Studio，界面如图 9-3 所示。

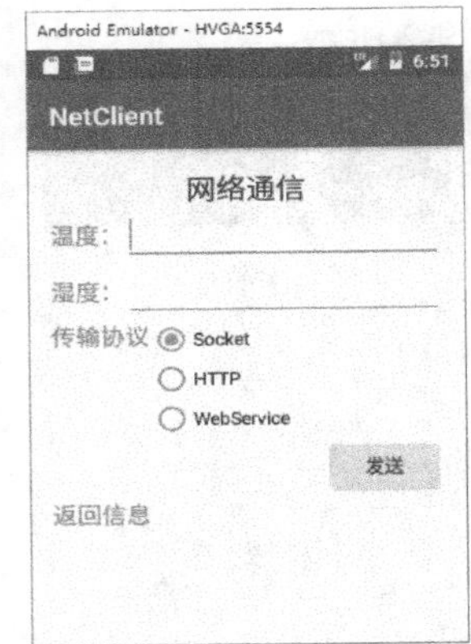

图 9-3　NetClient 项目的客户端界面

9.5.1　客户端界面

（1）主界面的布局文件为 activity_main.xml，代码如下：

```
<LinearLayout xmlns:android="http://schemas.android.com/apk/res/android"
    xmlns:tools="http://schemas.android.com/tools"
    android:id="@+id/LinearLayout1"
    android:layout_width="match_parent"
    android:layout_height="match_parent"
    android:orientation="vertical"
    android:paddingBottom="@dimen/activity_vertical_margin"
    android:paddingLeft="@dimen/activity_horizontal_margin"
    android:paddingRight="@dimen/activity_horizontal_margin"
    android:paddingTop="@dimen/activity_vertical_margin"
    tools:context=".MainActivity" >

    <TextView
        android:id="@+id/tv_title"
        android:layout_width="wrap_content"
        android:layout_height="wrap_content"
        android:layout_gravity="center"
        android:text="网络通信"
        android:textAppearance="?android:attr/textAppearanceLarge" />

    <LinearLayout
        android:layout_width="match_parent"
        android:layout_height="wrap_content" >

        <TextView
            android:id="@+id/textView1"
            android:layout_width="wrap_content"
            android:layout_height="wrap_content"
            android:text="温度："
            android:textAppearance="?android:attr/textAppearanceMedium" />

        <EditText
            android:id="@+id/etTemperature"
            android:layout_width="wrap_content"
            android:layout_height="wrap_content"
            android:layout_weight="1"
            android:ems="10"
```

```
            android:inputType="numberDecimal" >

            <requestFocus />
        </EditText>
    </LinearLayout>

    <LinearLayout
        android:layout_width="match_parent"
        android:layout_height="wrap_content" >

        <TextView
            android:id="@+id/textView2"
            android:layout_width="wrap_content"
            android:layout_height="wrap_content"
            android:text="湿度："
            android:textAppearance="?android:attr/textAppearanceMedium" />

        <EditText
            android:id="@+id/etHumidity"
            android:layout_width="wrap_content"
            android:layout_height="wrap_content"
            android:layout_weight="1"
            android:ems="10"
            android:inputType="numberDecimal" />
    </LinearLayout>

    <LinearLayout
        android:layout_width="match_parent"
        android:layout_height="wrap_content" >

        <TextView
            android:id="@+id/textView3"
            android:layout_width="wrap_content"
            android:layout_height="wrap_content"
            android:text="传输协议"
            android:textAppearance="?android:attr/textAppearanceMedium" />

        <RadioGroup
            android:id="@+id/radioGroup1"
            android:layout_width="wrap_content"
            android:layout_height="wrap_content" >

            <RadioButton
                android:id="@+id/radio0"
                android:layout_width="wrap_content"
                android:layout_height="wrap_content"
                android:checked="true"
                android:text="Socket" />

            <RadioButton
                android:id="@+id/radio1"
                android:layout_width="wrap_content"
                android:layout_height="wrap_content"
                android:text="HTTP" />

            <RadioButton
                android:id="@+id/radio2"
                android:layout_width="wrap_content"
```

```
                    android:layout_height="wrap_content"
                    android:text="WebService" />
            </RadioGroup>

        </LinearLayout>

        <Button
            android:id="@+id/btn_send"
            android:layout_width="wrap_content"
            android:layout_height="wrap_content"
            android:layout_gravity="right"
            android:text="发送" />

        <TextView
            android:id="@+id/tv_message"
            android:layout_width="wrap_content"
            android:layout_height="wrap_content"
            android:text="返回信息"
            android:textAppearance="?android:attr/textAppearanceMedium" />

    </LinearLayout>
```

（2）Activity 类的代码如下：

```
package com.example.administrator.netclient;

import android.support.v7.app.AppCompatActivity;
import java.io.BufferedReader;
import java.io.InputStream;
import java.io.InputStreamReader;
import java.io.OutputStream;
import java.io.OutputStreamWriter;

public class MainActivity extends AppCompatActivity implements OnClickListener {
    private Button btnSend;
    private EditText etTemperature;
    private EditText etHumudity;
    private TextView tvMessage;
    private RadioButton rb0;
    private RadioButton rb1;
    private RadioButton rb2;

    @Override
    protected void onCreate(Bundle savedInstanceState) {
        super.onCreate(savedInstanceState);
        setContentView(R.layout.activity_main);

        etTemperature = (EditText) this.findViewById(R.id.etTemperature);
        etHumudity = (EditText) this.findViewById(R.id.etHumidity);
        btnSend = (Button) this.findViewById(R.id.btn_send);
        tvMessage = (TextView) this.findViewById(R.id.tv_message);
        rb0 = (RadioButton) this.findViewById(R.id.radio0);
        rb1 = (RadioButton) this.findViewById(R.id.radio1);
        rb2 = (RadioButton) this.findViewById(R.id.radio2);

        btnSend.setOnClickListener(this);
    }

    @Override
    public void onClick(View v) {
```

```
            final String temperature = etTemperature.getText().toString();
            final String humudity = etHumudity.getText().toString();

            if (rb0.isChecked()) {
                // Socket，详见 9.5.2 节
            } else if (rb1.isChecked()) {
                // HTTP，详见 9.5.3 节
            } else if (rb2.isChecked()) {
                // WebService，详见 9.5.4 节
            }
        }
    }
```

注意：本项目需要访问互联网，因此在 AndroidManifest.xml 中设置访问网络的权限，代码如下：

```
<uses-permission android:name="android.permission.INTERNET" />
```

9.5.2 Socket 编程

首先，编写一个服务器端的 Socket 程序，开发工具为 Eclipse；然后，在上述客户端程序中添加 TCP 客户端代码，将数据通过 TCP 协议发送到服务器端进行保存。

说明：客户端开发工具为 Android Studio；Socket 服务器端开发工具为 Eclipse；HTTP 服务器端开发工具常用 Tomcat+浏览器（过去曾用 MyEclipse+Tomcat+浏览器）；WebService 服务器端开发工具为 MyEclipse+Tomcat+浏览器。

1．服务器端

创建一个名为 SocketServer 的普通 Java Application 项目，代码如下：

```
import java.io.BufferedReader;
import java.io.File;
import java.io.IOException;
import java.io.InputStreamReader;
import java.io.PrintWriter;
import java.io.RandomAccessFile;
import java.net.ServerSocket;
import java.net.Socket;
import java.net.SocketException;

public class SocketServer {
    private int port = 9807;    // 监听的端口号
    private ServerSocket serverSocket;

    public static void main(String[] args) throws IOException {
        new SocketServer();
    }

    public SocketServer() throws IOException {
        System.out.println("服务器启动，在端口" + port + "监听多个客户。");
        service();
    }

    public void service() {
        try {
            serverSocket = new ServerSocket(port);
        } catch (IOException e) {
            e.printStackTrace();
        }
```

```
        try {
            while (true) { // 循环监听，每个客户的请求都用一个新的线程处理
                new ServerThread(serverSocket.accept(), serverSocket).start();
            }
        } catch (SocketException e) {
            System.out.println("将退出服务器端程序，原因之一是被客户端强制中止。");
        } catch (IOException e) {
            e.printStackTrace();
        }
    }
}

class ServerThread extends Thread {
    Socket socket; // 保存与本线程相关的 Socket 对象
    ServerSocket serverSocket; // 保存 ServerSocket，退出时关闭它

    public ServerThread(Socket socket, ServerSocket serverSocket) { // 构造函数
        this.socket = socket; // 初始化 socket 变量
        this.serverSocket = serverSocket; // 初始化 serverSocket 变量
    }

    public void run() { // 线程主体
        try {
            System.out.println("接受新连接: " + socket.getInetAddress() + ":"
                    + socket.getPort());
            BufferedReader br = new BufferedReader(new InputStreamReader(
                    socket.getInputStream()));
            PrintWriter pw = new PrintWriter(socket.getOutputStream(), true);

            String txt = "";
            String line = null;
            while ((line = br.readLine()) != null) {
                if(line.length()==0){
                    // 收到一个空白行，表示数据结束
                    break;
                }
                txt += line;
            }
            // 将收到的数据写入文本文件
            writeTextFile("data_tcp.txt",txt);
            pw.println("Received by tcp: " + txt);
        } catch (IOException e) {
            e.printStackTrace();
        } finally {
            if (socket != null) {
                try {
                    System.out
                            .println("客户" + socket.getInetAddress() + "断开连接。");
                    socket.close(); // 结束与该客户的连接
                } catch (IOException e) {
                    e.printStackTrace();
                }
            }
        }
    }

    private static void writeTextFile(String fileName, String txt) {
```

```
            RandomAccessFile raf;
            try {
                // 写入文本文件。在实际应用中，一般写入数据库
                raf = new RandomAccessFile(new File("d:/" + fileName), "rw");
                raf.seek(raf.length());
                raf.writeBytes(txt + "\r\n");
                raf.close();
                System.out.println("数据已保存到文件中:" + txt);
            } catch (Exception e) {
                e.printStackTrace();
            }
        }
}
```

2. 客户端

TCP 客户端的代码如下：

```
private void sendMsgByTcp(String temperature, String humudity) {
    String host = "10.30.12.59"; // 服务器的地址
    int port = 9807; // 请求的端口号，与服务器的监听端口相同
    Socket socket;
    String data = "t=" + temperature + ", h=" + humudity + ";\r\n\r\n"; // 后接一个空白行，表示数据结束
    try {
        socket = new Socket(host, port);// 建立连接
        OutputStream socketOut = socket.getOutputStream();
        InputStream socketIn = socket.getInputStream();
        BufferedReader br = new BufferedReader(new InputStreamReader(
                socketIn));// 获得输入流
        PrintWriter pw = new PrintWriter(socketOut, true); // 获得输出流

        pw.println(data);

        String txt = "";
        String line;
        while ((line = br.readLine()) != null) {
            txt += line;
        }
        socket.close();

        Message msg = new Message();
        Bundle b = new Bundle();
        b.putString("txt", txt);
        msg.setData(b);
        showMessage.sendMessage(msg);
    } catch (Exception e) {
        e.printStackTrace();
    }
}
```

由于多线程的原因，子线程与主线程进行通信要通过 Handler()方法实现。

```
@SuppressLint("HandlerLeak")
Handler showMessage = new Handler() {
    // 子线程与主线程进行通信
    @Override
    public void handleMessage(Message msg) {
        super.handleMessage(msg);
        Bundle b = msg.getData();
        String line = b.getString("txt");
```

```
            tvMessage.setText(line);
        }
    };
```

调用 showMessage()方法的代码如下：

```
@Override
public void onClick(View v) {
    final String temperature = etTemperature.getText().toString();
    final String humudity = etHumudity.getText().toString();

    if (rb0.isChecked()) {
        // Socket，客户端必须采用多线程
        new Thread(new Runnable() {
            @Override
            public void run() {
                sendMsgByTcp(temperature, humudity);
            }
        }).start();
    } else if (rb1.isChecked()) {
        // HTTP，详见 9.5.3 节
    } else if (rb2.isChecked()) {
        // WebService ，详见 9.5.4 节
    }
}
```

3．运行

（1）在 Windows 系统中运行 Java Application 服务器端，显示结果如下：

```
服务器启动，在端口 9807 监听多个客户。
```

（2）在 Android 虚拟设备中运行客户端。请注意，Android 虚拟设备必须能够访问 host 指定的服务器 IP 地址，本例的服务器 IP 地址为“10.30.12.59”。

（3）检查服务器中的“d:/data_tcp.txt”文件，该文件包含客户端发送数据的记录。

运行结果如图 9-4 所示。

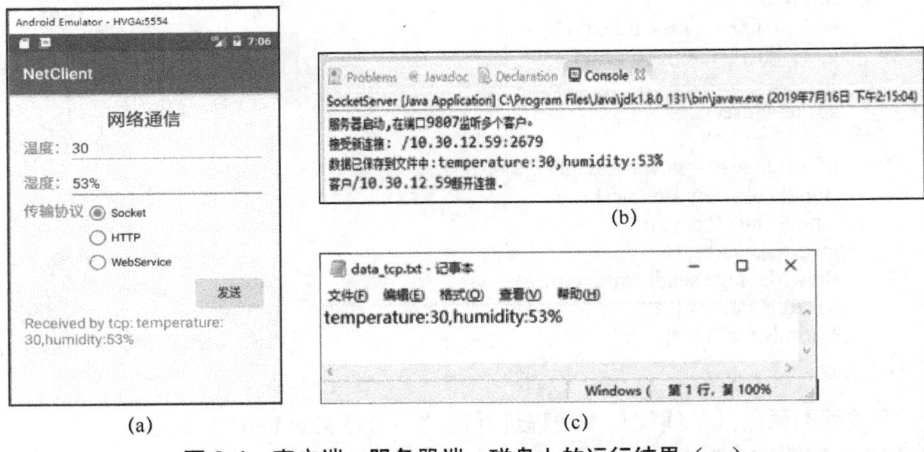

图 9-4 客户端、服务器端、磁盘上的运行结果（一）

9.5.3 HTTP 编程

本节采用 HTTP 编程，完成与 Socket 编程相同的功能。

首先，使用记事本工具编写一个 Web 服务器端程序；然后，在客户端程序中编写 HTTP 的客户端代码，并将 HTTP 格式的数据（即 POST 的数据）通过 HTTP 协议发送到服务器端进行保存。

1．服务器端

先安装 Tomcat，然后用记事本工具编写一个 Web 服务器端程序（http_server.jsp），存放路径为"Tomcat 安装路径\webapps\ROOT"。代码如下：

```
<%@ page language="java" contentType="text/html; charset=UTF-8" pageEncoding="UTF-8"%>
<%@ page import="java.io.*,javax.servlet.*,javax.servlet.http.*"%>
<html>
<body>
HTTP Server
<%
    String param1 = request.getParameter("param1");
    String param2 = request.getParameter("param2");
    if(param1!=null && param2!=null){
        String txt = "temperature:" + param1 + ", humidity:" + param2 + ";";
        out.print("Received by http:" + txt);
        RandomAccessFile raf;
        try {
            // 写入文本文件。在实际应用中，一般写入数据库
            raf = new RandomAccessFile(new File("d:/data_http.txt"), "rw");
            raf.seek(raf.length());
            raf.writeBytes(txt + "\r\n");
            raf.close();
            System.out.println("数据已保存到文件中:" + txt);
        } catch (Exception e) {
            e.printStackTrace();
        }
    }
%>
</body>
</html>
```

2．客户端

HTTP 客户端的代码如下：

```
private void sendMsgByHttp(String temperature, String humudity) {
    String host = "10.30.12.59"; // 服务器的地址
    int port = 8080; // 请求的端口号，与服务器的监听端口相同
    String path = "/Http_Server.jsp";
    URLConnection conn;

    try {
        OutputStreamWriter wr = null;
        BufferedReader rd = null;
        try {
            // 准备要发送的数据，数据的格式为 key1=value1&key2=value2
            // 对特殊符号（如空格）和中文需要编码，然后才能发送
            String data = "param1=" + temperature;
            data += "&param2=" + humudity;

            // 建立连接
            String str = "http://" + host + ":" + port + path;
            URL url = new URL(str);
            conn = url.openConnection();

            conn.setDoOutput(true); //将数据写入 URL 连接
```

```
                // 提交用户数据
                wr = new OutputStreamWriter(conn.getOutputStream());
                wr.write(data);
                wr.flush();
                System.out.println("向" + str + "发送数据" + data);

                // 获得网页的返回结果
                rd = new BufferedReader(new InputStreamReader(
                        conn.getInputStream()));
                String txt = "";
                String line = "";
                while ((line = rd.readLine()) != null) {
                    txt += line;
                }
                Message msg = new Message();
                Bundle b = new Bundle();
                b.putString("txt", txt);
                msg.setData(b);
                showMessage.sendMessage(msg);
                // tvMessage.setText(line);
            } catch (Exception e) {
                e.printStackTrace();
            } finally {
                wr.close();

            }
        } catch (Exception e) {
            e.printStackTrace();
        }
}
```

调用 showMessage()方法的代码如下：

```
@Override
public void onClick(View v) {
    final String temperature = etTemperature.getText().toString();
    final String humudity = etHumudity.getText().toString();

    if (rb0.isChecked()) {
        // Socket，在 9.5.2 节中已完成
    } else if (rb1.isChecked()) {
        // HTTP，客户端必须采用多线程
        new Thread(new Runnable() {
            @Override
            public void run() {
                sendMsgByHttp(temperature, humudity);
            }
        }).start();
    } else if (rb2.isChecked()) {
        // WebService，详见 9.5.4 节
    }
}
```

3．运行

（1）运行“Tomcat 安装路径\bin\startup.bat”文件。

（2）打开浏览器，输入 URL，测试 Web 服务器端能否正常运行。代码如下：

```
http://10.30.12.59:8080/Http_Server.jsp
```

（3）在 Android 虚拟设备中运行客户端，输入温度与湿度两个参数值，单击“发送”按钮。观察客户端接收信息的情况，以及检查 Web 服务器中的“d:/data_http.txt”文件，该文件包含客户端发送数据的记录。

运行结果如图 9-5 所示。

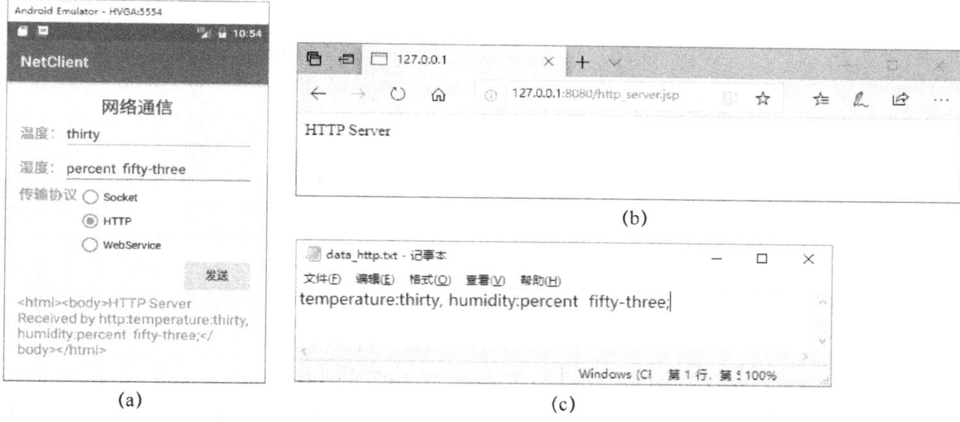

图 9-5　客户端、服务器端、磁盘上的运行结果（二）

9.5.4　WebService 编程

本节采用 WebService 编程，完成与 Socket 编程相同的功能。

首先，在 MyEclipse 环境中，使用 Java EE 编写一个 Web 服务器程序；然后，在客户端程序中编写 WebService 的客户端代码，并将 WebService 格式的数据（即 SOAP 数据）通过 HTTP 协议发送到服务器端进行保存。

Web 服务器需要 axis2.war 的支持，文件参见项目素材。

1．服务器端

在 Eclipse 环境中，导入 axis2.war，并将项目命名为 WsServer，如图 9-6 所示。

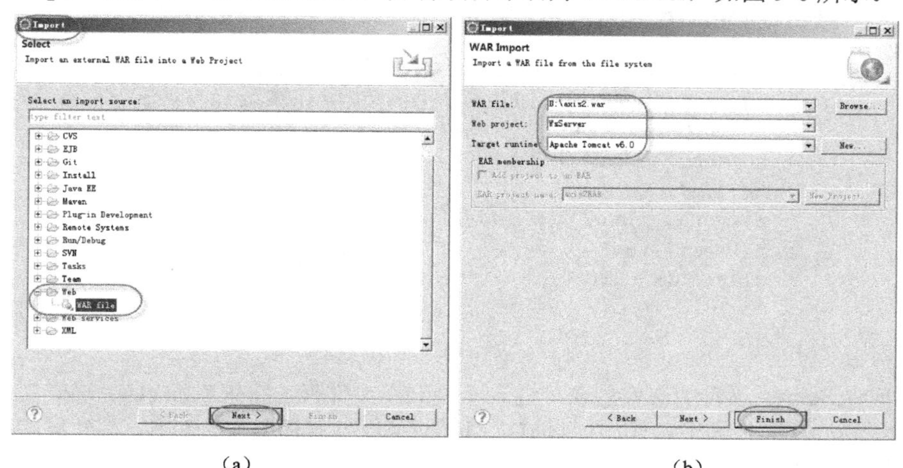

图 9-6　导入 axis2.war 并命名项目

将 WsServer 项目添加到 Tomcat 服务器，并通过地址测试该项目能否正常运行，代码如下：

```
http://127.0.0.1:8080/WsServer/
```

正常的运行结果如图 9-7 所示。

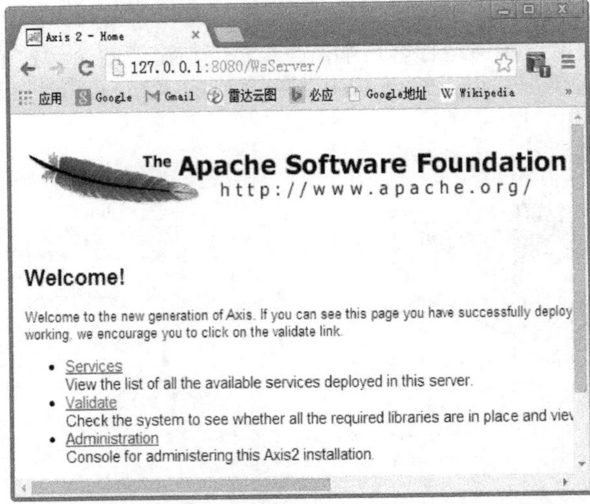

图 9-7　WsServer 项目运行结果

在本例中，对 axis2.war 不做任何配置，而是将一个简单的 POJO（Plain Ordinary Java Object）发布为 WebService，即将 pojo 中所有的 public 方法自动发布为 WebService 方法，代码如下：

```
import java.io.File;
import java.io.RandomAccessFile;

public class MyServer {
    public String saveData(String humudity, String temperature) {
        // WebService 的功能为将收到的数据保存到文件中
        String txt = "t=" + temperature + "; h=" + humudity + ";";
        writeTextFile("data_ws.txt", txt);
        return "Received by ws:" + txt;
    }

    private static void writeTextFile(String fileName, String txt) {
        RandomAccessFile raf;
        try {
            // 写入文本文件。在实际应用中，一般写入数据库
            raf = new RandomAccessFile(new File("d:/" + fileName), "rw");
            raf.seek(raf.length());
            raf.writeBytes(txt + "\r\n");
            raf.close();
            System.out.println("数据已保存到文件中:" + txt);
        } catch (Exception e) {
            e.printStackTrace();
        }
    }
}
```

打开 WsServer 项目的文件列表，进入“build”文件夹，将生成的.class 格式文件复制到“WEB-INF/pojo”文件夹中，刷新项目后，项目的文件列表如图 9-8 所示。重新启动项目，然后访问以下地址：

```
http://127.0.0.1:8080/WsServer/services/MyServer?wsdl
```

这样，pojo 便发布为 WebService，运行效果如图 9-9 所示。

注意：每次修改“pojo”源代码文件后，都要将.class 格式文件重新复制到“pojo”文件夹中，并刷新项目，然后再重新启动项目。

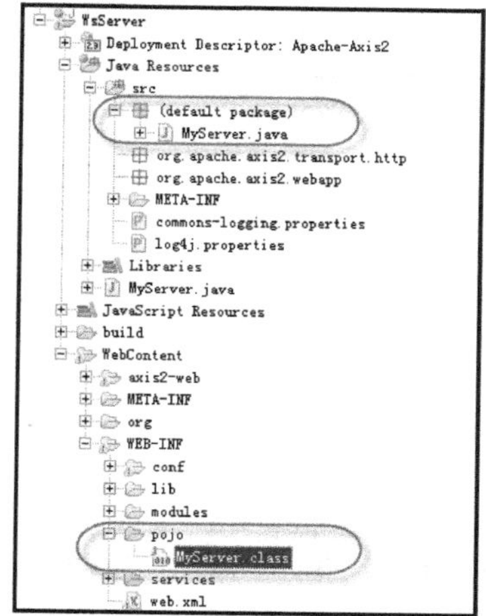

图 9-8　WebService 项目文件列表

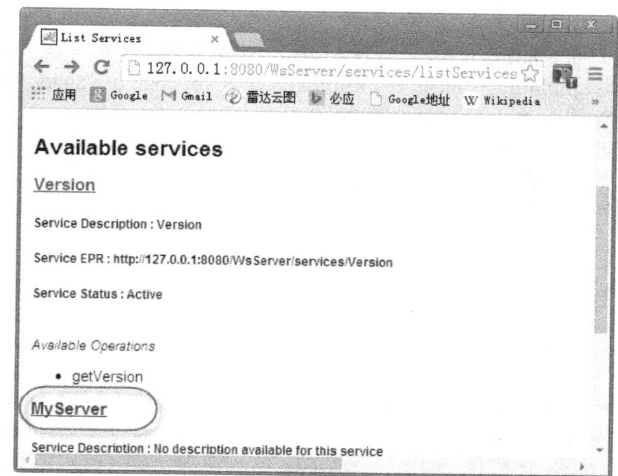

图 9-9　pojo 发布为 WebService

单击图 9-9 中的超链接（MyServer），将显示 MyServer 的 WSDL 定义，代码如下：

```
<wsdl:definitions xmlns:wsdl="http://schemas.xmlsoap.org/wsdl/" xmlns:ns1="http://org.apache.axis2/xsd" xmlns:ns="http://ws.apache.org/axis2" xmlns:wsaw="http://www.w3.org/2006/05/addressing/wsdl" xmlns:http="http://schemas.xmlsoap.org/wsdl/http/" xmlns:xs="http://www.w3.org/2001/XMLSchema" xmlns:mime="http://schemas.xmlsoap.org/wsdl/mime/" xmlns:soap="http://schemas.xmlsoap.org/wsdl/soap/" xmlns:soap12="http://schemas.xmlsoap.org/wsdl/soap12/" targetNamespace="http://ws.apache.org/axis2">
<wsdl:types>
<xs:schema attributeFormDefault="qualified" elementFormDefault="unqualified" targetNamespace="http://ws.apache.org/axis2">
<xs:element name="saveData">
<xs:complexType>
<xs:sequence>
<xs:element minOccurs="0" name="humudity" nillable="true" type="xs:string"/>
<xs:element minOccurs="0" name="temperature" nillable="true" type="xs:string"/>
</xs:sequence>
</xs:complexType>
</xs:element>
<xs:element name="saveDataResponse">
<xs:complexType>
<xs:sequence>
<xs:element minOccurs="0" name="return" nillable="true" type="xs:string"/>
</xs:sequence>
</xs:complexType>
</xs:element>
</xs:schema>
</wsdl:types>
<wsdl:message name="saveDataRequest">
<wsdl:part name="parameters" element="ns:saveData"/>
</wsdl:message>
<wsdl:message name="saveDataResponse">
<wsdl:part name="parameters" element="ns:saveDataResponse"/>
</wsdl:message>
<wsdl:portType name="MyServerPortType">
<wsdl:operation name="saveData">
<wsdl:input message="ns:saveDataRequest" wsaw:Action="urn:saveData"/>
<wsdl:output message="ns:saveDataResponse" wsaw:Action="urn:saveDataResponse"/>
</wsdl:operation>
```

```
</wsdl:portType>
<wsdl:binding name="MyServerSoap11Binding" type="ns:MyServerPortType">
<soap:binding transport="http://schemas.xmlsoap.org/soap/http" style="document"/>
<wsdl:operation name="saveData">
<soap:operation soapAction="urn:saveData" style="document"/>
<wsdl:input>
<soap:body use="literal"/>
</wsdl:input>
<wsdl:output>
<soap:body use="literal"/>
</wsdl:output>
</wsdl:operation>
</wsdl:binding>
<wsdl:binding name="MyServerSoap12Binding" type="ns:MyServerPortType">
<soap12:binding transport="http://schemas.xmlsoap.org/soap/http" style="document"/>
<wsdl:operation name="saveData">
<soap12:operation soapAction="urn:saveData" style="document"/>
<wsdl:input>
<soap12:body use="literal"/>
</wsdl:input>
<wsdl:output>
<soap12:body use="literal"/>
</wsdl:output>
</wsdl:operation>
</wsdl:binding>
<wsdl:binding name="MyServerHttpBinding" type="ns:MyServerPortType">
<http:binding verb="POST"/>
<wsdl:operation name="saveData">
<http:operation location="saveData"/>
<wsdl:input>
<mime:content type="application/xml" part="parameters"/>
</wsdl:input>
<wsdl:output>
<mime:content type="application/xml" part="parameters"/>
</wsdl:output>
</wsdl:operation>
</wsdl:binding>
<wsdl:service name="MyServer">
<wsdl:port name="MyServerHttpSoap11Endpoint" binding="ns:MyServerSoap11Binding">
<soap:address location="http://192.168.1.101:8080/WsServer/services/
MyServer.MyServerHttpSoap11Endpoint/"/>
</wsdl:port>
<wsdl:port name="MyServerHttpSoap12Endpoint" binding="ns:MyServerSoap12Binding">
<soap12:address location="http://192.168.1.101:8080/WsServer/services/
MyServer.MyServerHttpSoap12Endpoint/"/>
</wsdl:port>
<wsdl:port name="MyServerHttpEndpoint" binding="ns:MyServerHttpBinding">
<http:address location="http://192.168.1.101:8080/WsServer/services/
MyServer.MyServerHttpEndpoint/"/>
</wsdl:port>
</wsdl:service>
</wsdl:definitions>
```

上述代码表示 MyServer 发布成功。

2．客户端

（1）将天气预报项目中的两个源代码文件（WebServiceCall、WebServiceCallBack）及第

三方 ksoap2.jar 包复制到本项目。

客户端的代码如下：

```
private void sendMsgByWs(String temperature, String humudity) {
        String url = "http://192.168.1.101:8080/WsServer/services/MyServer";
        String namespace = "http://ws.apache.org/axis2";
        String methodName = "saveData";
        HashMap<String, String> params = new HashMap<String, String>();
        params.put("humudity", humudity);
        params.put("temperature", temperature);

        WebServiceCall.callWebService(url, namespace, methodName, params,
                webServiceCallBack, false);
}
```

（2）完善 WebServiceCallBack 类中 callBack()方法的代码，其作用与 Handler()方法相似，代码如下：

```
WebServiceCallBack webServiceCallBack = new WebServiceCallBack() {
        @Override
        public void callBack(SoapObject result) {
                if (result != null) {
                        String msg = "";
                        for (int i = 0; i < result.getPropertyCount(); i++) {
                                msg += result.getProperty(i) + "\r\n";
                        }
                        tvMessage.setText(msg);
                } else {
                        Toast.makeText(MainActivity.this, "获取 WebService 数据错误",
                                        Toast.LENGTH_SHORT).show();
                }
        }
};
```

（3）调用 WebServiceCallBack()方法的代码如下：

```
@Override
public void onClick(View v) {
        final String temperature = etTemperature.getText().toString();
        final String humudity = etHumudity.getText().toString();

        if (rb0.isChecked()) {
                // Socket，详见 9.5.2 节
        } else if (rb1.isChecked()) {
                // HTTP，详见 9.5.3 节
        } else if (rb2.isChecked()) {
                // WebService，不必用多线程，因为在方法内部采用回调技术
                sendMsgByWs(temperature, humudity);
        }
}
```

3．运行

运行步骤和结果与 9.5.3 节相似，此处不再赘述。

9.6　实训：完善天气预报项目

为本章的天气预报项目增加以下功能。

- 显示我国所有省、直辖市、自治区、特别行政区的列表，供用户选择。
- 若用户选择我国的省、直辖市、自治区、特别行政区后，且该省级行政单位下包含地

级行政区（如地级市），请显示相关的地级行政区列表，供用户选择。

- 显示用户选择的地级行政区的天气预报。

9.7 实训：词典项目

利用网络编程中的 WebService 技术，开发一个名为 Dictionary 的项目，实现英汉单词互译的功能。

本项目的界面与运行效果如图 9-10 所示。

（a）

（b）

图 9-10 Dictionary 项目界面与运行效果

（1）本项目只有一个界面，包括三个 TextView 控件，一个 EditText 控件，一个 Button 控件。主界面的布局文件为 activity_main.xml，请扫描下方的二维码进行查看。

（2）在 Android 系统中访问 WebService 需要第三方 jar 包，本项目使用的第三方 jar 包为 ksoap2-android-assembly-2.5.4-jar-with-dependencies.jar，将其复制到项目的 libs 文件夹中。

（3）编写 WebServiceCall 类。利用 ksoap2 提供的功能，可以编写一个调用 WebService 的 WebServiceCall 类，请扫描下方的二维码查看 WebServiceCall.java 代码。

（4）编写 WebServiceCallBack 接口。如果想调用 WebServiceCall 类，还需要一个 WebServiceCallBack 接口，其作用与 Handler()方法类似，请扫描下方的二维码查看 WebServiceCallBack.java 代码。

（5）请扫描下方的二维码，浏览 MainActivity.java 代码。

activity_main.xml

WebServiceCall.java

WebServiceCallBack.java

MainActivity.java

9.8 作业

1．网络编程主要有哪几种？请分别叙述各自的特点。

2．本章介绍的天气预报服务是通过 WebService 提供的，WebService 有哪些方法？其功能分别是什么？

3．网络编程与多线程有什么关系？

4．在天气预报项目中，WebServiceCallBack 接口的作用是什么？

第10章 嵌入式开发：串口编程——读卡器项目

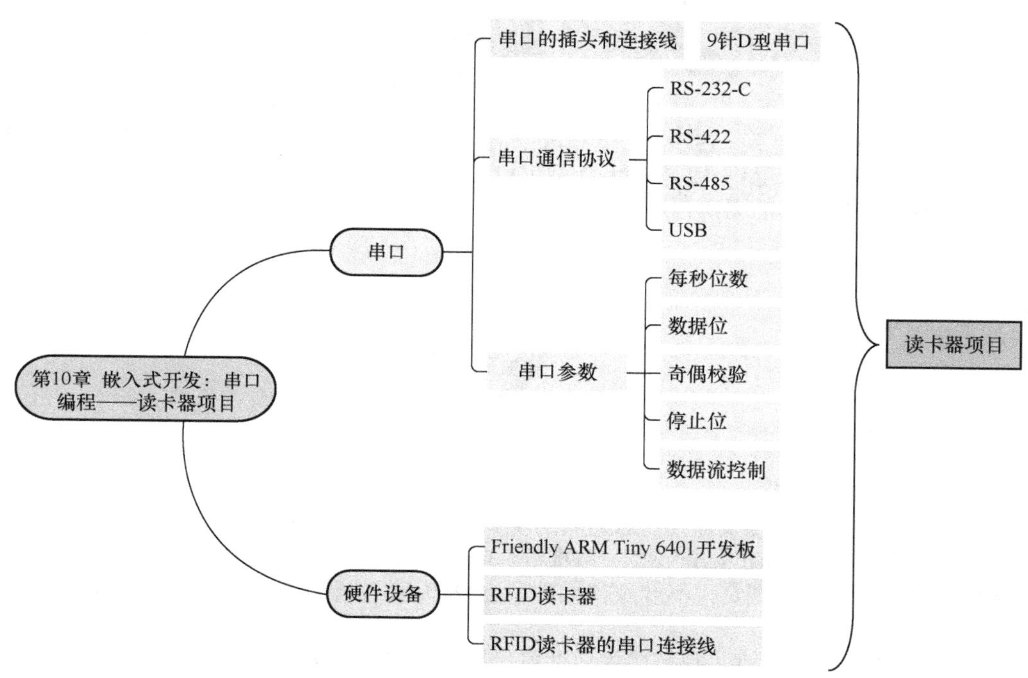

本章知识要点思维导图

Android 系统的嵌入式开发主要研究 Android 硬件设备、传感器、执行器之间的交互行为。通过传感器获取相关信息，经由装载嵌入式系统的 Android 硬件设备处理，最终由执行器控制相关设备。

其中，传感器包括条码阅读器、RFID 读卡器、温度传感器、湿度传感器、光电传感器、火焰传感器、可燃气体检测探头等。

执行器包括继电器开关、执行机构等。

本章介绍的程序源代码适用于 Friendly ARM Tiny 6401 开发板和 Friendly ARM 210 开发板。

10.1 需求分析

开发一个读卡器项目，用于读取 RFID 卡的卡号。

10.2 串口介绍

串行接口（Serial Port）简称“串口”，其特点是数据按顺序逐位传送，通信线路简单，只需一对传输线就可以实现双向通信，从而极大地降低了成本。

1．串口的插头和连接线

传统的串口标准为 22 根线，采用标准 25 针 D 型插头/插座，后来简化为 9 针 D 型插头/插座，25 针 D 型插头/插座已经被淘汰。台式计算机一般有一个或两个串口，在计算机主机箱体背后可以看到 9 针 D 型串口，如图 10-1 所示。

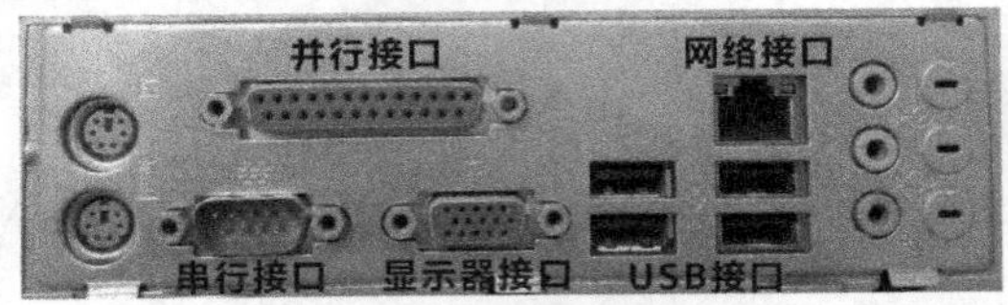

图 10-1 台式计算机的串口及串口连接线

笔记本电脑一般不配备串口，但是可以通过 USB-串口转换线引出串口，USB-串口转换线的一端是 USB 接口，另一端是 9 针 D 型串口，如图 10-2 所示。使用 USB-串口转换线时需要在计算机上安装相应的驱动程序。

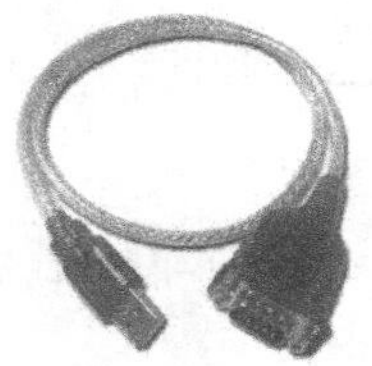

图 10-2 USB-串口转换线

2．串口通信协议

串口按照电气标准及通信协议进行分类，可分为 RS-232-C 标准、RS-422 标准、RS-485 标准、USB （Universal Serial Bus）标准等。RS-232-C 标准、RS-422 标准与 RS-485 标准只对接口的电气特性进行规定，不涉及接插件、电缆和通信协议。USB 标准是新型接口标准，主要用于高速传输数据。

（1）RS-232-C 标准，也被称为标准串口，是一种较常用的串口。它的全名为“数据终端设备（DTE）和数据通信设备（DCE）之间串行二进制数据交换接口技术标准”。它的最大传送距离为 15m，最高传输速率为 20kb/s。RS-232-C 标准是为点对点（即只用一对收、发设备）通信而设计的，主要用于本地设备之间进行通信。

（2）RS-422 标准。为改进 RS-232-C 标准的通信距离短、速率低的缺点，RS-422 标准定义了一种平衡通信接口，将传输速率提高到 10Mb/s，传输距离延长到 4000ft（速率低于 100kb/s 时），并允许在一条平衡总线上最多连接 10 个接收器。RS- 422 标准是一种单机发送、多机接收的单向、平衡传输规范，被命名为 TIA/EIA-422-A 标准。

（3）RS-485 标准。为扩展应用范围，EIA 于 1983 年在 RS-422 标准的基础上制定了 RS-485 标准，该标准新增了多点、双向通信能力，即允许多个发送器连接到同一条总线上；此外，还增加了发送器的驱动能力和冲突保护特性，扩展了总线共模范围，被命名为 TIA/EIA-485-A 标准。

（4）USB 标准，即通用串行总线标准，是计算机中广泛应用的接口规范。在计算机主板上，USB 端口为 4 针接口，中间的 2 针用于传输数据，两边的 2 针用于向外部设备供电。USB 接口速度快、连接简单、不需要外接电源、支持热插拔。USB 标准下的传输速度为 12Mbps，

USB 2.0 标准下的传输速度可以达到 480Mbps。采用 USB 标准的电缆，其最大长度为 5m，电缆内含 4 条线，即 2 条信号线和 2 条电源线，可提供+5V 电压，采用 USB 标准，最多可以串接 127 台设备。

3. 串口参数

为了使串口能够正确通信，必须对串口的参数进行设置。如图 10-3 所示为使用 Windows 系统的超级终端设置串口通信的参数。

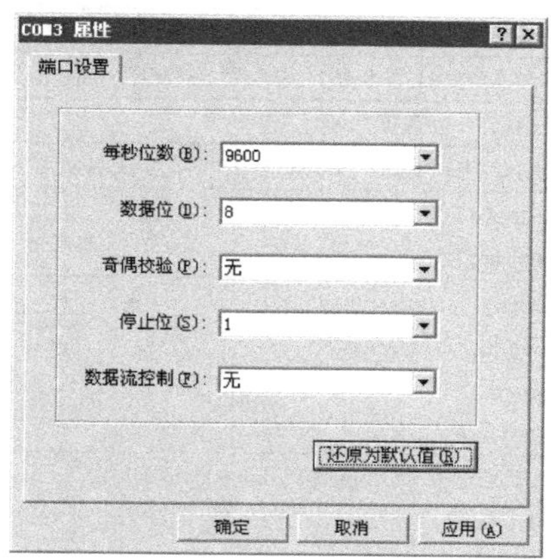

图 10-3　使用 Windows 系统的超级终端设置串口通信的参数

如图 10-3 所示的参数均为默认值，开发人员通常会按照此参数进行设置，下面对各参数进行介绍。

（1）每秒位数。每秒位数即波特率，表示每秒传输的比特数，最常用的传输速率为 9600b/s，即每秒传输 1200B。

（2）数据位。在实际的通信过程中，计算机发送数据包的数据位不一定是 8 位，数据位参数需要根据实际需求确定。例如，在串口通信中，数据位可以选择 5 位、6 位、7 位或 8 位。

（3）奇偶校验。奇偶校验用于对串口通信进行简单检错。奇偶校验的主要校验内容包括：偶校验、奇校验、标记、空格等。用户也可以不使用奇偶校验，而使用其他方式进行数据校验。

（4）停止位。停止位用于对数据包进行标记，表明数据结束，停止位可以取 1 位、1.5 位或 2 位。

（5）数据流控制。通过串口传输数据时，由于计算机的处理速度或其他因素的影响，会发生丢失数据的现象，数据流控制用于解决此类问题。通过控制发送数据的速度，确保数据不会出现丢失现象。但是，目前几乎很少使用数据流控制，而是改由上层软件负责处理数据丢失问题。

10.3　实验设备

10.3.1　硬件设备

硬件设备如下。

（1）Friendly ARM Tiny 6401 开发板一块。

（2）RFID 读卡器一个。

（3）RFID 读卡器的串口连接线一条。

10.3.2　Friendly ARM Tiny 6410 简介

Friendly ARM Tiny 6410 开发板是一款以 ARM 11 芯片（三星 S3C6410）为主处理器的嵌

入式核心板，该 CPU 基于 ARM1176JZF-S 核设计，内部集成了强大的多媒体处理单元，支持 MPEG4、 H.264/H.263 等格式的视频文件的编码和解码，可同时输出至液晶显示屏和电视机屏幕进行显示；它还带有 3D 图形硬件加速器，帮助 OpenGL ES 1.1 & 2.0 加速渲染；另外，它还支持 2D 图形图像的翻转、平滑缩放等操作。Friendly ARM Tiny 6410 核心板的资源见表 10-1，Friendly ARM Tiny 6410 底板的资源见表 10-2。

表 10-1　Friendly ARM Tiny 6410 核心板的资源

项　目	说　明
CPU	三星 S3C6410，ARM1176JZF-S 核，主频为 533MHz，最高频率为 667MHz
DDR RAM	256MB DDR RAM，32 位数据总线
FLASH 存储	标配 256MB SLC NAND Flash，掉电非易失，可选 1GB SLC NAND Flash 或 2GB MLC NAND Flash
接口资源	2×60 pin 2.0mm space DIP connector 2×30 pin 2.0mm space DIP connector
主板资源	4×User Leds (Green) 10 pin 2.0mm space Jtag connector Reset button on board Supply Voltage from 2~6V
操作系统支持	Linux2.6.38 + Qtopia-2.2.0 + Qtopia4-Phone + QtE-4.7.0 Windows CE.NET 6.0(R3) Android 2.3.4 Ubuntu-0910

表 10-2　Friendly ARM Tiny 6410 底板的资源

接　口	说　明
三个 LCD 接口	接显示器
四线电阻触摸屏接口	
100MB 标准网络接口	有线网络
标准 DB9 五线串口	四路 TTL 串口
Mini USB 2.0 接口	
USB Host 1.1	
3.5mm 音频输入、输出口	
标准 TV-OUT 接口	电视视频接口
SD 卡座	
红外线接收器	
SDIO2 接口	可接 SD Wi-Fi
蜂鸣器	
I2C-EEPROM	
AD 可调电阻	
八个中断式按键	可以直接操作

10.3.3　Friendly ARM Tiny 6410 的串口编程

Friendly ARM Tiny 6410 开发板共有四个串口，其中，一个串口用于调试，用户可以使用剩余的三个串口，这三个串口的名称为：

```
/dev/s3c2410_serial1
/dev/s3c2410_serial2
/dev/s3c2410_serial3
```

1. 串口函数库

Friendly ARM Tiny 6401 开发板提供了一个串口函数库，即 libfriendlyarm-hardware.so。在

Friendly ARM Tiny 6401 开发板上进行串口编程必须使用串口函数库。该函数库的使用方法如下。

（1）定位到 Android 应用程序的目录文件夹，在该文件夹中依次创建 libs→armeabi 文件夹，然后将 libfriendlyrm-hardware.so 库文件复制到 armeabi 文件夹中。

（2）在 src 文件夹中创建 com.friendlyarm.AndroidSDK 包，然后在该包内创建一个 HardwareControler 类，代码如下：

```
package com.friendlyarm.AndroidSDK;

import android.util.Log;

public class HardwareControler {

    /* LED */
    static public native int setLedState(int ledID, int ledState);
    /* PWM */
    static public native int PWMPlay(int frequency);
    static public native int PWMStop();
    /* ADC */
    static public native int readADC();
    /* I2C */
    static public native int openI2CDevice();
    static public native int writeByteDataToI2C(int fd, int pos, byte byteData);
    static public native int readByteDataFromI2C(int fd, int pos);
    /*  通用接口  */
    /* Serial Port */
    static public native int openSerialPort(String devName, long baud,
            int dataBits, int stopBits);
    static public native int write(int fd, byte[] data);
    static public native int read(int fd, byte[] buf, int len);
    static public native int select(int fd, int sec, int usec);
    static public native void close(int fd);
    static {
        try {
            System.loadLibrary("friendlyarm-hardware");
        } catch (UnsatisfiedLinkError e) {
            Log.d("HardwareControler",
                    "libfriendlyarm-hardware library not found!");
        }
    }
}
```

最终，Android 程序的目录架构如图 10-4 所示。串口函数库的接口说明见表 10-3。

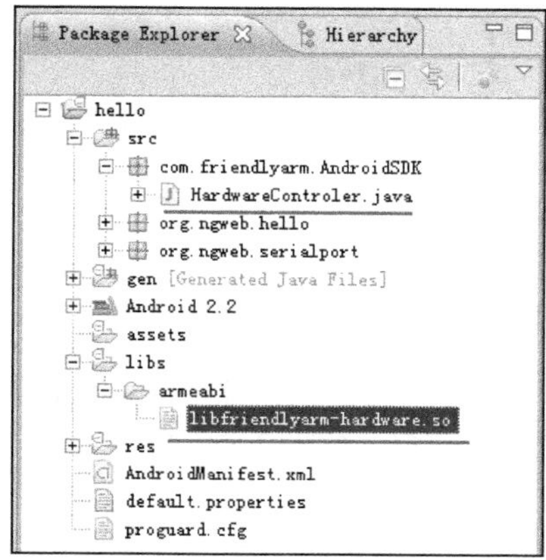

图 10-4　Android 程序的目录架构

表 10-3　串口函数库的接口说明

接口名称	参数与返回值说明		功能说明
int openSerialPort (String devName, long baud, int dataBits, int stopBits)	参数说明	（1）devName：串口设备文件名，可选的值包括/dev/s3c2410_serial1、/dev/s3c2410_serial2、/dev/s3c2410_serial3、/dev/ttyUSB0、/dev/ttyUSB1、/dev/ttyUSB2、/dev/ttyUSB3 （2）baud：波特率 （3）dataBits：数据位（取值范围是 5~8 位，一般用 8 位） （4）stopBits：停止位（取值范围是 1~2 位，一般用 1 位）	打开指定的串口设备，并返回文件描述符
	返回值说明	成功打开串口时，将返回串口的文件描述符，用该描述符可进行 read、write、select 等操作，如果打开失败，则返回 -1	
int write(int fd, byte[] data)	参数说明	（1）fd：要写入数据的文件描述符 （2）data：要写入的数据	向打开的设备或文件中写数据
	返回值说明	成功返回写入的字节数，出错返回-1	
int read(int fd, byte[] buf,	参数说明	（1）fd：要读出数据的文件描述符 （2）buf：存储数据的缓冲区 （3）len：要读取的字节数	从打开的设备或文件中读取数据
	返回值说明	成功返回读取的字节数，出错返回-1，如果在调用 read 前已到达文件末尾，则本次调用 read 后返回 0	
int select(int fd, int sec,int usec)	参数说明	（1）fd：要查询的文件描述符 （2）sen：阻塞等待数据的时间（单位：s） （3）usec：阻塞等待数据的时间（单位：ns，1ms=1000ns）	查询当前打开的设备或文件是否有数据可读
	返回值说明	如果 fd 有数据可读，则返回 1；如果没有数据可读，则返回 0；出错时返回-1	
void close(int fd)	参数说明	fd：要关闭的文件描述符	关闭指定的文件描述符
	返回值说明	无	

2．串口的写入

串口的写入过程如下。

（1）打开串口。

（2）写入数据。

（3）关闭串口。

3．串口的读取

读取串口的过程比较复杂，这是因为读取串口时需要多线程技术的帮助，具体过程如下。

（1）打开串口。

（2）在主线程中开启一个线程。

（3）在线程中循环检测串口是否收到数据。

（4）如果检测到有数据，则读取，并通过回调机制将数据返回给主线程。

（5）继续读取下一条数据。

（6）当应用程序关闭时，关闭串口。

10.3.4　RFID 读卡器的串口通信协议

RFID 读卡器的串口通信协议由 RFID 读卡器设备制造商提供，不同设备制造商生产的 RFID 读卡器可能有所不同，本书只摘取其中的一部分。

1．RFID 读卡器的串口参数

RFID 读卡器的串口参数可以根据通信的需要由软件设置，RFID 读卡器的串口参数及默认值见表 10-4。

表 10-4　RFID 读卡器的串口参数

参　数	描　述	默 认 值
每秒位数	可选: 9600 b/s、19200 b/s、38400 b/s、57600 b/s、1152000 b/s	9600（单位省略）
数据位	固定: 8 位	8（单位省略）
停止位	固定: 1 位	1（单位省略）
奇偶校验	可选: 奇校验、偶检验、无	无
数据流控制	固定: 无	无

2. 数据包格式

数据包是在主机和读卡器之间传输的数据帧，数据包有两大类：命名包和返回包。命令包由主机发送到读卡器，格式见表 10-5；返回包由读卡器返回给主机，格式见表 10-6。

表 10-5 命令包格式

STX	STATION ID	DATA LENGTH	CMD	DATA [0..N]	BCC	ETX

表 10-6 返回包格式

STX	STATION ID	DATA LENGTH	STATUS	DATA[0..N]	BCC	ETX

命名包和返回包各字节的含义见表 10-7。

表 10-7　命令包和返回包各字节的含义

字　段	长　度	描　述	备　注
STX	1	起始字节，表示一个数据包开始。其值固定为 0x02	
STATION ID	1	设备地址，多机通信时必须使用，读卡器收到数据包后判断包内的地址与自身预设的地址是否相符，只有相符才会响应	0x00 是在单机模式下使用的特殊地址。读卡器会响应任何带 0 地址的数据包（不进行地址判断）
DATALENGTH	1	数据包中数据字节的长度。包括 CMD/STATUS 和 DATA field，但不包括 BCC	LENGTH= 字节数 (CMD/STATUS + DATA[0..N])
CMD	1	命令字：由一个命令字节组成。详细说明见后续章节	可以参照命令表，该字节只在命令包中使用
STATUS	1	返回状态字节：由读卡器返回主机的状态	该字节只在返回包中使用
DATA[0..N]	0~255	该数据流的长度与命令字有关。有部分命令不需要附加数据	如写入的数据、读取的数据等
BCC	1	校验字节，用于字节（除 STX 和 ETX 外）的异或校验	
ETX	1	终止字节，表示一个数据包结束。其值固定为 0x03	

3. 命令简介

RFID 读卡器支持 32 条命令，表 10-8 展示了部分主要命令。

表 10-8　RFID 读卡器的部分主要命令

命 名 字 节	名　称	说　明
0x20	MF_ Read	集成了寻卡，防冲突，选卡，验证密码，读卡等操作，一个命令完成读卡的操作
0x21	MF_ Write	集成了寻卡，防冲突，选卡，验证密码，写卡等操作，一个命令完成写卡的操作
0x22	MF_ InitVal	集成了寻卡，防冲突，选卡，验证密码等操作，一个命令完成块值初始化的操作
0x23	MF_Decrement	集成了寻卡，防冲突，选卡，验证密码，块值减等操作，一个命令完成块减值的操作
0x24	MF_Increment	集成了寻卡，防冲突，选卡，验证密码，块值加等操作，一个命令完成块值加的操作

续表

命名字节	名　　称	说　　明
0x25	MF_GET_SNR	集成了寻卡，防冲突，选卡等操作，一个命令完成读取卡片序列号的操作
0x80	SetAddress	设置读卡器地址
0x81	SetBaudrate	设置通信波特率

例如，对于读卡号操作，必须分两步进行：首先，向读卡器发送一条读卡命令（卡必须事先放在读卡器之上）；然后，读卡器执行读卡操作，并将读到的卡号作为读卡命令的结果返回给发送方。

4．读卡号操作的数据包传输过程

下面简单介绍读卡号操作的数据包传输过程。

（1）Android 系统向读卡器发送数据帧（有下画线的部分为数据）。

发送数据：

```
02 00 03 25 26 00 00 03
```

（2）读卡器返回读卡的结果（有下画线的部分为数据）。

返回数据：

```
02 00 06 00 00 0D 80 01 56 DC 03
```

对上述命名包和返回包的每个字段进行分析，详细情况见表 10-9。

表 10-9　读卡号操作的命名包和返回包的字段分析

字　　段	发送数据	返回数据
STX	02：起始字节	02：起始字节
STATION ID	00：设备地址	00：设备地址
DATA LENGTH	03：3 字节	06：6 字节
CMD	25：读取卡片序列号	不适用
STATUS	不适用	00：表示成功（OK）
DATA	26：IDLE 模式，一次只对一张卡操作 00：不需要执行 halt 指令	00：监测到一张卡 0D 80 01 56：卡芯片号为 0D800156
BCC	00：校验位	DC：校验位
ETX	03：结束字节	03：结束字节

10.3.5　串口小助手

为了更好地理解设备的串口通信协议，提高程序调试水平，建议读者使用“串口小助手”软件分析传输的数据包，测试设备是否正常工作。“串口小助手”连接读卡器时的运行界面如图 10-5 所示。

（a）

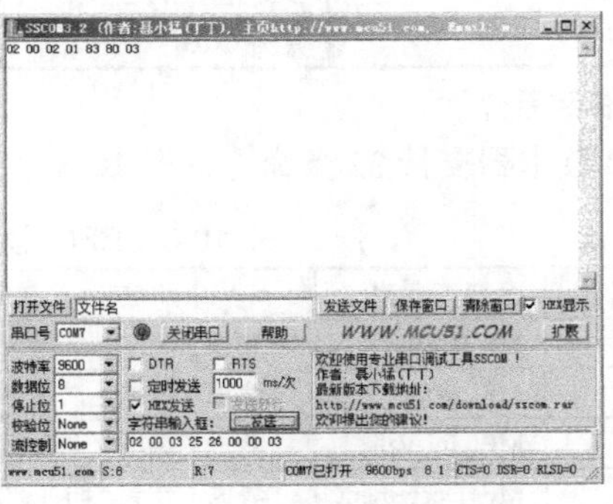

（b）

图 10-5　“串口小助手”连接读卡器时的运行界面

硬件连接：使用串口连接线（两端均为母头）连接台式计算机的串口与设备的串口，如果设备需要电源，则应当为设备供电。如果使用笔记本电脑，则还需要一条 USB-串口转换线，并安装相应的驱动程序。

在装有 Windows 操作系统的计算机上运行“串口小助手”软件，打开与读卡器连接的串口，显示串口号为 COM7。将发送数据和接收数据均设置为十六进制，然后输入读卡号操作的数据包（02 00 03 25 26 00 00 03），单击“发送”按钮，会立即接收到读卡器返回的数据，在图 10-5 中，左侧为有卡时的返回数据，右侧为无卡时的返回数据。

10.4 实施

10.4.1 连接设备

（1）设备与配件。本章的读卡器项目会用到以下设备和配件。

- Friendly ARM Tiny 6410 开发板一块（含电源线一条，USB 线一条）
- 读卡器一个（含电源的串口连接线一条）

（2）接线方式。按照如图 10-6 所示的连接方式接线。图中靠下的位置有两条线，左侧的线是 Friendly ARM Tiny 6410 开发板的电源线，必须使用配套的电源线；右侧的线是 USB 线，将它连接到开发程序用的计算机上。图中靠上的位置有一条线，这是与读卡器配套的串口连接线，并且该线带有电源线，该线右侧的支线连接到 USB 接口，用于向读卡器供电。

（3）项目的运行界面如图 10-7 所示。

图 10-6　读卡器的接线方式

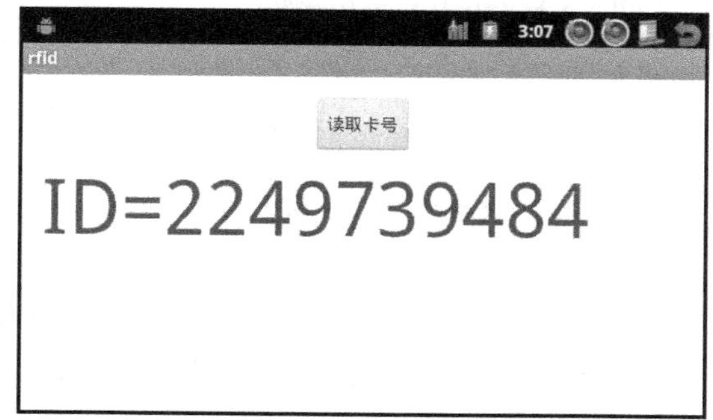

图 10-7　读卡器项目的运行界面

10.4.2 实例代码

读卡器项目的主界面只用到两个控件：一个 Button 控件和一个 TextView 控件。

（1）主界面布局文件的代码如下：

```
<LinearLayout xmlns:android="http://schemas.android.com/apk/res/android"
    xmlns:tools="http://schemas.android.com/tools"
    android:id="@+id/LinearLayout2"
    android:layout_width="match_parent"
    android:layout_height="match_parent"
    android:gravity="center_horizontal"
    android:orientation="vertical"
    android:paddingBottom="@dimen/activity_vertical_margin"
    android:paddingLeft="@dimen/activity_horizontal_margin"
    android:paddingRight="@dimen/activity_horizontal_margin"
    android:paddingTop="@dimen/activity_vertical_margin"
    tools:context=".MainActivity" >
```

```
    <Button
        android:id="@+id/btn_read_rfid"
        android:layout_width="wrap_content"
        android:layout_height="wrap_content"
        android:text="读取卡号" />

    <TextView
        android:id="@+id/tv_rfid"
        android:layout_width="wrap_content"
        android:layout_height="wrap_content"
        android:layout_gravity="left"
        android:text="ID="
        android:textSize="64sp" />

</LinearLayout>
```

（2）Activity 类的功能是给“读取卡号”按钮设置一个单击事件的处理程序，用户单击“读取卡号”按钮后，系统将发出一条读卡指令，该功能被封装在 SerialPort 类中，代码如下：

```
package com.example.administrator.rfid;

import org.ngweb.serial.SerialPort;
import android.support.v7.app.AppCompatActivity;
import android.os.Bundle;
import android.view.Menu;
import android.view.View;
import android.view.View.OnClickListener;
import android.widget.Button;
import android.widget.TextView;

public class AppCompatActivity extends Activity {
    public static Button btnReadRfid;
    public static TextView tvRfid;
    private SerialPort serialPort = new SerialPort();

    @Override
    protected void onCreate(Bundle savedInstanceState) {
        super.onCreate(savedInstanceState);
        setContentView(R.layout.activity_main);

        tvRfid = (TextView) this.findViewById(R.id.tv_rfid);

        btnReadRfid = (Button) this.findViewById(R.id.btn_read_rfid);
        btnReadRfid.setOnClickListener(new OnClickListener() {

            @Override
            public void onClick(View arg0) {
                serialPort.sendReadCmd();
            }
        });
    }

    @Override
    public void onResume() {
        super.onResume();
        serialPort.open();
        serialPort.read();
    }
```

```
    @Override
    public void onPause() {
        super.onPause();
        serialPort.close();
    }
}
```

（3）为了封装读卡器的数据包，设计一个数据类，用来表示将要发送或接收的数据包，代码如下：

```
package com.example.administrator.rfid;

public class RfidData {
    // 数据包结构如下:
    // byte stx = 0x02;
    // byte stationId = 0x00;
    // byte dataLength;
    // byte cmdStatus;
    // byte[] data;
    // byte bcc;
    // byte etx = 0x03;

    private byte[] data;

    public static RfidData getReadRfidCmd() {
        RfidData data = new RfidData((byte) 0x25, new byte[] { 0x26, 0x00 });
        return data;
    }

    public RfidData(byte[] bytes) {
        data = bytes;
    }

    public RfidData(byte cmd, byte[] bytes) {
        data = new byte[bytes.length + 6];
        data[0] = 0x02; // 起始字节，固定为 0x02
        data[1] = 0x00; // 设备地址，固定为 0x00
        data[2] = (byte) (bytes.length + 1); // 数据长度
        data[3] = cmd; // 命令字节
        System.arraycopy(bytes, 0, data, 4, bytes.length); // 数据
        data[data.length - 2] = getBcc(); // ECC
        data[data.length - 1] = 0x03; // 结束字节，固定为 0x03
    }

    public byte getBcc() {
        byte c = 0;
        for (int i = 1; i < data.length - 2; i++) {
            c ^= data[i];
        }
        return c;
    }

    public boolean checkBcc() {
        byte c = getBcc();
        return c == data[data.length - 2];
    }

    public byte[] getData() {
        return data;
```

```
        }

        public String toString() {
                StringBuffer sb = new StringBuffer();

                for (int i = 0; i < data.length; i++) {
                        String b = "00" + Integer.toHexString(data[i]);
                        b = b.substring(b.length() - 2);
                        sb.append(b);
                        sb.append(" ");
                }

                return sb.toString();
        }

        public long getRfidNo() {
                long l = 0;

                for (int i = 5; i < data[2] + 3; i++) {
                        l = l << 8;
                        l = l | (data[i] & 0x000000FF);
                }

                return l;
        }
}
```

（4）串口操作（包括打开串口、关闭串口，读数据和写数据等）的代码被封装在 SerialPort 类中，代码如下：

```
package com.example.administrator.rfid;

import com.example.administrator.rfid.RfidData;
import android.annotation.SuppressLint;
import android.os.Bundle;
import android.os.Handler;
import android.os.Message;
import android.util.Log;
import com.example.administrator.rfid.MainActivity;
import com.friendlyarm.AndroidSDK.HardwareControler;

public class SerialPort {
        int dev;
        boolean isRunning = false;

        public void open() {
                dev = HardwareControler.openSerialPort("/dev/s3c2410_serial3", 9600, 8,
                                1);
                Log.v("RFID", "open=" + dev);
        }

        public void close() {
                isRunning = false;
                if (dev != -1) {
                        HardwareControler.close(dev);
                        dev = -1;
                        Log.v("RFID", "close=" + dev);
                }
        }
```

```
public void sendReadCmd() {
        RfidData data = RfidData.getReadRfidCmd();
        HardwareControler.write(dev, data.getData());
        Log.v("RFID", "send=" + data.toString());
}

@SuppressLint("HandlerLeak")
Handler handleRfid = new Handler() {
        @Override
        public void handleMessage(Message m) {
                Bundle data = m.getData();
                String id = data.getString("id");
                MainActivity.tvRfid.setText("ID=" + id);
        }
};

int maxLen = 11;

public void read() {
        new Thread(new Runnable() {
                @Override
                public void run() {
                        readRfid();
                }
        }).start();
}

private void readRfid(){
        isRunning = true;
        byte[] buffer = new byte[maxLen];
        int bufLen = 0;
        while (isRunning) {
                int ready = 0;
                ready = HardwareControler.select(dev, 0, 100000);
                if (ready == 1) {
                        // 有数据可读
                        byte[] buf = new byte[maxLen];
                        int len = HardwareControler.read(dev, buf, maxLen);
                        Log.v("RFID",
                                        "get(" + len + ")="
                                                        + new RfidData(buf).toString());
                        for (int i = 0; i < len; i++) {
                                buffer[bufLen + i] = buf[i];
                        }
                        bufLen += len;
                        if (bufLen == 11) {
                                // 这样的数据才有效

                                Log.v("RFID",
                                                "buffer=" + new RfidData(buffer).toString());

                                RfidData rfid = new RfidData(buffer);
                                long id = rfid.getRfidNo();

                                Log.v("RFID", "id=" + id);

                                Bundle b = new Bundle();
                                b.putString("id", Long.toString(id));
```

```
                    Message m = new Message();
                    m.setData(b);

                    handleRfid.sendMessage(m);

                    bufLen = 0;
                }
            }
        }
    }
}
```

10.5 实训：完善读卡器项目

针对读卡器项目，完成以下需求。

- 将读取的卡号上传到服务器的数据库中，并与留存的一组卡号进行比对，比对成功时，返回持卡人的姓名，否则返回空值。

提示：为了简化有关服务器的编程工作，在服务器端可以不使用数据库管理系统，而用一个二维数组储存卡号和持卡人的姓名。

10.6 作业

1．什么是串口？串口参数有哪些？默认值是什么？

2．在 Windows 系统中，串口命名的规律是什么？在 Linux 系统中，串口命名的规律是什么？在 Linux 系统中，如何查看对应的设备文件名？

3．向串口写数据时是否需要多线程技术的支持？从串口读数据时是否需要多线程技术的支持？为什么？

4．从串口读数据时需要注意哪些问题？这些问题应如何解决？

第 11 章

综合实训——诗词赏析项目

本章将引入综合实训——诗词赏析项目，旨在通过完整的实训项目巩固之前所学知识。

11.1 项目介绍

11.1.1 项目概述

请设计一款诗词赏析 App，用户使用该 App 可以欣赏唐、宋两个朝代共九位著名诗人的风采以及他们的代表作品。在精美的页面布局的映衬下，让用户品味唐代大诗人李白的《望庐山瀑布》、宋代文学家苏轼的《题西林壁》等。在该 App 中，每首诗词均配有诗人的图片、诗词原文、诗词译文、诗词赏析及诗词朗诵配音的 mp3 格式文件。

11.1.2 开发工具

开发工具及 IDE 搭建要求：JDK8、Android SDK19、Android Studio2.2.2、Java 编程语言。

11.1.3 界面设计

诗词赏析项目有三个主要界面：引导界面、登录界面、主界面（包括主界面左侧的侧拉菜单和主界面右侧的 FrameLayout），分别如图 11-1~图 11-5 所示。

图 11-1　引导界面

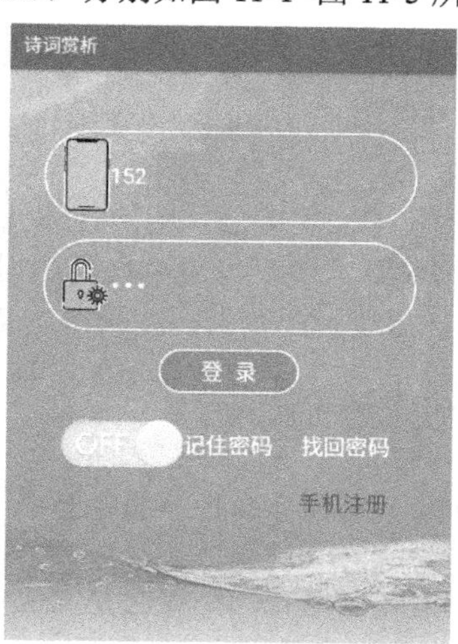

图 11-2　登录界面

图 11-3　主界面

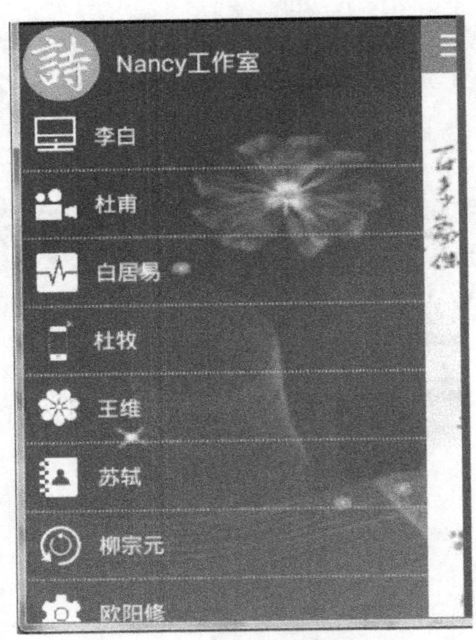

图 11-4　主界面左侧的侧拉菜单

图 11-5　主界面右侧的 FrameLayout

11.2　需求分析与功能分析

11.2.1　需求分析

传统的阅读模式主要基于纸质载体，用户通常会购买图书或从图书馆借阅书籍，这种方式一般会增加用户的经济投入，并且在购买和借阅环节需要一定的时间成本。随着电子设备的研发与普及，在线阅读应运而生，这种模式突破了传统的阅读模式，用户可以从应用商店下载免费的 App，随时随地进行阅读，极大地方便了生活和学习。

本项目的主界面为侧拉菜单风格，采用 android.support.v4.widget.SlidingPaneLayout 布局，将主界面分为左、右两部分：左侧采用一个 ListView 控件构成的侧拉菜单，右侧为显示内容的主框架。

11.2.2　功能分析

1．引导界面

引导界面是诗词赏析项目的第一个界面，用户打开 App 后，屏幕会显示倒计时 5 秒，之后会自动跳转到登录界面。倒计时采用 CountDownTimer()方法实现，设置延迟 5 秒；自动跳转采用线程的 Handler 机制中的 postDelayed()方法实现。

2．登录界面

在登录界面中，可以实现以下四项功能。

（1）登录功能。

① 如果用户输入的手机号没有储存在程序的容器中，用户输入手机号和密码，单击“登录”按钮后，屏幕提示“手机号不存在！”。

② 如果用户输入的手机号已经储存在程序的容器中，则用户具备了登录的基本条件，用户输入手机号和密码，单击“登录”按钮后，程序开始判断用户的手机号与密码是否匹配。

③ 若用户的手机号和密码不匹配，则屏幕提示“密码不匹配！”。

④ 若用户的手机号和密码匹配，则屏幕提示“登录成功！”，然后跳转到主界面。

（2）是否记住密码功能。用户登录之后，提供是否记住密码功能，以便用户再次使用程序时快捷登录。是否记住密码功能采用 ImageView 控件：准备两张图片，图片内容分别为英文单词“ON”和“OFF”，设置 ImageView 控件的单击属性为 True，默认状态下显示“OFF”，表示系统忽略手机号与密码；当用户单击按钮切换到“ON”时，表示系统将记住当前的手机号与密码。该功能主要采用 SharedPreferences 技术，用于存储手机号与密码，以及开关的状态。

（3）手机注册功能。新用户可以使用手机号注册，“用户注册”对话框如图 11-6 所示。

如果用户从未注册过，则必须先注册新用户，否则登录时会出现提示“没有这样的用户存在，请先注册新用户！”。

（4）找回密码功能。如果用户忘记登录密码，可以使用找回密码功能，在如图 11-7 所示的“找回密码”对话框中输入手机号，单击“发送”按钮，等待短信，找回原密码。

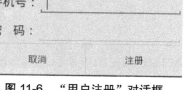

图 11-6　“用户注册”对话框

图 11-7　“找回密码”对话框

提醒：手机注册功能和找回密码功能需要借助自定义对话框实现，而且对话框采用透明风格。

3. 主界面

本项目采用 android.support.v4.widget.SlidingPaneLayout 布局，主要分为左、右两部分。

（1）左侧的侧拉菜单其布局文件为 activity_main_left.xml，采用垂直线性布局，上半部分是项目的 LOGO 及个性化标志；下半部分是一个 ListView 控件，采用 Menu 类封装图片与对应的菜单项。

（2）右侧的 FrameLayout 布局文件为 activity_main_right.xml，采用垂直线性布局，分为上、下两部分。

上半部分为水平线性布局，包括两个控件（ImageView 与 TextView），其中 ImageView 控件用于后控制 SlidingPaneLayout 面板的打开与关闭，单击 ImageView 控件可实现此功能。TextView 控件用于显示菜单项对应的文字，单击 ImageView 控件可实现此功能。

下半部分为相对布局，id 为 mainContent，当用户单击左侧的侧拉菜单项后，mainContent 可以呈现加载的 Fragment。这里需要说明：用户单击左侧的侧拉菜单项后，菜单项的对应代码会继承 Fragment 类，所以各菜单项对应的布局就会覆盖原来的 mainRight 对应的屏幕内容。

4. 修改密码

本项目具备修改密码的功能，修改密码界面如图 11-8 所示，该功能的实现原理如下。

在“请输入原密码”的 TextView 控件上增加了一个 setOnFocusChangeListener 监听器。如果用户的原密码输入有误，则会出现提示“原密码输入不正确！”。

用户输入原密码与新密码后，单击“确认修改”按钮，就可以修改 Set 容器中存储的与手机号对应的密码，并且覆盖 SharedPreferences 中存储的信息，删除“data/data/包名/shared_pref”文件夹中生成的 xml 文件。

修改密码成功后，会自动返回登录界面，请用户重新登录。

用户修改密码时需要注意，注册时使用的手机号不能变更，因为要根据该手机号获取原来的密码。

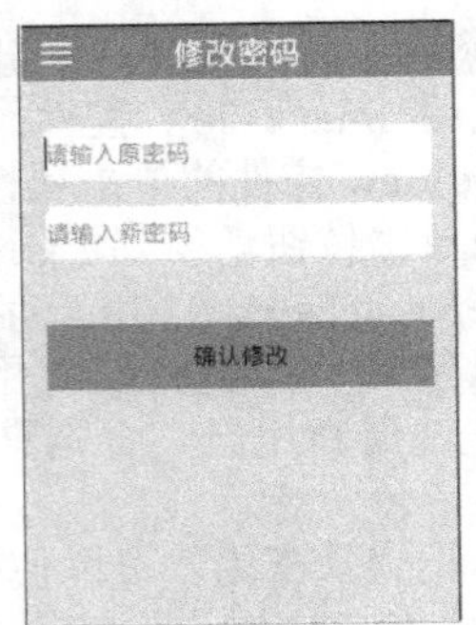

图 11-8 修改密码界面

5. 注销登录

本项目具备注销登录功能。在左侧的侧拉菜单中，单击最后的菜单项，可以由主界面返回登录界面，请用户重新登录。

11.2.3 功能模块设计

根据功能分析，绘制诗词赏析项目的整体模块架构图，如图 11-9 所示。

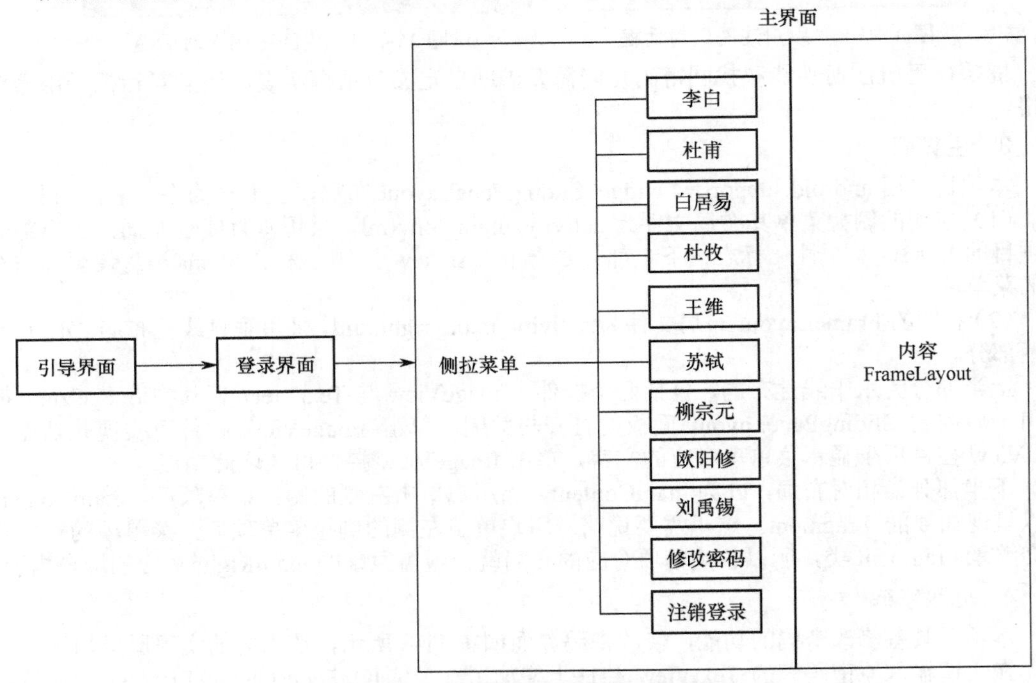

图 11-9 诗词赏析项目的整体模块架构图

11.3 实施

11.3.1 数据设计

诗词欣赏项目主要采用容器类完成数据的组织，感兴趣的读者也可以采用数据库 SQLite 完成数据的组织。

1．注册信息的存储

为了避免注册信息重复存储，建议采用 Set 容器存储 User 类。User 类封装了两个属性：phone（手机号）和 pwd（密码）。

采用 UserSet(LinkedHashSet)类存储对 User 类的操作，主要包括四种方法，即 initUser()方法（初始用户添加）、findByPhone (String phone)方法（根据手机号查找用户）、selectAll()方法（查询显示所有注册过的用户信息）和 userCount()方法（用户数统计）。

2．诗词信息的存储

采用 List 容器存储 ShiCi 类。Shici 类封装了七个属性，分别为 img（诗人图片）、author（诗人姓名）、title（诗名）、content（诗词内容）、yiwen（诗词译文）、shangxi（诗词赏析）和 langsong（诗词朗诵配音文件的名称）。

采用 ShiCiList(LinkedList)类存储对 ShiCi 类的操作，主要包括三种方法，即 add(ShiCi shici)方法（添加诗词）、findByAuthor(String author)方法（根据诗人姓名查找作品）和 SelectAll()方法（查询所有诗词信息）。

3．各类的 Java 代码

（1）User 类（用户类）的代码如下：

```
package com.example.administrator.scsx;

public class User {
    String phone;   //手机号
    String pwd;   //密码

    public User() {
        super();
    }

    public User(String phone, String pwd) {
        super();
        this.phone = phone;
        this.pwd = pwd;
    }

    public String getPhone() {
        return phone;
    }

    public void setPhone(String phone) {
        this.phone = phone;
    }

    public String getPwd() {
        return pwd;
    }

    public void setPwd(String pwd) {
        this.pwd = pwd;
    }
}
```

（2）UserSet 类（用户管理类）的代码如下：

```
package com.example.administrator.scsx;

import java.util.LinkedHashSet;
import java.util.Set;
```

```
public class UserSet{
    Set<User> set = new LinkedHashSet();

    public void initUser(){
        set.add(new User("152","123"));
        set.add(new User("189","123"));
    }

    public User findByPhone(String phone) {
        User user = new User();
        int i = 0;
        for (User u : set) {
            if (phone.equals(u.getPhone())) {
                user = u;
                break;
            }
            i += 1;
        }
        if (i==set.size())
            return null;
        else
            return user;
    }

    public void selectAll() {
        for (User u : set) { // 这种写法较简洁
            System.out.println(u.getPhone() + "," + u.getPwd());
        }
    }

    public int userCount() {
        return set.size();
    }
}
```

（3）ShiCi 类（诗词类）的代码如下：

```
package com.example.user.scsx;

public class ShiCi {
    int img;   //诗人图片
    String title;   //诗名
    String author;   //诗人姓名
    String music;   // 诗词朗诵配音文件的名称
    String content; //诗词内容
    String yiwen;      //诗词译文
    String shangxi; //诗词赏析

    public ShiCi() {
    }

    public ShiCi(int img, String author,String title,String content,String shangxi,String yiwen,String music) {
        this.author = author;
        this.content = content;
        this.img = img;
        this.shangxi = shangxi;
        this.title = title;
        this.yiwen = yiwen;
```

```
        this.music = music;
    }

    public String getAuthor() {
        return author;
    }

    public void setAuthor(String author) {
        this.author = author;
    }

    public String getContent() {
        return content;
    }

    public void setContent(String content) {
        this.content = content;
    }

    public int getImg() {
        return img;
    }

    public void setImg(int img) {
        this.img = img;
    }

    public String getMusic() {
        return music;
    }

    public void setMusic(String music) {
        this.music = music;
    }

    public String getShangxi() {
        return shangxi;
    }

    public void setShangxi(String shangxi) {
        this.shangxi = shangxi;
    }

    public String getTitle() {
        return title;
    }

    public void setTitle(String title) {
        this.title = title;
    }

    public String getYiwen() {
        return yiwen;
    }

    public void setYiwen(String yiwen) {
        this.yiwen = yiwen;
    }
}
```

（4）ShiCiList 类（诗词管理类）的 Java 代码如下：

```
package com.example.user.scsx;

import java.util.LinkedList;
import java.util.List;

import static android.R.id.list;

public class ShiCiList {
    List<ShiCi> list = new LinkedList<>();

    public void initShiCi() {
        list.add(new ShiCi(R.drawable.libai, "李白", "望庐山瀑布",
                "日照香炉生紫烟，遥看瀑布挂前川。\r\n 飞流直下三千尺，疑是银河落九天。",
                "香炉峰在阳光的照射下生起紫色烟霞，远远望见瀑布似白色绢绸悬挂在山前。
高崖上飞腾直落的瀑布好像有几千尺，让人恍惚以为银河从天上泻落到人间。",
                "这是诗人李白五十岁左右隐居庐山时写的一首风景诗。这首诗形象地描绘了庐
山瀑布雄奇壮丽的景色，反映了诗人对祖国大好河山的无限热爱。\r\n\r\n 首句“日照香炉生紫烟”中的
“香炉”指庐山的香炉峰，此峰地处庐山的西北方位，形状尖圆，像座香炉。由于瀑布飞泻，水气蒸腾而
上，在烈日的照耀下，仿佛有座顶天立地的香炉冉冉升起了团团紫烟。一个“生”字把烟云冉冉上升的
景象写活了。此句为瀑布设置了雄奇的背景，也为下文直接描写瀑布渲染了气氛。\r\n\r\n 次句“遥看瀑
布挂前川”中的“遥看瀑布”四字照应了诗名《望庐山瀑布》。“挂前川”指瀑布像一条巨大的白练从悬
崖直挂到前面的河流上。“挂”字化动为静，惟妙惟肖地勾勒出遥望中的瀑布。\r\n\r\n 诗的前两句从大处
着笔，概写望中全景：山顶紫烟缭绕，山间白练悬挂，山下激流奔腾，构成一幅绚丽壮美的图景。\r\n\r\n
第三句“飞流直下三千尺”是从近处细致地描写瀑布。“飞流”表现瀑布凌空而出，喷涌飞泻。“直下”
既写出岩壁的陡峭，又写出水流之急。“三千尺”极力夸张，描绘山的高峻。\r\n\r\n 诗人觉得这种写法还
没有把瀑布的雄奇气势表现得淋漓尽致，于是又写道“疑是银河落九天”，使人怀疑这 “飞流直下”的
瀑布就是从九天倾泻而下的银河。一个“疑”字，用得空灵活泼，若真若幻，引人遐想，增添了瀑布的
神奇色彩。\r\n\r\n 这首诗成功地运用了比喻、夸张和想象手法，构思奇特，语言生动形象、洗炼明快。
苏轼十分赞赏这首诗，说“帝遣银河一脉垂，古来唯有谪仙词。” “谪仙”就是李白。《望庐山瀑布》的
确是状物写景和抒情的范例。",
                "001.mp3"));
    //其他八首诗词内容，详见项目素材文件
        }

    public void printShiCi(){
        for(ShiCi sc:list){
            System.out.println(sc.getTitle());
            System.out.println(sc.getAuthor());
            System.out.println(sc.getContent());
        }
    }
}
```

（5）在 Menu 类（侧拉菜单类）中，封装了 img（图片）和 name（菜单名称）两个属性，这两个属性用于填充自定义布局文件 lst_layout.xml 中的相应控件。Menu 类的代码如下：

```
package com.example.user.scsx;

public class Menu {
    int img;
    String name;

    public Menu() {
    }

    public Menu(int img, String name) {
```

```
            this.img = img;
            this.name = name;
        }

        public int getImg() {
            return img;
        }

        public void setImg(int img) {
            this.img = img;
        }

        public String getName() {
            return name;
        }

        public void setName(String name) {
            this.name = name;
        }
}
```

11.3.2　界面实现

在界面实现过程中，需要设计引导界面、登录界面、主界面、修改密码界面、注销登录界面、侧拉菜单、“用户注册”对话框和“找回密码”对话框的布局文件。

1．引导界面

引导界面的自定义布局文件为 activity_yindao.xml，代码如下：

```
<?xml version="1.0" encoding="utf-8"?>
<RelativeLayout xmlns:android="http://schemas.android.com/apk/res/android"
    xmlns:tools="http://schemas.android.com/tools"
    android:id="@+id/activity_yidao"
    android:layout_width="match_parent"
    android:layout_height="match_parent"
    android:paddingBottom="@dimen/activity_vertical_margin"
    android:paddingLeft="@dimen/activity_horizontal_margin"
    android:paddingRight="@dimen/activity_horizontal_margin"
    android:paddingTop="@dimen/activity_vertical_margin"
    android:background="@drawable/yd_bg"
    tools:context="com.example.user.scsx.YindaoActivity">

    <TextView
        android:text="倒计时"
        android:layout_width="wrap_content"
        android:layout_height="wrap_content"
        android:layout_marginRight="11dp"
        android:layout_marginEnd="11dp"
        android:layout_marginTop="11dp"
        android:textSize="30sp"
        android:textColor="#FF0000"
        android:id="@+id/tvTime"
        android:layout_alignParentTop="true"
        android:layout_alignParentRight="true"
        android:layout_alignParentEnd="true" />
</RelativeLayout>
```

2. 登录界面

登录界面的自定义布局文件为 activity_login.xml，代码如下：

```
<?xml version="1.0" encoding="utf-8"?>
<LinearLayout xmlns:android="http://schemas.android.com/apk/res/android"
    xmlns:app="http://schemas.android.com/apk/res-auto"
    xmlns:tools="http://schemas.android.com/tools"
    android:id="@+id/activity_login"
    android:layout_width="match_parent"
    android:layout_height="match_parent"
    android:paddingBottom="@dimen/activity_vertical_margin"
    android:paddingLeft="@dimen/activity_horizontal_margin"
    android:paddingRight="@dimen/activity_horizontal_margin"
    android:orientation="vertical"
    android:background="@drawable/dl_bg2"
    android:paddingTop="@dimen/activity_vertical_margin"
    tools:context="com.example.user.scsx.LoginActivity">

    <EditText
        android:layout_width="500dp"
        android:layout_height="100dp"
        android:layout_gravity="center"
        android:layout_marginTop="50dp"
        android:inputType="phone"
        android:ems="10"
        android:background="@drawable/line1"
        android:drawableLeft="@drawable/phone"
        android:paddingLeft="10dp"
        android:hint="手机号"
        android:textSize="30sp"
        android:textColor="#FFFFFF"
        android:textColorHint="#FFFFFF"
        android:id="@+id/etphone" />

    <EditText
        android:layout_width="500dp"
        android:layout_height="100dp"
        android:layout_gravity="center"
        android:layout_marginTop="30dp"
        android:inputType="numberPassword"
        android:ems="10"
        android:background="@drawable/line1"
        android:drawableLeft="@drawable/pwd"
        android:paddingLeft="10dp"
        android:hint="密码"
        android:textSize="30sp"
        android:textColor="#FFFFFF"
        android:textColorHint="#FFFFFF"
        android:id="@+id/etpwd" />

    <Button
        android:text="登　录"
        android:layout_width="180dp"
        android:layout_height="80dp"
        android:background="@drawable/line2"
        android:layout_marginTop="50dp"
        android:layout_gravity="center"
        android:textSize="30sp"
```

```
            android:textColor="#FFFFFF"
            android:id="@+id/btnLogin" />

        <LinearLayout
            android:orientation="horizontal"
            android:layout_width="500dp"
            android:layout_height="wrap_content"
            android:layout_gravity="center"
            android:layout_marginTop="80dp">

            <ImageView
                android:layout_width="wrap_content"
                android:layout_height="wrap_content"
                app:srcCompat="@drawable/off"
                android:clickable="true"
                android:id="@+id/ivOnOff" />

            <TextView
            android:text="记住密码"
            android:layout_width="wrap_content"
            android:layout_height="wrap_content"
            android:textSize="30sp"
            android:textColor="#FFFFFF"
            android:layout_gravity="center"
            android:id="@+id/textView" />

            <TextView
                android:text="找回密码"
                android:layout_width="wrap_content"
                android:layout_height="wrap_content"
                android:textSize="30sp"
                android:textColor="#FFFFFF"
                android:layout_gravity="center"
                android:layout_marginLeft="60dp"
                android:id="@+id/tvFind" />
        </LinearLayout>

        <TextView
            android:text="手机注册"
            android:layout_width="wrap_content"
            android:layout_height="wrap_content"
            android:textSize="30sp"
            android:textColor="#FF0000"
            android:layout_gravity="center"
            android:layout_marginTop="20dp"
            android:layout_marginLeft="120dp"
            android:id="@+id/tvRegister" />

    </LinearLayout>
```

3. 主界面

主界面的布局文件为 activity_main.xml，采用 RelativeLayout 方式，其中加入 android.support.v4.widget.SlidingPaneLayout 布局。屏幕分为左、右两部分，即左侧布局和右侧布局。

在主界面中，需要实现以下两项功能。

- 用户在手机屏幕中从左向右拖动，可以弹出左侧的侧拉菜单。
- 用户单击右侧布局左上角的图片（ImageView 控件），也可以打开、折叠左侧的侧拉菜单。

主界面的整体框架设计如图 11-10 所示。

图 11-10 主界面的整体框架设计

activity_main.xml 的代码如下：

```
<?xml version="1.0" encoding="utf-8"?>
<RelativeLayout xmlns:android="http://schemas.android.com/apk/res/android"
    xmlns:tools="http://schemas.android.com/tools"
    android:id="@+id/activity_main"
    android:layout_width="match_parent"
    android:layout_height="match_parent"
    tools:context="com.example.user.scsx.MainActivity">

    <android.support.v4.widget.SlidingPaneLayout
        android:layout_width="match_parent"
        android:layout_height="match_parent"
        android:id="@+id/slidingPaneLayout1">

        <include layout="@layout/activity_main_left"/>
        <include layout="@layout/activity_main_right"/>
    </android.support.v4.widget.SlidingPaneLayout>
</RelativeLayout>
```

（1）主界面的左侧布局文件为 activity_main_left.xml，采用垂直线性布局，内含一个水平线性布局和一个 ListView 控件。

其中，水平线性布局包括一个 Logo 和一个 TextView 控件。

activity_main_left.xml 的代码如下：

```
<?xml version="1.0" encoding="utf-8"?>
<LinearLayout xmlns:android="http://schemas.android.com/apk/res/android"
    xmlns:app="http://schemas.android.com/apk/res-auto"
    android:orientation="vertical"
    android:background="@drawable/bg_left_menu"
    android:layout_width="230dp"
    android:layout_height="match_parent">

    <LinearLayout
        android:orientation="horizontal"
        android:layout_width="match_parent"
        android:layout_height="wrap_content">

        <ImageView
```

```
            android:layout_width="90dp"
            android:layout_height="100dp"
            app:srcCompat="@drawable/logo"
            android:id="@+id/imageView" />
        <TextView
            android:text="Nancy 工作室"
            android:layout_width="wrap_content"
            android:layout_height="wrap_content"
            android:textSize="22sp"
            android:textColor="#FFFFFF"
            android:layout_gravity="center"
            android:id="@+id/textView" />
    </LinearLayout>

    <ListView
        android:layout_width="230dp"
        android:layout_height="match_parent"
        android:divider="#10FF0000"
        android:dividerHeight="20dp"
        android:id="@+id/lvMenu"/>
</LinearLayout>
```

（2）主界面的右侧布局文件为 activity_main_right.xml，采用垂直线性布局，内含一个水平线性布局和一个相对布局（id=mainContent）。

其中，水平线性布局包括一个 ImageView 控件和一个 TextView 控件，ImageView 控件可以单击，用于打开、关闭侧拉菜单，TextView 控件用于显示左侧的各菜单名。

activity_main_right.xml 的代码如下：

```
<?xml version="1.0" encoding="utf-8"?>
<LinearLayout xmlns:android="http://schemas.android.com/apk/res/android"
    xmlns:app="http://schemas.android.com/apk/res-auto"
    android:orientation="vertical"
    android:background="@drawable/main_bg"
    android:layout_width="match_parent"
    android:layout_height="match_parent">

    <LinearLayout
        android:orientation="horizontal"
        android:layout_width="match_parent"
        android:background="#56ABE4"
        android:layout_height="50dp">

        <ImageView
            android:layout_width="wrap_content"
            android:layout_height="wrap_content"
            app:srcCompat="@drawable/ic_top_bar_category"
            android:id="@+id/ivRightTop" />

        <TextView
            android:id="@+id/tvMenuname"
            android:layout_width="wrap_content"
            android:layout_height="wrap_content"
            android:text="菜单名"
            android:textSize="28sp"
            android:textColor="#FFFFFF" />
    </LinearLayout>

    <RelativeLayout
        android:id="@+id/mainContent"
```

```
        android:layout_width="match_parent"
        android:layout_height="match_parent">

    </RelativeLayout>
</LinearLayout>
```

4. 修改密码界面

修改密码界面的布局文件为 activity_change_password.xml，代码如下：

```
<LinearLayout xmlns:android="http://schemas.android.com/apk/res/android"
    xmlns:tools="http://schemas.android.com/tools"
    android:id="@+id/LinearLayout1"
    android:layout_width="match_parent"
    android:layout_height="match_parent"
    android:orientation="vertical"
    android:paddingBottom="@dimen/activity_vertical_margin"
    android:paddingLeft="@dimen/activity_horizontal_margin"
    android:paddingRight="@dimen/activity_horizontal_margin"
    android:paddingTop="@dimen/activity_vertical_margin"
    android:background="#DCDCDC"
    tools:context=".ChangPassword" >

    <EditText
        android:id="@+id/etPwd1"
        android:layout_width="match_parent"
        android:layout_height="40dp"
        android:layout_marginTop="20dp"
        android:background="@drawable/round2"
        android:hint="请输入原密码"
        android:ems="10" >

        <requestFocus />
    </EditText>

     <EditText
        android:id="@+id/etPwd2"
        android:layout_width="match_parent"
        android:layout_height="40dp"
          android:layout_marginTop="20dp"
        android:background="@drawable/round2"
        android:hint="请输入新密码"
        android:ems="10" >

    </EditText>

     <Button
          android:id="@+id/btnUpdate"
          android:layout_width="match_parent"
          android:layout_height="50dp"
            android:layout_marginTop="50dp"
               android:background="#1E90FF"
          android:text="确认修改" />

</LinearLayout>
```

5. 注销登录界面

注销登录界面的布局文件为 activity_cancel_login.xml，代码如下：

```
<RelativeLayout xmlns:android="http://schemas.android.com/apk/res/android"
    xmlns:tools="http://schemas.android.com/tools"
```

```
    android:layout_width="match_parent"
    android:layout_height="match_parent"
    android:paddingBottom="@dimen/activity_vertical_margin"
    android:paddingLeft="@dimen/activity_horizontal_margin"
    android:paddingRight="@dimen/activity_horizontal_margin"
    android:paddingTop="@dimen/activity_vertical_margin"
    tools:context=".CancelLogin" >

    <TextView
        android:layout_width="wrap_content"
        android:layout_height="wrap_content"
        android:text="@string/cancel_login" />"

</RelativeLayout>
```

6. 侧拉菜单

侧拉菜单（ListView）对应的自定义布局文件为 lst_layout.xml，代码如下：

```
<?xml version="1.0" encoding="utf-8"?>
<LinearLayout xmlns:android="http://schemas.android.com/apk/res/android"
    xmlns:app="http://schemas.android.com/apk/res-auto"
    android:layout_width="230dp"
    android:layout_height="50dp"
    android:orientation="horizontal"
    android:weightSum="1">

    <ImageView
        android:layout_width="50dp"
        android:layout_height="match_parent"
        app:srcCompat="@drawable/ic_launcher"
        android:layout_marginLeft="10dp"
        android:id="@+id/ivimg" />

    <TextView
        android:id="@+id/tvname"
        android:layout_width="wrap_content"
        android:layout_height="wrap_content"
        android:layout_gravity="left"
        android:text="诗人姓名"
        android:textSize="28sp"
        android:textColor="#FFFFFF" />
</LinearLayout>
```

7. “用户注册”对话框

“用户注册”对话框对应的自定义布局为 dialog_reg.xml，代码如下：

```
<?xml version="1.0" encoding="utf-8"?>
<LinearLayout xmlns:android="http://schemas.android.com/apk/res/android"
    android:layout_width="match_parent"
    android:layout_height="match_parent"
    android:orientation="vertical" >

    <LinearLayout
        android:layout_width="match_parent"
        android:layout_height="wrap_content" >

        <TextView
            android:id="@+id/textView1"
            android:layout_width="wrap_content"
            android:layout_height="wrap_content"
```

```
            android:text="手机号："
            android:textAppearance="?android:attr/textAppearanceMedium" />

        <EditText
            android:id="@+id/etPhoneReg"
            android:layout_width="wrap_content"
            android:layout_height="wrap_content"
            android:layout_weight="1"
            android:ems="10"
            android:inputType="phone" >

            <requestFocus />
        </EditText>
    </LinearLayout>

    <LinearLayout
        android:layout_width="match_parent"
        android:layout_height="wrap_content"
        android:layout_marginTop="5dp" >

        <TextView
            android:id="@+id/textView1"
            android:layout_width="wrap_content"
            android:layout_height="wrap_content"
            android:text="密　码："
            android:textAppearance="?android:attr/textAppearanceMedium" />

        <EditText
            android:id="@+id/etPwdReg"
            android:layout_width="wrap_content"
            android:layout_height="wrap_content"
            android:layout_weight="1"
            android:ems="10"
            android:inputType="text" >
        </EditText>
    </LinearLayout>

</LinearLayout>
```

8．“找回密码”对话框

“找回密码”对话框对应的自定义布局为 dialog_find.xml，代码如下：

```
<?xml version="1.0" encoding="utf-8"?>
<LinearLayout xmlns:android="http://schemas.android.com/apk/res/android"
    android:layout_width="match_parent"
    android:layout_height="match_parent"
    android:orientation="vertical" >

    <LinearLayout
        android:layout_width="match_parent"
        android:layout_height="wrap_content" >

        <TextView
            android:id="@+id/textView1"
            android:layout_width="wrap_content"
            android:layout_height="wrap_content"
            android:text="手机号："
            android:textAppearance="?android:attr/textAppearanceMedium" />
```

```
        <EditText
            android:id="@+id/etPhoneFind"
            android:layout_width="wrap_content"
            android:layout_height="wrap_content"
            android:layout_weight="1"
            android:ems="10"
            android:inputType="phone" >

            <requestFocus />
        </EditText>
    </LinearLayout>
</LinearLayout>
```

11.3.3　Java 代码

1. 引导界面

引导界面的 YindaoActivity.java 代码如下：

```
package com.example.user.scsx;

import android.content.Intent;
import android.os.CountDownTimer;
import android.os.Handler;
import android.support.v7.app.AppCompatActivity;
import android.os.Bundle;
import android.widget.TextView;

public class YindaoActivity extends AppCompatActivity {
    final long SPLASH_LENGTH=5000;
    Handler handler=new Handler();
    TextView tvTime;

    @Override
    protected void onCreate(Bundle savedInstanceState) {
        super.onCreate(savedInstanceState);
        setContentView(R.layout.activity_yindao);

        tvTime= (TextView) this.findViewById(R.id.tvTime);
        /** 倒计时 5 秒*/
        CountDownTimer timer = new CountDownTimer(5 * 1000, 1000) {
            @Override
            public void onTick(long millisUntilFinished) {
                tvTime.setText(millisUntilFinished / 1000 + "秒");
            }

            @Override
            public void onFinish() {

            }
        }.start();

        handler.postDelayed(new Runnable() {//延迟 5 秒后，跳转到 LoginActivity
            @Override
            public void run() {
                Intent intent=new Intent(YindaoActivity.this,LoginActivity.class);
                startActivity(intent);
                finish();
            }
```

```
            },SPLASH_LENGTH);
        }
}
```

2．登录界面

登录界面的 LoginActivity.java 代码如下：

```
package com.example.user.scsx;

import android.annotation.SuppressLint;
import android.app.Activity;
import android.app.AlertDialog;
import android.app.Notification;
import android.app.NotificationManager;
import android.content.DialogInterface;
import android.content.Intent;
import android.content.SharedPreferences;
import android.support.v7.app.AppCompatActivity;
import android.os.Bundle;
import android.view.LayoutInflater;
import android.view.View;
import android.view.Window;
import android.view.WindowManager;
import android.widget.Button;
import android.widget.EditText;
import android.widget.ImageView;
import android.widget.TextView;
import android.widget.Toast;

import static com.example.user.scsx.R.drawable.phone;

public class LoginActivity extends AppCompatActivity {
    EditText etphone, etpwd;
    Button btnLogin;
    public static UserSet set = new UserSet();
    public static String inputphone, inputpwd;
    ImageView ivOnOff;
    boolean clickState=false;
    int clickNumber=0;
    TextView tvFind,tvRegister;
    User user=new User();

    @Override
    protected void onCreate(Bundle savedInstanceState) {
        super.onCreate(savedInstanceState);
        setContentView(R.layout.activity_login);

        set.initUser();
        set.selectAll();

        etphone = (EditText) this.findViewById(R.id.etphone);   //按 Alt+Enter 组合键
        etpwd = (EditText) this.findViewById(R.id.etpwd);

        btnLogin = (Button) this.findViewById(R.id.btnLogin);
        btnLogin.setOnClickListener(new View.OnClickListener() {
            @Override
            public void onClick(View v) {
                inputphone = etphone.getText().toString();
                inputpwd = etpwd.getText().toString();
```

```
                if (set.findByPhone(inputphone) == null) {
                    Toast.makeText(LoginActivity.this, "手机号不存在！", Toast.LENGTH_
LONG).show();
                } else {
                    if (inputpwd.equals(set.findByPhone(inputphone).getPwd())) {
                        Toast.makeText(LoginActivity.this, "登录成功！", Toast.LENGTH_
LONG).show();
                        Intent intent = new Intent(LoginActivity.this, MainActivity.class);
                        startActivity(intent);
                    } else {
                        Toast.makeText(LoginActivity.this, "密码不匹配！", Toast.LENGTH_
LONG).show();
                        etpwd.setText("");
                    }
                }
            }
        });

        ivOnOff = (ImageView) this.findViewById(R.id.ivOnOff);
        ivOnOff.setOnClickListener(new View.OnClickListener() {
            @Override
            public void onClick(View v) {
                clickNumber++;
                if(clickNumber%2!=0){
                    clickState=true;
                    ivOnOff.setImageResource(R.drawable.on);
                }else {
                    clickState=false;
                    ivOnOff.setImageResource(R.drawable.off);
                }
            }
        });

        tvFind = (TextView) this.findViewById(R.id.tvFind);
        tvFind.setOnClickListener(new View.OnClickListener() {
            @Override
            public void onClick(View v) {
                AlertDialog.Builder builder_Find = new AlertDialog.Builder(
                        LoginActivity.this);
                LayoutInflater inflater_Find = getLayoutInflater();
                final View layout_Find = inflater_Find.inflate(
                        R.layout.dialog_find, null);// 获取自定义布局
                builder_Find.setView(layout_Find);
                builder_Find.setIcon(R.drawable.key);// 设置标题图标
                builder_Find.setTitle("找回密码");// 设置标题内容
                final EditText etPhoneFind = (EditText) layout_Find
                        .findViewById(R.id.etPhoneFind);

                // "发送"按钮
                builder_Find.setPositiveButton("发送",
                        new DialogInterface.OnClickListener() {

                            @SuppressLint("NewApi")
                            @Override
                            public void onClick(DialogInterface arg0, int arg1) {
                                Toast.makeText(layout_Find.getContext(),
                                        "密码已发送", Toast.LENGTH_LONG).show();
```

```
                                Notification.Builder builder_Noti = new Notification.Builder(
                                        LoginActivity.this);
                                builder_Noti.setSmallIcon(R.drawable.sms); // 设置图标
                                builder_Noti.setContentTitle("找回密码"); // 设置标题

                                inputphone= etPhoneFind.getText().toString();
                                user = set.findByPhone(inputphone);
                                System.out.println(user.getPwd());

                                builder_Noti.setContentText("您好！密码是："
                                        + user.getPwd()); // //设置内容
                                builder_Noti.setWhen(System.currentTimeMillis());
                                // 设置启动时间

                                Notification msg = builder_Noti.build(); // 生成通知
                                msg.defaults = Notification.DEFAULT_SOUND;
                                //获取系统的 Notification 服务
                                NotificationManager    manager    =    (NotificationManager)
getSystemService(NOTIFICATION_SERVICE);
                                manager.notify(1, msg); // 发出通知
                            }
                        });

                AlertDialog dlg_Find = builder_Find.create();
                Window window_Find = dlg_Find.getWindow();
                WindowManager.LayoutParams lp_Find = window_Find
                        .getAttributes();
                lp_Find.alpha = 0.8f;
                window_Find.setAttributes(lp_Find);
                dlg_Find.show();
            }
        });

        tvRegister = (TextView) this.findViewById(R.id.tvRegister);
        tvRegister.setOnClickListener(new View.OnClickListener() {
            @Override
            public void onClick(View v) {
                AlertDialog.Builder builder = new AlertDialog.Builder(
                        LoginActivity.this);
                LayoutInflater inflater = getLayoutInflater();
                final View layout = inflater.inflate(R.layout.dialog_reg, null);// 获取自定义布局
                builder.setView(layout);
                builder.setIcon(R.drawable.userreg);// 设置标题图标
                builder.setTitle("用户注册");// 设置标题内容
                final EditText etPhoneReg = (EditText) layout
                        .findViewById(R.id.etPhoneReg);
                final EditText etPwdReg = (EditText) layout
                        .findViewById(R.id.etPwdReg);

                // “注册”按钮
                builder.setPositiveButton("注册",
                        new DialogInterface.OnClickListener() {

                            @Override
                            public void onClick(DialogInterface arg0, int arg1) {
                                String phoneReg, pwdReg;
                                phoneReg = etPhoneReg.getText().toString();
```

```
                        pwdReg = etPwdReg.getText().toString();
                        set.set.add(new User(phoneReg, pwdReg));
                        set.selectAll();
                        Toast.makeText(layout.getContext(),
                                "注册成功:" + phoneReg + "," + pwdReg,
Toast.LENGTH_LONG)

                                .show();
                        etPhoneReg.setText("");
                        etPwdReg.setText("");
                    }
                });

            // “取消”按钮
            builder.setNegativeButton("取消",
                    new DialogInterface.OnClickListener() {
                        @Override
                        public void onClick(DialogInterface arg0, int arg1) {
                            // 什么也不做
                        }
                    });

            AlertDialog dlg = builder.create();
            Window window = dlg.getWindow();
            WindowManager.LayoutParams lp = window.getAttributes();
            lp.alpha = 0.8f;
            window.setAttributes(lp);
            dlg.show();
        }
    });
}

@Override
protected void onPause() { //离开屏幕前
    super.onPause();
    if(clickState==true){
        SharedPreferences sharedPreferences=getSharedPreferences("info",Activity.MODE_PRIVATE);
        SharedPreferences.Editor editor=sharedPreferences.edit();
        editor.putString("phone",inputphone);
        editor.putString("pwd",inputpwd);
        editor.putBoolean("clickState",clickState);
        editor.commit();
    }
}

@Override
protected void onResume() { //当屏幕来到前台时
    super.onResume();
    SharedPreferences
sharedPreferences=getSharedPreferences("info",Activity.MODE_PRIVATE);
    etphone.setText(sharedPreferences.getString("phone",null));
    etpwd.setText(sharedPreferences.getString("pwd",null));
    clickState= sharedPreferences.getBoolean("clickState",false);
    if(clickState==true)
        ivOnOff.setImageResource(R.drawable.on);
}
}
```

在登录界面中，有一个 ImageView 控件，用户单击该控件，可以使开关在“OFF”和“ON”

状态之间进行切换，如图 11-11 所示。

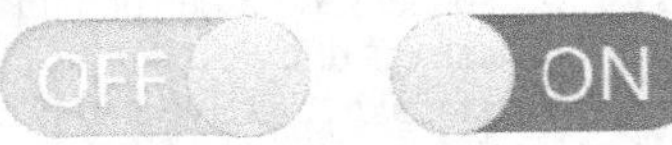

图 11-11　在“OFF”和“ON”状态之间进行切换

3．主界面右侧 Fragment 调用的类

主界面右侧的 Fragment 调用侧拉菜单的前 9 个菜单项对应类的代码分别为 Fragment1Activity.java ~ Fragment9Activity.java，这 9 个类的设计思路和调用方式都是一样的，下面以 Fragment1Activity.java 为例进行介绍，代码如下：

```
package com.example.user.scsx;

import android.media.AudioManager;
import android.media.MediaPlayer;
import android.os.Environment;
import android.support.annotation.Nullable;
import android.support.v4.app.Fragment;
import android.os.Bundle;
import android.view.LayoutInflater;
import android.view.View;
import android.view.ViewGroup;
import android.widget.ImageView;
import android.widget.TextView;
import java.io.IOException;

public class Fragment1Activity extends Fragment {
    ImageView ivImg,imgMusic;
    TextView tvTitle,tvContent;
    int numClicked=0;
    MediaPlayer player=new MediaPlayer();

    @Nullable
    @Override
    public View onCreateView(LayoutInflater inflater, @Nullable ViewGroup container,
                             @Nullable Bundle savedInstanceState) {
        View view=inflater.inflate(R.layout.activity_fragment1,container,false);

        ivImg= (ImageView) view.findViewById(R.id.ivImg);
        imgMusic= (ImageView) view.findViewById(R.id.imgMusic);
        tvTitle= (TextView) view.findViewById(R.id.tvTitle);
        tvContent= (TextView) view.findViewById(R.id.tvContent);

        ivImg.setImageResource(MainActivity.img_sc);
        tvTitle.setText(MainActivity.title_sc);
        tvContent.setText(MainActivity.content_sc+"\r\n\n"+" 译 文： \r\n\n"+MainActivity.yiwen_sc+
"\r\n\n"+"赏析：\r\n\n"+MainActivity.shangxi_sc);

        imgMusic.setOnClickListener(new View.OnClickListener() {
            @Override
            public void onClick(View v) {
                numClicked++;
                if(numClicked%2!=0){
                    imgMusic.setImageResource(R.drawable.stop);
                    String url="file://"+ Environment.getExternalStorageDirectory().toString()
                            +"/Music/"+MainActivity.filename_sc;
                    player.setAudioStreamType(AudioManager.STREAM_MUSIC);
                    try {
```

```
                    player.reset();
                    player.setDataSource(url);
                    player.prepare();
                    player.start();
                } catch (IOException e) {
                    e.printStackTrace();
                }
            }else{
                if(player.isPlaying()){
                    player.stop();   //停止播放
                    imgMusic.setImageResource(R.drawable.play);
                }
            }
        }
    });

    return view;
    }
}
```

在此界面中，有一个 ImageView 控件，该控件在默认状态下显示的图片为 play.png（“开始播放”按钮），用户单击该控件，可以切换到 stop.png（“停止播放”按钮），如图 11-12 所示。上述操作用于控制诗词朗诵配音文件的播放与停止。

图 11-12　“开始播放”按钮与“停止播放”按钮

4．ListView 适配器填充类

ListView 适配器填充类的 MenuAdapter.java 代码如下：

```
package com.example.user.scsx;

import android.content.Context;
import android.support.annotation.NonNull;
import android.view.LayoutInflater;
import android.view.View;
import android.view.ViewGroup;
import android.widget.ArrayAdapter;
import android.widget.ImageView;
import android.widget.TextView;

import java.util.List;

public class MenuAdapter extends ArrayAdapter{
    int resId;
    List<Menu> list;

    public MenuAdapter(Context context, int resource, List objects) {
        super(context, resource, objects);
        this.resId=resource;
        this.list=objects;
    }

    @NonNull
    @Override
    public View getView(int position, View convertView, ViewGroup parent) {
        Menu menu=list.get(position);
```

```
            View view= LayoutInflater.from(getContext()).inflate(resId,parent,false);
            ImageView imageView= (ImageView) view.findViewById(R.id.ivimg);
            TextView textView= (TextView) view.findViewById(R.id.tvname);
            imageView.setImageResource(menu.getImg());
            textView.setText("    "+menu.getName());
            return view;
        }
}
```

5．主界面

主界面的 MainActivity.java 代码如下：

```
package com.example.user.scsx;

import android.content.Intent;
import android.support.v4.app.Fragment;
import android.support.v4.widget.SlidingPaneLayout;
import android.support.v7.app.AppCompatActivity;
import android.os.Bundle;
import android.view.View;
import android.widget.AdapterView;
import android.widget.ImageView;
import android.widget.ListView;
import android.widget.TextView;

import java.util.ArrayList;
import java.util.List;

public class MainActivity extends AppCompatActivity {
    List<Menu> menuList = new ArrayList<>();
    int[] arr_img = new int[]{R.drawable.icon1, R.drawable.icon2, R.drawable.icon3,
            R.drawable.icon4, R.drawable.icon5, R.drawable.icon6,
            R.drawable.icon7, R.drawable.icon8, R.drawable.icon9,
            R.drawable.icon10, R.drawable.icon11};
    String[] arr_name = new String[]{"李白", "杜甫", "白居易", "杜牧", "王维", "苏轼",
            "柳宗元", "欧阳修", "刘禹锡", "修改密码", "注销登录"};
    ListView lvMenu;
    ImageView ivRightTop;
    SlidingPaneLayout slidingPaneLayout1;
    TextView tvMenuname;
    ShiCiList scl=new ShiCiList();
    ShiCi sc;
    public static int img_sc;
    public static String title_sc, filename_sc, content_sc,yiwen_sc,shangxi_sc;
    Fragment[]  arr_frag=new  Fragment[]{new  Fragment1Activity(),new  Fragment2Activity(),new
Fragment3Activity(),
            new Fragment4Activity(),new Fragment5Activity(),new Fragment6Activity(),
            new Fragment7Activity(),new Fragment8Activity(),new Fragment9Activity(),
            new ChangePassword(),new CancelLogin()};

    @Override
    protected void onCreate(Bundle savedInstanceState) {
        super.onCreate(savedInstanceState);
        setContentView(R.layout.activity_main);

        PermisionUtils.verifyStoragePermissions(this);

        scl.initShiCi();
        scl.printShiCi();
```

```
        lvMenu = (ListView) this.findViewById(R.id.lvMenu);
        for (int i = 0; i < arr_img.length; i++) {
            Menu menu = new Menu(arr_img[i], arr_name[i]);
            menuList.add(menu);
        }

        MenuAdapter adapter = new MenuAdapter(MainActivity.this, R.layout.lst_layout,
                menuList);
        lvMenu.setAdapter(adapter);
        tvMenuname = (TextView) this.findViewById(R.id.tvMenuname);
        lvMenu.setOnItemClickListener(new AdapterView.OnItemClickListener() {
            @Override
            public void onItemClick(AdapterView<?> parent, View view, int position, long id) {
                switch (position) {
                    case 0:
                    case 1:
                    case 2:
                    case 3:
                    case 4:
                    case 5:
                    case 6:
                    case 7:
                    case 8:
                        getSupportFragmentManager().beginTransaction().replace(R.id.mainContent,
                                arr_frag[position]).commit();
                        sc = scl.list.get(position);
                        img_sc = sc.getImg();
                        title_sc = sc.getTitle();
                        filename_sc = sc.getMusic();
                        content_sc = sc.getContent();
                        yiwen_sc = sc.getYiwen();
                        shangxi_sc = sc.getShangxi();
                        break;
                    case 9:
                        getSupportFragmentManager().beginTransaction().replace(R.id.mainContent,
                                arr_frag[position]).commit();
                        break;
                    case 10:
                        Intent intent=new Intent(MainActivity.this,YindaoActivity.class);
                        startActivity(intent);
                        finish();
                        break;
                }
                tvMenuname.setText(menuList.get(position).getName());
            }
        });

        slidingPaneLayout1 = (SlidingPaneLayout) this.findViewById(R.id.slidingPaneLayout1);
        ivRightTop = (ImageView) this.findViewById(R.id.ivRightTop);
        ivRightTop.setOnClickListener(new View.OnClickListener() {
            @Override
            public void onClick(View v) {
                if (slidingPaneLayout1.isOpen())
                    slidingPaneLayout1.closePane();
                else
                    slidingPaneLayout1.openPane();
            }
        });
    }
```

```
}
```

6．修改密码

修改密码的 ChangePassword.java 代码如下：

```
package com.example.administrator.scsx;

import android.support.v7.app.AppCompatActivity;
import android.content.Intent;
import android.content.SharedPreferences;
import android.os.Bundle;
import android.support.v4.app.Fragment;
import android.view.LayoutInflater;
import android.view.View;
import android.view.View.OnClickListener;
import android.view.View.OnFocusChangeListener;
import android.view.ViewGroup;
import android.widget.Button;
import android.widget.EditText;
import android.widget.Toast;

public class ChangePassword extends Fragment{
    private View view;
    EditText etPwd1,etPwd2;
    Button btnUpdate;
    String pwd1,pwd2;

    @Override
    public View onCreateView(LayoutInflater inflater, ViewGroup container,
            Bundle savedInstanceState) {
        view = inflater.inflate(R.layout.activity_change_password, null);
        etPwd1=(EditText)view.findViewById(R.id.etPwd1);
        etPwd1.setOnFocusChangeListener(new View.OnFocusChangeListener(){

            @Override
            public void onFocusChange(View v, boolean hasFocus) {
                if(hasFocus){

                }else
                {
                    pwd1=etPwd1.getText().toString();
                    if(!LoginActivity.pwd.equals(pwd1))
                        Toast.makeText(getActivity(), "原密码输入不正确！", 0).show();
                }
            }});

        etPwd2=(EditText)view.findViewById(R.id.etPwd2);
        btnUpdate=(Button)view.findViewById(R.id.btnUpdate);
        btnUpdate.setOnClickListener(new OnClickListener(){

            @Override
            public void onClick(View arg0) {
                pwd2=etPwd2.getText().toString();
                //LoginActivity.userSet.set.remove(new
User(LoginActivity.phone,LoginActivity.pwd));
                /*for (User u : LoginActivity.userSet.set) {
                    if (LoginActivity.userSet.findByPhone(LoginActivity.phone) != null)
                        LoginActivity.userSet.set.remove(u);
                }*/
```

```
                    LoginActivity.userSet.set.add(new User(LoginActivity.phone,pwd2));
                    SharedPreferences                                    share                              =
getActivity().getSharedPreferences("info",Activity.MODE_PRIVATE);
                    SharedPreferences.Editor editor = share.edit();
                    editor.putString("phone", LoginActivity.phone);
                    editor.putString("pwd", pwd2);
                    editor.commit();
                    Toast.makeText(getActivity(), "修改成功", 0).show();
                    Intent intent = new Intent(getActivity(),LoginActivity.class);
                    startActivity(intent);
                    getActivity().finish();
                }}});
        return view;
    }

    @Override
    public void onActivityCreated(Bundle savedInstanceState) {
        super.onActivityCreated(savedInstanceState);
    }
}
```

7．注销登录

注销登录的 CancelLogin.java 代码如下：

```
package com.example.administrator.scsx;

import android.os.Bundle;
import android.support.v7.app.AppCompatActivity;
import android.support.v4.app.Fragment;
import android.view.LayoutInflater;
import android.view.Menu;
import android.view.View;
import android.view.ViewGroup;

public class CancelLogin extends Fragment{
    private View view;

    @Override
    public View onCreateView(LayoutInflater inflater, ViewGroup container,
            Bundle savedInstanceState) {
        view = inflater.inflate(R.layout.activity_cancel_login, null);
        return view;
    }

    @Override
    public void onActivityCreated(Bundle savedInstanceState) {
        super.onActivityCreated(savedInstanceState);
    }
}
```

11.4　运行测试

运行诗词赏析项目，在模拟器中观察运行结果是否与图 11-1~图 11-5 一致。

提醒：在本项目中，由于 9 首诗词的朗诵配音文件储存在 SD 卡中，因此在测试前，需要在 AndroidManifest.xml 文件中开通访问 SD 卡的权限。

参 考 文 献

[1] 孙浏毅. Java 宝典[M]. 北京：电子工业出版社, 2009.

[2] 郭霖. 第一行代码: Android（第 2 版）[M]. 北京：人民邮电出版社, 2016.

[3] 毕小朋. 精通 Android Studio[M]. 北京：清华大学出版社, 2016.

[4] 王翠萍. Android Studio 应用开发实战详解[M]. 北京：人民邮电出版社, 2017.

[5] 欧阳燊. Android Studio 开发实战：从零基础到 App 上线[M]. 北京：清华大学出版社, 2017.